Sustainable Civil Infrastructures

Sustainable Infrastructure impacts our well-being and day-to-day lives. The infrastructures we are building today will shape our lives tomorrow. The complex and diverse nature of the impacts due to weather extremes on transportation and civil infrastructures can be seen in our roadways, bridges, and buildings. Extreme summer temperatures, droughts, flash floods, and rising numbers of freeze-thaw cycles pose challenges for civil infrastructure and can endanger public safety. We constantly hear how civil infrastructures need constant attention, preservation, and upgrading. Such improvements and developments would obviously benefit from our desired book series that provide sustainable engineering materials and designs. The economic impact is huge and much research has been conducted worldwide. The future holds many opportunities, not only for researchers in a given country, but also for the worldwide field engineers who apply and implement these technologies. We believe that no approach can succeed if it does not unite the efforts of various engineering disciplines from all over the world under one umbrella to offer a beacon of modern solutions to the global infrastructure. Experts from the various engineering disciplines around the globe will participate in this series, including: Geotechnical, Geological, Geoscience, Petroleum, Structural, Transportation, Bridge, Infrastructure, Energy, Architectural, Chemical and Materials, and other related Engineering disciplines.

More information about this series at http://www.springer.com/series/15140

Hugo Rodrigues · Amr Elnashai
Editors

Advances and Challenges in Structural Engineering

Proceedings of the 2nd GeoMEast International Congress and Exhibition on Sustainable Civil Infrastructures, Egypt 2018 – The Official International Congress of the Soil-Structure Interaction Group in Egypt (SSIGE)

Springer

Editors
Hugo Rodrigues
Polytechnic Institute of Leiria
Leiria, Portugal

Amr Elnashai
Division of Research
University of Houston
Houston, TX, USA

ISSN 2366-3405 ISSN 2366-3413 (electronic)
Sustainable Civil Infrastructures
ISBN 978-3-030-01931-0 ISBN 978-3-030-01932-7 (eBook)
https://doi.org/10.1007/978-3-030-01932-7

Library of Congress Control Number: 2018957131

This Springer imprint is published by the registered company Springer Nature Switzerland AG
The registered company address is: Gewerbestrasse 11, 6330 Cham, Switzerland

Contents

About the Editors

Hugo Rodrigues Senior Lecturer at the School of Technology and Management, Polytechnic Institute of Leiria, Portugal, teaching several topics related with structural analysis, building pathology and rehabilitation. He received his Ph.D. in Civil Engineering from the University of Aveiro in 2012. He has experience in Seismic analysis, having participated as principal investigator and team member several research and projects, specialized consultancy studies ordered by several public institutions and companies regarding the assessment of seismic risk. His major research interests are building rehabilitation, structural health monitoring and seismic safety including experimental and numerical activities. Research member of the RIsks and Sustainability in Construction (RISCO). He has co-authored more than 150 publications in top-tier peer-reviewed journals and national and international conferences.

Amr Elnashai Fellow of the British Royal Academy of Engineering, Amr Elnashai is the Dean of Engineering at the Pennsylvania State University, and the Harold and Inge Marcus Chair of Engineering. He was previously head of the Department of Civil and Environmental Engineering at the University of Illinois at Urbana-Champaign, USA, and the Bill and Elaine Hall Endowed Professor in the same department. He was Director of the NSF multi-institution interdisciplinary Mid-America Earthquake Center (2004–2009). He was also Director of the NSF Network for

Earthquake Engineering Simulations (NEES) Laboratory at Illinois (2002–2009). He obtained his BSc from Cairo University, followed by MSc and Ph.D. from Imperial College, University of London. Before joining the University of Illinois in June 2001, he was Professor of Earthquake Engineering and Head of Division at Imperial College (1985–2001). He has been Visiting Professor at the University of Surrey, UK, since 1997. Other Visiting Professor appointments include the University of Tokyo, the University of Southern California and the European School for Advanced Studies in Reduction of Seismic Risk, Italy.

He is founder and co-editor of the Journal of Earthquake Engineering and Editorial Board Member of several other journals, a Member of the drafting panel of the European design codes, past Chair of the UK Earthquake Engineering Association, UK delegate to and past Senior Vice-President of the European Association of Earthquake Engineering and a Member of the Council of the UK Institution of Structural Engineers. He is the winner of the Imperial College Unwin Prize for the best Ph.D. thesis in Civil and Mechanical Engineering (1984), the Oscar Faber Medal for best paper in the Institution of Structural Engineering and two best paper medals from the International Association of Tall Buildings, Los Angeles.

His technical interests are multi-resolution distributed analytical simulations, network analysis, large-scale hybrid testing and field investigations of the response of complex networks and structures to earthquakes. He has produced more than 250 research publications, comprising over 140 refereed journal papers and many conference papers, keynote and prestige lectures (including the Nathan Newmark Distinguished Lecture), research reports, three books and several chapters, magazine articles and earthquake field investigation reports. He has supervised 45 doctoral and over 100 Master of Science theses. Many of his students hold significant positions in industry, academia and government in over 14 countries.

Retrofitting Gravity Load Designed R.C Frames Using FRP

Mohamed I. S. Elmasry[1(✉)], Melad Belal Said[1,2], and Essam A. Elkordy[2]

[1] Structural Engineering, Graduate Studies, Arab Academy for Science and Technology (AASTMT), Alexandria, Egypt
elmasryi@aast.edu

[2] Structural Engineering, Alexandria University (AU), Alexandria, Egypt
esmelkordi@yahoo.com

Abstract. Many existing worldwide Reinforced Concrete (RC) structures, such as non-ductile RC frames, were designed for gravity loads only during the 1950s through 1970s or earlier. Due to variations in the identification of seismic active zones by national codes, such structures may not satisfy the current design requirements, especially when lying in a recently identified seismic active zone. This is because such structures, as a result of poorer reinforcement detailing, may generally do not possess the adequate ductility and strength needed to withstand an expected earthquake. Consequently, older RC frames may undergo substantial damage during earthquakes. One of the main damage aspects in such case is clear cracks around and within the beam-column connections. This is the case where the failure of beam-column joints is governed by bond and shear failure mechanisms which are usually brittle. This may be attributed to inadequate shear reinforcement in the beam-column joints region., This paper presents the first results of an experimental campaign performed – at the Laboratory of Materials & Structures of the Arab Academy And Technology (EGYPT) – with the aim to investigate the seismic performance of RC beam-columns joints strengthened with FRP. The experimental program includes testing specimens realized to be representative of existing exterior beam-column subassemblies with inadequate seismic details. A technical solution was selected in order to improve the joint seismic behaviour, and the performance of the proposed strengthening system is investigated in this paper. To this aim, two as built beam-column joints have been tested, one strengthened specimen and a reference unstrengthened one. Test results provide useful information for the adopted strengthening systems in terms of strength, ductility and energy dissipation capacity. Results indicate that the proposed strengthening technique was successful in adding up to 50% of the beam column joint capacity. The results are encaging to apply this technique on existing gravity load designed buildings.

1 Introduction

Old buildings have many factors that can affect their integrity such as being subjected to earthquakes, or not being designed in the first place to endure lateral loads. This could lead to structural failures. This the case with knowing that the cost of

H. Rodrigues and A. Elnashai (Eds.): GeoMEast 2018, SUCI, pp. 1–13, 2019.
https://doi.org/10.1007/978-3-030-01932-7_1

demolishing and rebuilding is very high. Nevertheless, the economic costs become measure due to delaying works and projects for individuals, companies and the government. Due to all what have been mentioned, it would be preferable to reinforce the building instead of demolishing it. However, to repair a building, one needs first to figure out the weak parts that need to be strengthened in a way that doesn't require the building to be evacuated if possible (Anagnostopoulos *et al.* 1996).

Beam-column connections have been diagnosed as potentially vulnerable components when Reinforced Concrete (R.C.) frame buildings are subjected to seismic loads, Generally, older buildings are normally designed only for gravity loads. Such Buildings when subjected to seismic loading may be subjected to damage, and the most affected parts are the structural beam- column joints, and particularly, the corner joints or the exterior joints (Wernli 2004). The interior joints are also affected due to seismic loads, but the damage is relatively smaller in magnitude. In addition, the accurate seismic assessment of beam-column subassemblies has been an important objective for many research groups over the years. Experimental and analytical research on different aspects of these members by various research groups have provided numerous design and assessment techniques with little consensus amongst these research groups.

In this paper, the feasibility of a novel strengthening solution using Fiber Reinforced Polymers (FRP) in increasing the seismic performance of exterior RC beam-column joints is investigated. An experimental program was organized at the Arab Academy For Science and Technology which includes several as-built specimens representative of existing beam-column joint subassemblies with inadequate seismic details that may exist in gravity load designed buildings. Of these specimens, the first one was strengthened by using FRP, while the second one was not strengthened. The two as-built specimens where loaded till failure in a step by step manner to identify the gain in the beam- column joint capacity after strengthening while anther specimen was used as a reference benchmark.

2 Problem Definition and Research Objectives

Prior to modern codes that include detailed information for designing structures under seismic loads, older RC frame structures were designed only for gravity loads. This is even the case when considering that the seismic active zones varied in the national codes by time, such that some older structures are presently lying in active seismic zones, though this was not the case in the past. Such existing RC frame structures, though performing well under conventional gravity load case, could lead to questionable structural performance under earthquakes. In most cases, those structures are vulnerable to any moderate or major earthquake, and thus need immediate assessment and retrofitting to avoid a sudden full or partial collapse mechanism bringing considerable losses in human lives and economic assets (Elmasry *et al.* 2016).

In this paper, objectives include developing an effective rehabilitation to strengthen beam-column joints in older structures to improve their seismic performance in terms of lateral strength and serviceability. Accordingly, An experimental program is prepared for validating the suggested technique. By interpreting results and studying the resulting advantages versus disadvantages, it is clearly shown that the proposed technique improves considerably the strength of the beam-column joints by using CFRP sheets (Elmasry *et al.* 2016).

3 Literature Review

3.1 Beam Column Joints

The functional requirement of a joint, which is the zone of intersection of beams and columns, is to enable the adjoining members to develop and sustain their ultimate capacity. The demand on this finite size element is always severe especially under seismic loading. The joints should have adequate strength and stiffness to resist the internal forces induced by the framing members. The beam-column joint in a frame is defined as the portion of the column within the depth of the deepest beam that frames into the column. In a moment resisting frame, three types of joints can be identified. an interior joint, an exterior joint and a corner joint as shown in Fig. 1. When four beams frame into the vertical faces of a column, the joint is called an interior joint. In contrast, when one beam frames into a vertical face of the column and two other beams frame from perpendicular directions into the joint, then the joint is called as an exterior joint. When two beams frame, into two adjacent vertical faces of a column, then the joint is called as a corner joint. The severity of forces and demands on the performance of these joints calls for greater understanding of their seismic behaviour. These forces develop complex mechanisms involving bond and shear within the joint (ACI 2002).

(A) Interior *(B) Exterior* *(C) Corner*

Fig. 1. Beam column joint connections types

3.1.1 Joints Behavior Under Seismic Loading

In two-dimensional (2D) frame joints subjected to earthquake loadings, beams and columns experience flexure and shear loadings. Figure 2 shows the forces that could be expected to develop in a 2D frame joint subjected to earthquake and gravity loading. Moreover, when recent seismically designed frames are subjected to moderate and severe earthquake loading, it is expected that the beams will develop flexural strength at the joint while columns will develop moments that approach the yield moment. Meanwhile, in older frames, shear failure of beams and columns or flexural yielding of columns may preclude beams achieving yielding flexural strength. Figure 2(A) shows the expected loads and resultants at the perimeter of the joint region. The load distribution, as shown in Fig. 2(B), can result in severe loading within the joint region. In addition, the moment reversal in the beams and columns results in large shear forces within the joint (Nilanjan 2007).

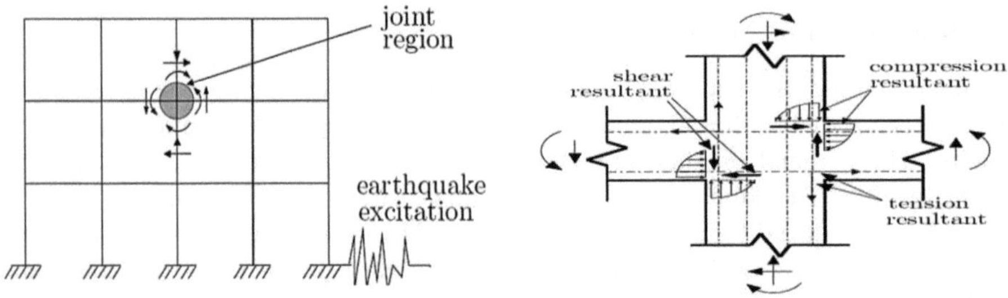

(A) *Building frame with joint region* (B) *Detailed view of joint region*

Fig. 2. Loads in the beam-column joint region

4 Fibre Reinforced Polymer (FRP)

Fibre Reinforced Polymer (FRP) composites comprise fibres of high tensile strength within a polymer matrix such as epoxy. FRP composites are used in a lot of applications such as aircraft, helicopters, space-craft, satellites, ships, submarines, automobiles, chemical processing equipment, sporting goods and civil infrastructure (Liyoung *et al.* (2002); Abhishek (2012) and Sreelatha (2013)). In general, one of the advantages of using FRP products would be strengthening of the existing or new RC structures with the possibility of application without disturbing the existing functionality of the structure (Liyoung *et al.* (2002) and Sreelatha (2013)). In addition, FRP composites had proved to be extremely useful for strengthening of RC structures against both normal and seismic loads as stated in previous research as shown earlier. Moreover, most of the elements of a structure can be applicably strengthened with FRP composite materials. Currently, this method has been applied to strengthen structural elements as columns, beams, walls, slabs (Nikita *et al.* (2015), Sasmal *et al.* (2011) and Obaidat *et al.* (2010)). This means in fact that FRP composites can take up the majority of the forces developed in a structure as long as they are transmitted by the strengthened element to the composite one as tensile stresses. Furthermore, strengthening with externally bonded FRP fabric has shown to be applicable to many kinds of structures. The use of external FRP reinforcement may be classified in: (i) Flexural strengthening, (ii) Improving the ductility of compression members, and (iii) Shear strengthening. Furthermore, Carbon Fibre-Reinforced Polymer (CFRP), Carbon Fibre-Reinforced Plastic (CRP), or Carbon Fibre-Reinforced Thermoplastic (CFRTP) are extremely strong and light FRP which contain carbon fibres. This is the case where carbon fibres give CFRP its strength and rigidity in terms of increasing the ultimate stress and elastic modulus respectively. Unlike isotropic materials like steel and aluminium, CFRP has directional strength properties. The properties of CFRP depend on the layouts of the carbon fibre and the proposition of the carbon fibres relative to the polymer. Advantages for CFRP include high tensile strength, high strength to weight ratio, low weight to volume ratio, excellent fatigue behaviour, and quicker application (Naveeena and Ranjitham (2016)). Thus, CFRP composites are able to strengthen beam column joints in terms of the shear capacity and ductility.

5 Experimental Program

The ongoing experimental program includes several RC beam-column joints that are subjected to forces (variable load) applied by setting two opposite forces at the specimen beam tip by keeping the column fixed at the specimen top and bottom ends. Two as-built specimens where loaded till failure in a step by step manner to identify the gain in the beam-column joint capacity after strengthening while anther specimen was used as reference benchmark.

6 Design of Specimens

Exterior as-built beam-column joints were designed for the experimental campaign, characterized by the amounts of longitudinal steel reinforcement in the beams and columns. Figure 3 depicts the geometry and the rebar configuration of the strengthened and unstrengthened specimens. For the unstrengthened specimen, the structural dimensions of the studied beam-column joint, as shown in Fig. 3a, are summarized in having a vertical column with a cross section of 300 mm × 300 mm intersecting a horizontal beam with a 300 mm × 300 mm cross section. The height of the specimen is 1300 mm and the beams extend by 500 mm before and after the joint. The longitudinal reinforcement of the column is 8 T16 rebar as shown in Fig. 3, and the shear reinforcement is 8 mm diameter stirrups with a spacing of 75 mm to avoid beam shear failure and to direct the stresses concentrations to the connection zones. The beam steel reinforcement is 4 T16 rebar as shown in Fig. 3, and the shear reinforcement is 8 mm diameter stirrups with a spacing of 75 mm. The reason for having excessive shear reinforcement in the beam and the column is to enforce the shear cracks to be initiated within the beam-column joint zone rather than in the beam or the column. The cubic strength of concrete is 20 MPa which is typical in GLD older structures, while the yield strength of the steel reinforcement is 360 MPa.

In addition, the strengthened specimen is shown in Fig. 4. The concrete dimensions and reinforcement details of this retrofitted specimen are the same as the unstrengthened specimen. However, in this specimen, CFRP sheets diagonally reinforce the beam- column joint. In addition, webbed built up angles and covering plates, as shown in Fig. 5, are used as a bearing for the CFRP sheets to get wrapped around the columns at the beam-column connection. The thickness of the CFRP sheet is considered 0.13 mm. The modulus of elasticity (E) for CFRP is taken as 24000 MPa in the longitudinal direction and 18581 MPa in the transversal sides.

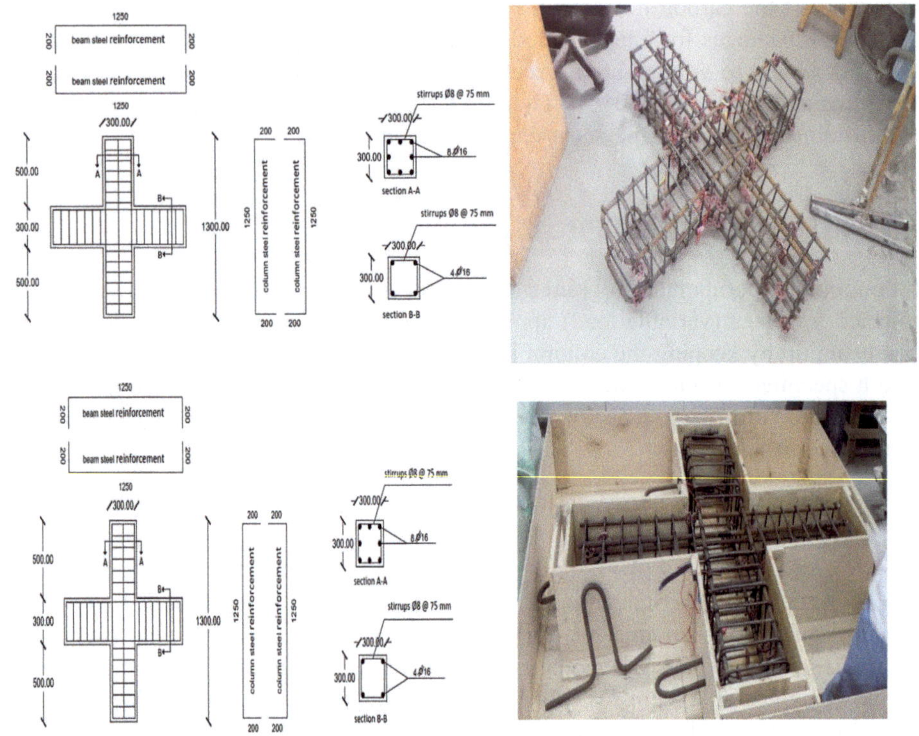

Fig. 3. The geometry and the rebar configuration of set of joints (a) The geometry and the rebars configuration of set of unstrengthened

7 Materials

The Concrete cubic strength for all concrete specimens was evaluated after 28 day and reported in Table 1. In order to replicate the GLD buildings, the concrete cubic strength was assumed 20 MPa to infer the strength quality by the time of construction of such buildings.

Table 1. Mechanical properties of concrete for test specimens.

Cubic sample	Compressive strength (fc) (MPa)
Unstrengthened	20
Strengthened	20

8 Test Set-up

The loading setup is shown in Fig. 6. The bottom and top of the column is considered constrained in three dimensions X, Y, Z using fixation steel jacket. Incrementally increasing concentrated loads are assumed to act on the specimen at the two cantilevering ends in opposite directions, to induce the resulting moments on the connection. A steel plate as shown in Fig. 7 is added at the expected loading location in order to avoid stress concentration problems. This provides a more even stress distribution over the loading area. There are ten strain gauges in the beam-column specimens that are distributed and fixed to the specimens as show in Fig. 8.

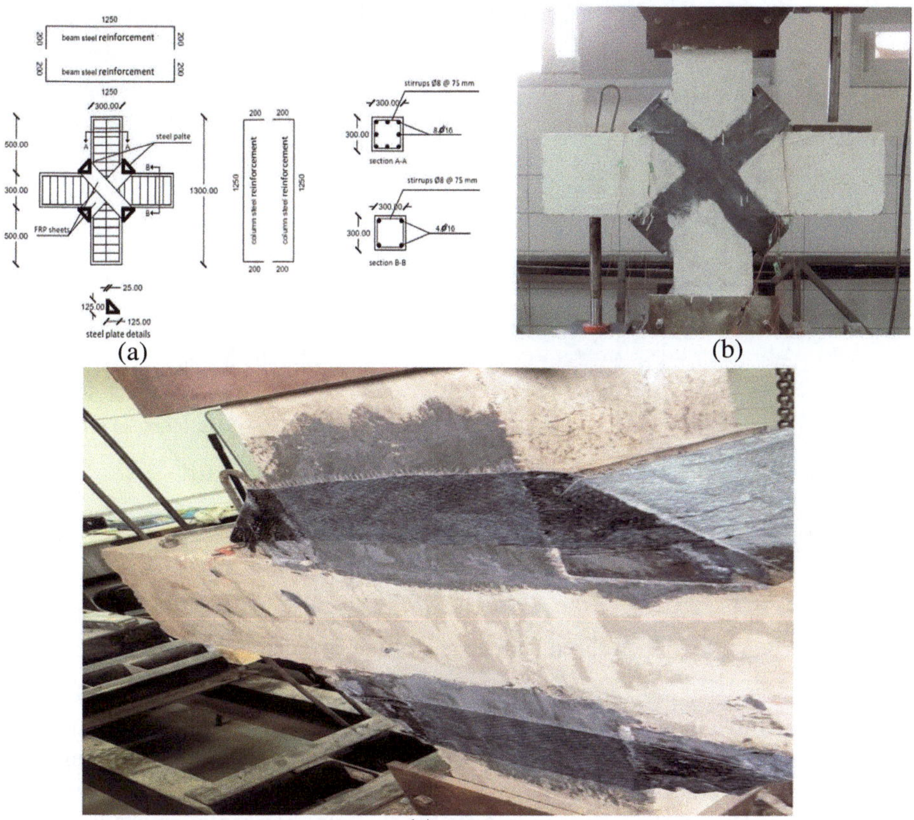

Fig. 4. The geometry and the rebars configuration of set of strengthened (a) The rebar in strengthened specimen (b) FRP fixation in front strengthened specimen (c) The back view of the strengthened specimen with FRP

9 Experimental Result

Crack patterns and failure modes of the unstrengthened specimen are shown in Fig. 9. Figure 10 shows the crack patterns in the strengthened specimen. The load & displacement curve for the unstrengthened cases shown in Fig. 11. Figure 12 shows the load & displacement curve for the strengthened specimen.

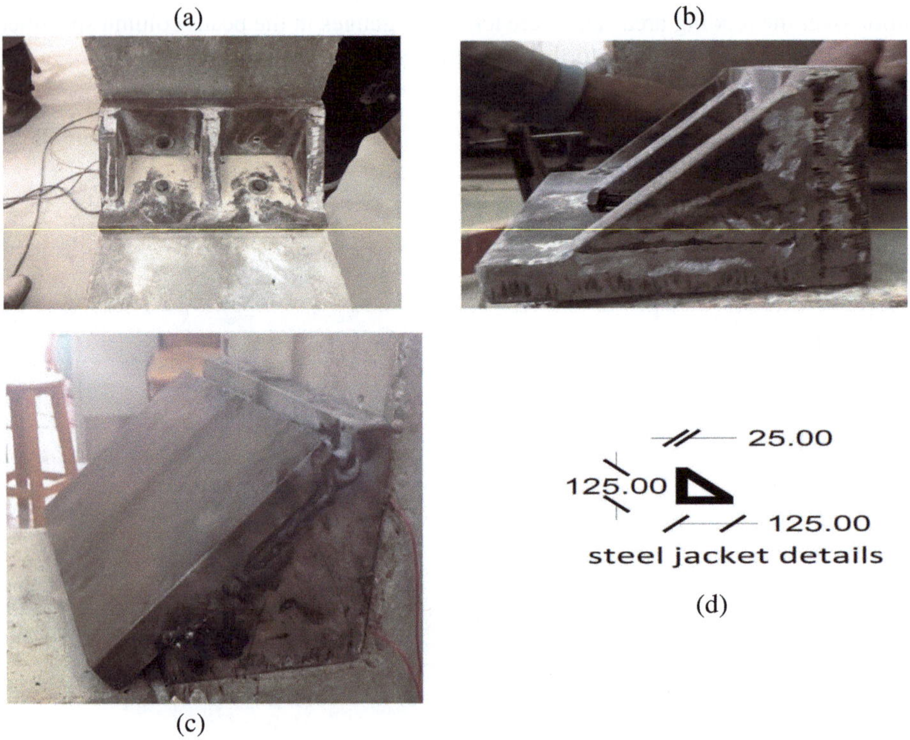

Fig. 5. Webbed built up angles and covering plates

From the above mentioned figures, it is clear that the cracks were concentrated for the unstrengthened case that showed the lowest strength, while in unstrengthened specimen, a lot of cracks appeared, some in the beam, some in the column as well as cracks in joint. The cracks were concentrated in the joint horizontal cracks in the form of vertical and diagonal cracks. In the strengthened specimen when using the strengthening technique by CFRP sheets, the cracks appeared rather in the beam and the joint. The cracks in the strengthened joint are less than the case as the unstrengthened joint. In spite of the fact that the cracks propagation in the strengthened specimen are bigger, yet the ductility was in the unstrengthened specimen less than the case of the strengthened specimen, which showed higher deflections. It is clear that the strengthened specimen showed the highest strength. The unstrengthened specimen

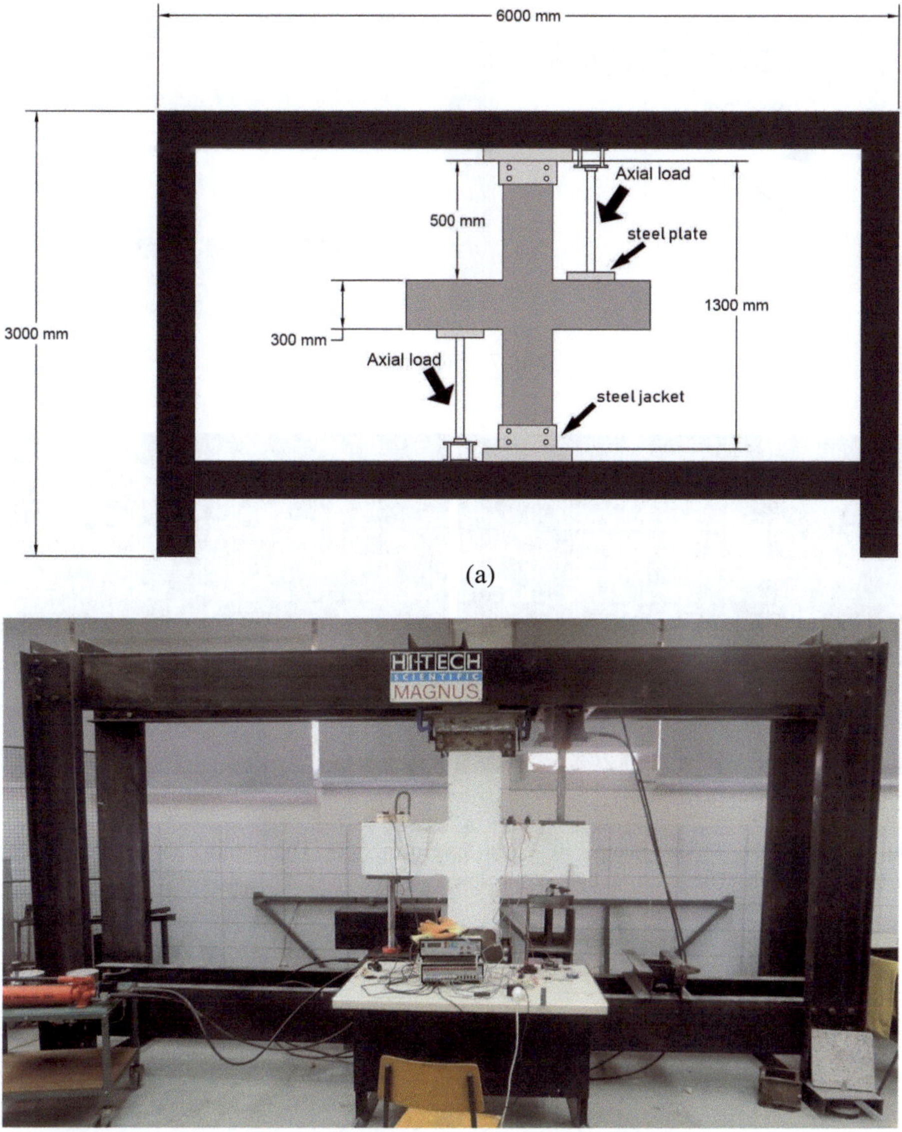

(a)

(b)

Fig. 6. (a) The schematic Test setup (b) Reis contrast Test setup

Fig. 7. Steel plate.

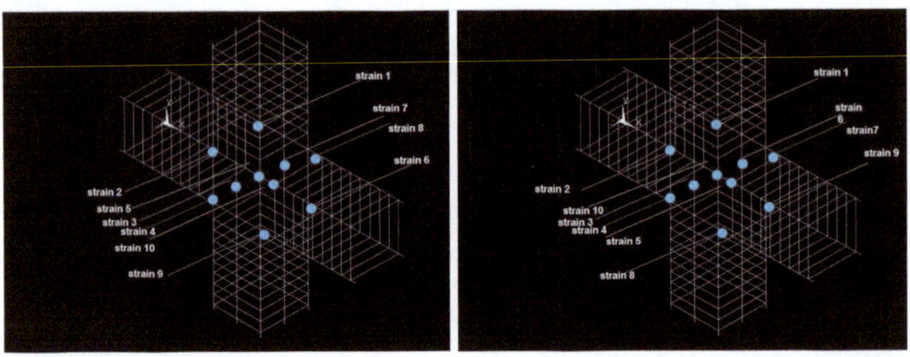

(a) (b)

Fig. 8. Show the strain gauge in each specimen. (a) The strain gauge in specimen unstrengthened (b) The strain gauge in specimen strengthened

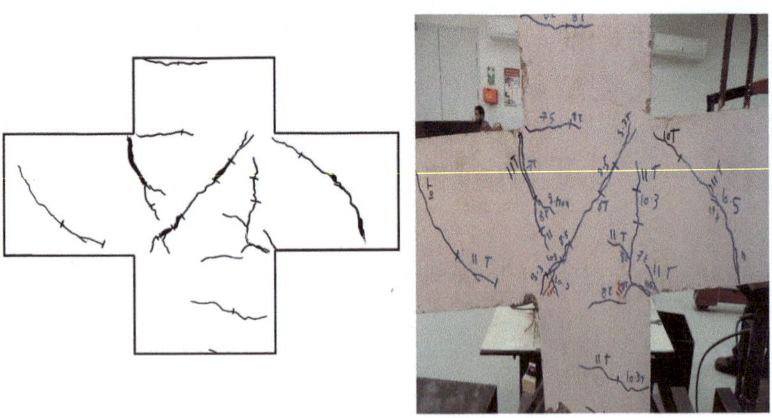

Fig. 9. The crack pattern of the unstrengthened specimen

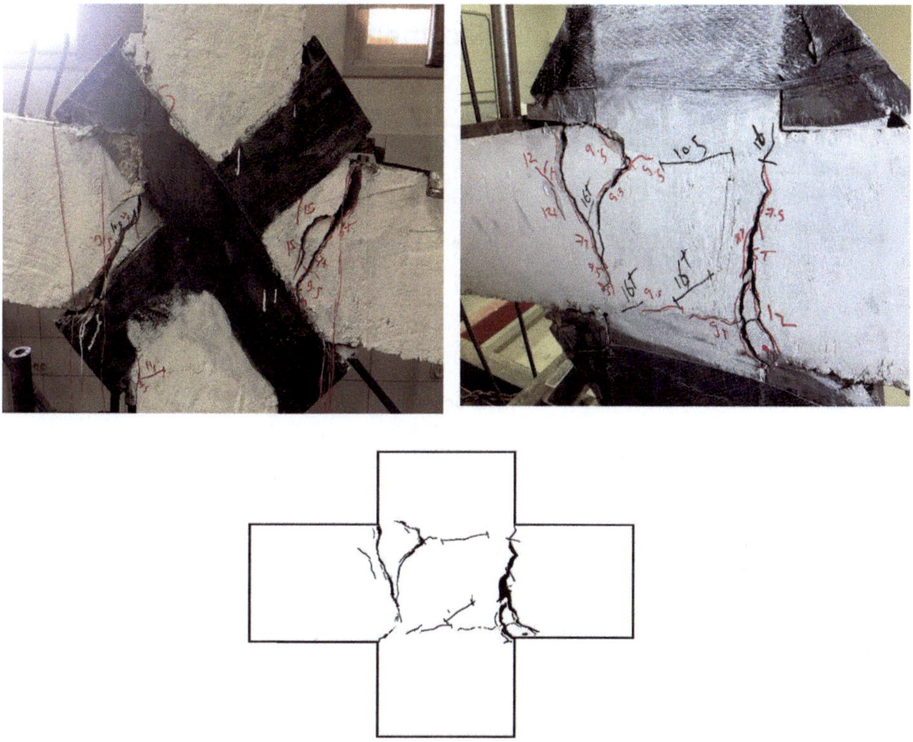

Fig. 10. The crack pattern of strengthened specimen

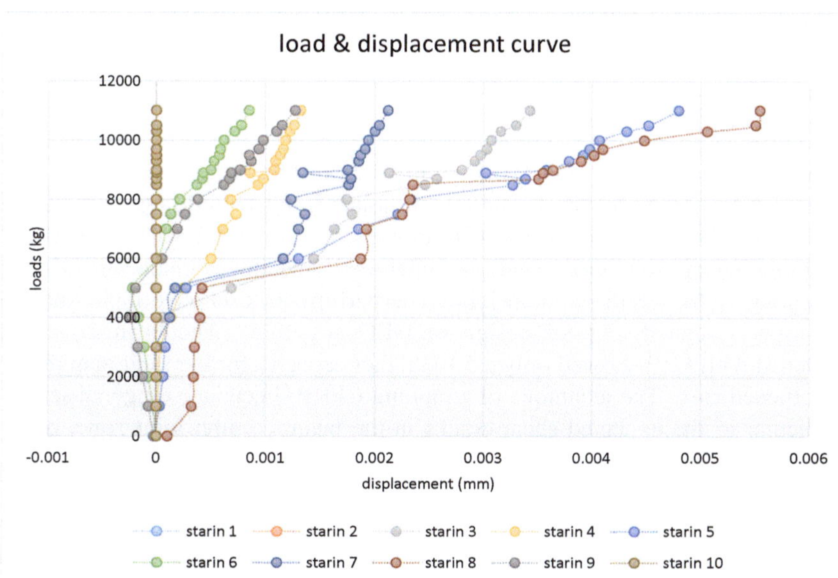

Fig. 11. The load & displacement curve for the unstrengthened specimen.

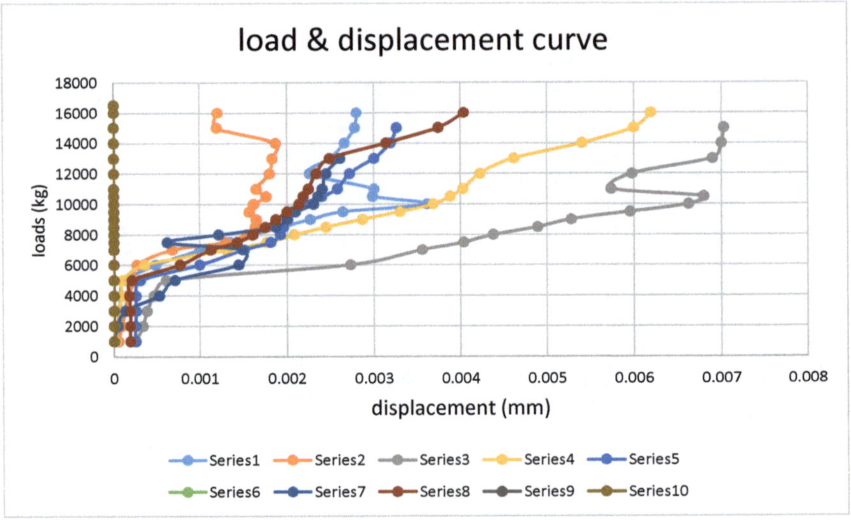

Fig. 12. Show the load & displacement curve for the strengthened specimen

failed when the axial load in both directions was equal to 110 KN (11 TON), and the strengthened specimen failed when the axial load in both directions was equal to 165 KN (16.5 TON).it is also noticed that the unstrengthened specimen failed in the beam-column joint, but the strengthened specimen failed rather in the beam. In summary, and as clear from this experimental study, it can be concluded that the proposed technique was successful in transferring the cracks from the beam-column joint to the beam through using diagonal wrapped CFRP sheets, this technique can be easily applied on older facades without varying or destroying the architecture layout.

10 Conclusions

An experimental study was undergone and shown in this paper to identify a suitable strengthening technique for the beam-column joints in GLD buildings. The results for the exterior RC beam-column joints were discussed. Two Specimens were tested, i.e., an unstrengthened specimen, and an upgraded specimen characterized by the strengthening of the beam -column joint. The performed tests have highlighted some critical aspects related to poor design of GLD structures. The beam-column joint strengthened with FRP showed only a 50% higher capacity for the joint than that of the unstrengthened case. The technique of wrapping CFRP sheets in a diagonal setup to be perpendicular to the expected shear cracks in the beam-column joint zone, proved to show better ductile response.

References

Elmasry1, M.I.S., Abdelkader, A.M., Elkordy, E.A.: An analytical study of improving beam-column joints behavior under earthquake, Egypt (2016)

Liyoung, T., Adrian, M., Michael, B.: Seismic rehabilitation of beam column joints using FRP laminates. J. Earthq. Eng. **5**(1), 113–129 (2002)

Naveeena, N., Ranjitham, M.: Numerical study on retrofitting of beam column joint strengthened with CFRP. Int. Res. J. Eng. Technol. (IRJET) **3**, 914–920 (2016)

Sasmal, S, Ramanjaneyulu, K., Novak, B., Srinivas, V., Kumar, K.S., Korkowski, C.: Upgradation of gravity load designed subassemblages subjected to seismic typeloading. Compos. Struct. **93**, 1561–1573 (2011)

Anagnostopoulos, S.A.: Building pounding re-examined: how serious a problem is it ? Greece (1996)

Wernli, M.: Eismic evaluation of 32 industrial buildings by screening process and analysis, Canada (2004)

Recommendations for design of beam-column-joints in monolithic reinforced concrete structures. American Concrete Institute, ACI 352R-02, ACIASCE, Committee 352, Detroit (2002)

Nilanjan, M.: An analytical study of reinforced concrete beam-column joint behavior under seismic loading. Doctor of Philosophy, University of Washington (2007)

Condition Monitoring of Overhead Line Equipment (OHLE) Structures Using Ground-Bourne Vibrations from Train Passages

Chayut Ngamkhanong[1], Sakdirat Kaewunruen[1](✉), Rui Calçada[2], and Rodolfo Martin[3]

[1] School of Engineering, The University of Birmingham, Birmingham, UK
s.kaewunruen@bham.ac.uk
[2] Faculty of Engineering, University de Porto, Porto, Portugal
[3] Evoleo Technology Pty Ltd., Porto, Portugal

Abstract. The most modern railway systems have fully adopted clean energy for train and track operations. Trains or rolling stocks are powered by electricity through the overhead wire or the third rail on ground. Commonly, the overhead line equipment (OHLE), which supplies electric power to the trains, is widely adopted in new railway networks around the world since its system enables trains to operate smoothly while track inspectors can safely work on tracks. The OHLE is supported by mast structure, which is located at the lineside along the track. The mast structure is often made of steel structure built on mat or pile foundation. Due to the train passages, ground-bourne periodic forces may cause damage to the OHLE structure especially mast structure, connections and its foundation, which can lead to operational failure of train electrification. On this ground, the structural integrity of mast structures must be inspected regularly. In this study, the modal analysis is used in order to identify the mode shapes and natural frequencies of the mast structure. A mast structure with varying rotational soil stiffness is used to construct dynamic influential lines for soil-structure integrity prediction. Finite element model updating technique has been used to perform modal analysis and modal parameter identification. This paper presents the integrated numerical of three-dimensional mast structure considering soil-structure interaction to evaluate the condition of OHLE structures for maintenance planning. The outcome of this study will help civil and track engineers to effectively and efficiently inspect OHLE structures using ground borne vibrations from train passages.

1 Introduction

Nowadays, the extra capacity of passenger train is needed due to the sudden growth of population and journeys (RailCorp 2011). It was noticed that freight and passenger journeys have increased by 60% and 100%, respectively. To get the destination faster, electric train has become the efficient railway systems, which are allowed to run more frequent and quicker than conventional diesel train. Also, electric train is more environmental friendly as it emits less carbon than diesel counterpart. Instead of carrying the fuel to make the train runs, the power is supplied by the Overhead Line Equipment

H. Rodrigues and A. Elnashai (Eds.): GeoMEast 2018, SUCI, pp. 14–22, 2019.
https://doi.org/10.1007/978-3-030-01932-7_2

(OHLE), which consists of masts, gantries, wires found along electrified railways. The electric system can be failed and the power supply can be lost. As seen in literature (Shing and Wong 2008; Robinson and Bryan 2009; Taylor 2013), extreme environmental events such as snow storm, strong wind, earthquake etc., have affected railway track and OHLE, especially single mast structure (Fig. 1), which is very slender and flexible (Beagles et al. 2016). Hence, condition monitoring of railway track and its overhead line components is needed for maintenance planning (Ngamkhanong and Kaewunruen 2018). Besides, ground bourne vibration from train passage is also one of the serious concern as it affects property and cause annoyance to people in surrounding area (Suhairy 2000; Lopes et al. 2016; Connolly et al. 2016).

Fig. 1. Cantilever mast structure

Based on previous studies (Kouroussis et al. 2013; Zou et al. 2015; Mouzakis and Vogiatzis 2016; Zou et al. 2017; Vogiatzis and Mouzakis 2017), it can be seen that the effect of ground-bourne vibrations from train passages on building have been established. However, the responses of mast structures, which are located closer and along the railway track, have not been fully studied. Modal analysis has been used to study vibration mode of cantilever mast structure. Crossing phenomena was observed when support stiffness reduced as this structure has a sensitivity under vibrations (Ngamkhanong et al. 2017a). It was confirmed by previous studies that soil-structure interaction affected the overall response of the structure (Prum and Jiravacharadet 2012; NEHRP 2012).

Due to the increase of train speed, ground-bourne periodic forces may cause damage to the OHLE structure especially mast structure, connections and its foundation, which can lead to operational failure of train electrification. This study aims to present the integrated numerical study of three-dimensional cantilever mast structure considering soil-structure interaction to evaluate the condition of OHLE structures for maintenance planning. To monitor the OHLE condition, adaptation technique is used by monitoring the condition of single mast structure, which its response affects the overhead wire movement. At this stage, the movements of overhead contact wire are measured based on finite element modelling. It is noted that only finite element modelling is employedusing finite element package STRAND7 (G+D Computing

2001). Ground borne vibrations are computed by the classical formulation based on the semi-empirical model for predicting low frequency vibration on soft ground condition (Kurzeil 1979; Madshus et al. 1996). The frequencies of ground borne vibration between 0 Hz and 100 Hz are considered to cover all possible frequencies of ground vibration and the first-eight fundamental mode of mast vibration (Kouroussis et al. 2014; Kouroussis et al. 2015; Ngamkhanong and Kaewunruen 2018). This study presents the maintenance index which can be used for maintenance planning and inspection of support condition of mast and OHLE structures. Moreover, it should be noted that on-site measurement will be further conducted in the next step. The outcome of this study will help civil and track engineers to effectively and efficiently inspect OHLE structures using ground borne vibrations from train passages.

2 Methodology

2.1 Modelling

In this study, it is assumed that mast structure is placed 3.5 m far from railway track in perpendicular. Three-dimensional modelling of single mast structure is constructed using finite element package STRAND7 (G+D Computing 2001). The parametric study of the soil stiffness and train speed is conducted. It is assumed that translational stiffness (k_x, k_y) of support is fixed in all directions, while the rotational stiffness (k_{zz}) is varied from 1000 to 1000000 kNm/rad (fully fixed support condition). It is noted that the rotational stiffness is affected by the soil-structure interaction condition and the quality of support connection. The support stiffness can be decrease due to the connection failure such as broken bolt, yielding weld, improper design and construction etc. and soil erosion and degradation. The 2-D schematic load to structure with support stiffness is shown in Fig. 2. The typical H-section steel (Section area: 2.219×10^{-2} m^2, I_{zz}: 5.08×10^{-4} m^4, I_{xx}: 1.84×10^{-4} m^4) is used and connected to the cantilever which made of round steel to support the overhead contact wire. It should be noted that the interesting point is at the end of cantilever which is the location overhead contact wire. The steel used has the young modulus of 2×10^5 MPa, density of 7850 kg/m^3 and Poisson's ratio of 0.25.

2.2 Condition Monitoring

In general, the overhead contact wire forms in a zig-zag path (also called "stagger") above the track to avoid wearing a groove in the pantograph. Thus, the overhead wire displacement in transverse direction affects overhead line system which has limitation of sway movement. The allowable displacement is assumed as construction tolerances of contact wire. Hence, 50 mm construction tolerance of contact wire is used as the maximum displacement at the end of cantilever mast in transverse direction (RailCorp 2011). This study presents the adaptation technique for overhead line equipment monitoring. As the movement of mast structure can affect the movement of contact wire, the sensors should be placed in several positions in order to monitor the responses of mast structure affected by passing train and other environment impacts. However,

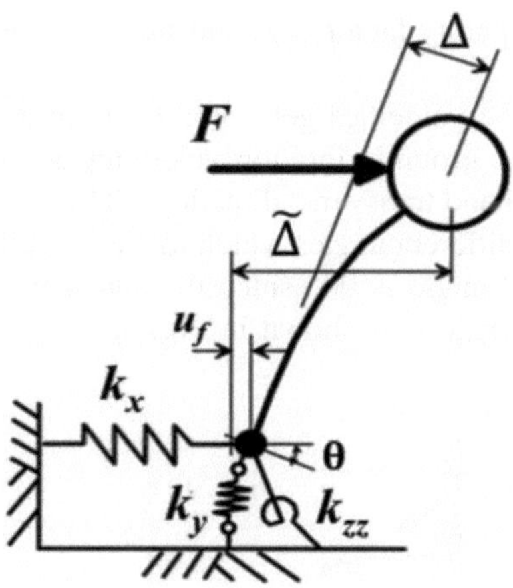

Fig. 2. Schematic load to structure with support stiffness

this study only demonstrates the finite element modelling instead of on site monitoring. It is recommended to monitor the OHLE condition on real site to compare with numerical results.

2.3 Ground Borne Vibration

It was noted that the fundamental modes of vibration of mast structure can be changed due to the reduction of support stiffness (Ngamkhanong et al. 2017b). Crossing phenomena was observed at the low frequency when the support stiffness was reduced. Hence, to cover all the fundamental modes of vibration, the frequency range between 1 and 100 Hz is considered. The train speed varies from 100 km/h to 300 km/h. When the train passes the mast structure, ground bourne vibrations are generated in both directions (longitudinal and transverse). The vibration velocities on the ground are based on several factors such as ground condition, train type and speed, distance from track to structure etc. The ground bourne velocity can be calculated using the formula by Madshus et al. (1996) as seen in Eq. 1

$$V = V_T F_S F_D F_R F_B \tag{1}$$

Where V_T is a train type specific vibration level, F_S is a speed factor, F_D is a distance factor, F_R is a track quality factor and F_B is a building amplification factor.

The vibration parameters are calculated based on the literature (Madshus et al. (1996)). The train type specific vibration level, (V_T) of 0.1 is used since it is assumed that the trains run along the track on soft ground. The speed factor (F_S) is calculated from the ratio between train speed (S) and the reference speed on standard track (S_0 = 70 km/h). The distance factor (F_D) can be calculated from the ratio between the centre of the track to the receiver and the reference distance from the centre of the tracks (20 m). It is assumed that the railway line considered is a single track and is assumed to be the same type of single story building based on the height of mast structure. Thus, the track quality factor (F_R) and building amplification factor (F_B) are 1.3 and 1.3.

This study considers the velocities generated by train at the different locations of running train on track. The ground vibration velocities are applied as harmonic excitations in both longitudinal and transverse directions. The vibration creation regions are formed on the track at the different angles which make the different vibration velocities. It is noted that the critical angle is 90° since this can generate the highest vibration velocities in transverse direction, as shown in Fig. 3.

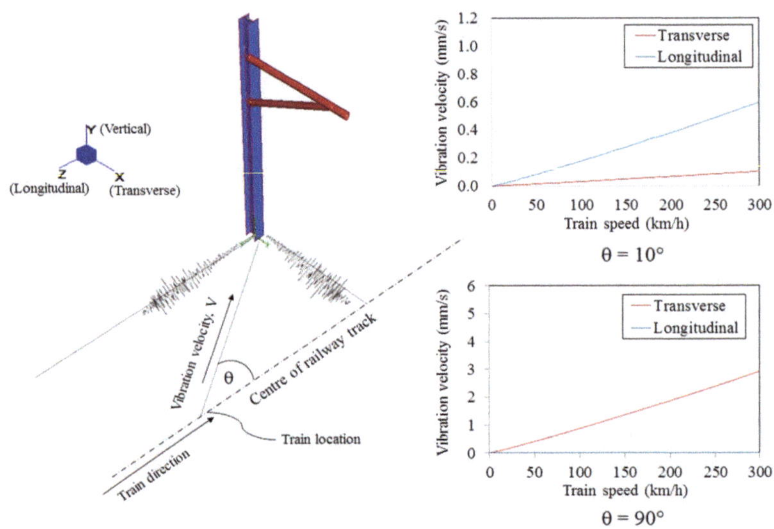

Fig. 3. Single mast structure modelling and ground vibration velocities from train passages

3 Results and Discussions

3.1 Maximum Displacement

The maximum displacements at the end of the cantilever, which is the location of overhead wire, in both directions are presented, as shown in Fig. 4. It should be noted that the direction concern is transverse as this is a sway direction of overhead wire.

At fully fixed support condition (Fig. 4a), it is clearly seen that the train speed plays a little role on transverse direction when angles of the train to the mast are in low range (less than 30°) but plays a significant role at higher angles. For 70–90 degrees, it should be noted that the maximum displacement in transverse direction increases nearly double and triple from the speed of 100 km/h to 200 km/h and 300 km/h, respectively. This can also be observed in case of lower support stiffness (Fig. 4b). It is clear that the transverse displacement is higher than longitudinal. About 42 mm is observed as a highest transverse displacement.

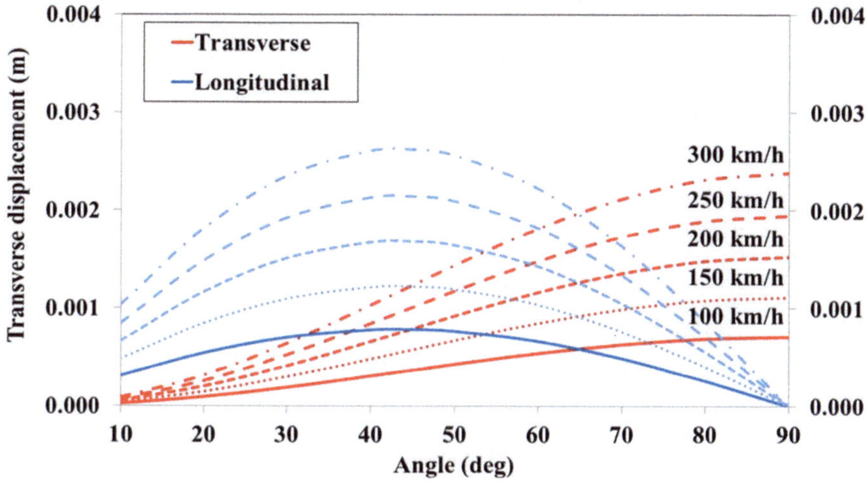

(a)

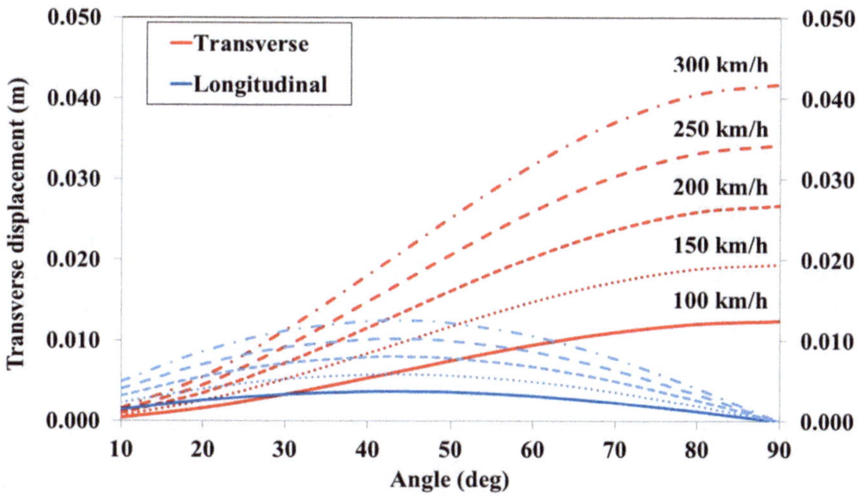

(b)

Fig. 4. Maximum displacements at the position of overhead contact wire on cantilever mast with support stiffness of (a) 1000000 kNm/rad (Fully fixed) (b) 1000 kNm/rad

3.2 Maintenance Index

The ratio between the overhead contact wire displacement and allowable displacement (50 mm) is indicated as maintenance index. It is assumed and noted that the allowable displacement used is the construction tolerances of contact stagger according to RailCorp (2011). As seen from Fig. 5, it is shown that mast structure with support stiffness of 1000 kNm/rad under the ground bourne vibration generated by 300 km/h train speed has the highest maintenance index (0.83). Even thought, all cases are under maintenance level, with support stiffness of 1000 kNm/rad, the maintenance index tends to reach the maintenance level. Thus, by linear extrapolation, the contact wire on single mast with support stiffness of 1000 kNm//rad needs to be under maintenance as the maintenance index equal to 1 at the train speed of 361 km/h.

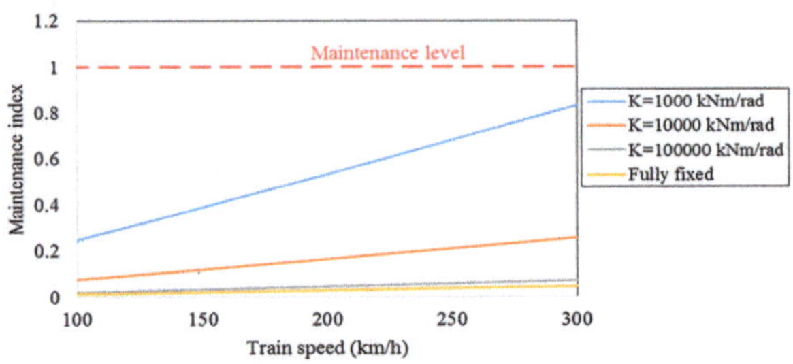

Fig. 5. Maintenance index

4 Conclusions

The overhead line equipment (OHLE), which supplies the electric power to the train, is one of the most vulnerable asset of the railway infrastructure. The electric system can be failed due to the large movement of overhead wire under harsh environment. The end of cantilever is a position concern as it holds the overhead contact wire. This study considers the variation of support stiffness of mast structure and vehicle speed from 100 to 300 km/h. It should be noted that support stiffness can be reduced by many reasons such as broken bolt, yielding weld, soil erosion and degradation. Thus the soil-structure connection should be carefully constructed. It is interesting that OHLE can be reached the maintenance level under the train speed of about 360 km/h in case of mast placed on poor support. The obtained results can be also used to detect the support condition and maintenance index for OHLE on real track site. The outcome of this study will help a better understanding of the critical responses and dynamic behaviour of cantilever mast structure and its support subjected to ground bourn vibrations in order to improve the further design standard of this structure. The finding of this study can also help improve inspection of OHLE structure. However, this is the first to study and monitor the condition of OHLE structure subjected to ground bourne vibrations from train passages, the onsite monitoring and measurement will be further conducted in order to verify the obtained results.

Acknowledgments. The authors are sincerely grateful to the European Commission for the financial sponsorship of the H2020-RISE Project No. 691135 "RISEN: Rail Infrastructure Systems Engineering Network", which enables a global research network that tackles the grand challenge of railway infrastructure resilience and advanced sensing in extreme environments (www.risen2rail.eu) (Kaewunruen et al. 2016).

References

Beagles, A., Fletcher, D., Peffers, M., Mak, P., Lowe, C.: Validation of a new model for railway overhead line dynamics. Proc. Inst. Civ. Eng. **169**, 339–349 (2016)

Connolly, D.P., Marecki, G.P., Kouroussis, G., Thalassinakis, I., Woodward, P.K.: The growth of railway ground vibration problems—a review. Sci. Total Environ. **568**, 1276–1282 (2016)

G+D Computing: Using Strand7: Introduction to the Strand7 finite element analysis system, Sydney, Australia (2001)

Kaewunruen, S., Sussman, J.M., Matsumoto, A.: Grand challenges in transportation and transit systems. Front. Built. Environ. **2**, 4 (2016)

Kouroussis, G., Van Parys, L., Conti, C., Verlinden, O.: Prediction of ground vibrations induced by urban railway traffic: an analysis of the coupling assumptions between vehicle, track, soil, and buildings. Int. J. Acoust. Vib. **18**(4), 163–172 (2013)

Kouroussis, G., Connolly, D.P., Verlinden, O.: Railway-induced ground vibrations—a review of vehicle effects. Int. J. Rail Transp. **2**(2), 69–110 (2014)

Kouroussis, G., Connolly, D.P., Alexandrou, G., Vogiatzis, K.: Railway ground vibrations induced by wheel and rail singular defects. Veh. Syst. Dyn. **53**(10), 1500–1519 (2015)

Kurzeil, L.G.: Ground borne noise and vibration from underground rail systems. J. Sound Vib. **66**, 363–371 (1979)

Lopes, P., Ruiz, J.F., Costa, P.A., Rodríguez, L.M., Cardoso, A.S.: Vibrations inside buildings due to subway railway traffic. Experimental validation of a comprehensive prediction model. Sci. Total Environ. **568**, 1333–1343 (2016)

Madshus, C., Bessason, B., Harvik, L.: Prediction model for low frequency vibration from high speed railways on soft ground. J. Sound Vib. **193**(1), 195–203 (1996)

Mouzakis, H., Vogiatzis, K.: Ground-borne noise and vibration transmitted from subway networks to a typical Athenian multi-storey reinforced concrete building. In: 23rd International Conference—ICSV 2016 (2016)

NEHRP Consultants Joint Venture: Soil-Structure Interaction for Building Structures (National Institute of Standards and Technology) (2012)

Ngamkhanong, C., Kaewunruen, S., Baniotopoulos, C., Papaelias, M.: Crossing phenomena in overhead line equipment (OHLE) structure in 3D space considering soil-structure interaction. In: IOP Conference Series: Materials Science and Engineering, vol. 245 (2017a)

Ngamkhanong, C., Kaewunruen, S., Baniotopoulos, C.: A review on modelling and monitoring of railway ballast. Struct. Monit. Maint., **4**(3), 195–220 (2017b). https://doi.org/10.12989/smm.2017.4.3.195

Ngamkhanong, C., Kaewunruen, S.: The effect of ground borne vibrations from high speed train on overhead line equipment (OHLE) structure considering soil-structure interaction. Sci. Total Environ. **627**, 934–941 (2018). https://doi.org/10.1016/j.scitotenv.2018.01.298

Prum, S., Jiravacharadet, M.: Effects of soil structure interaction on seismic response of buildings. In: International Conference on Advances in Civil Engineering for Sustainable Development (2012)

RailCorp: Design of Overhead Wiring Structures and Signal Gantries. Engineering Manual–Civil (2011)

Robinson, P., Bryan, C.: Network Rail Electrical Power Reliability Study. Network Rail, Milton Keynes (2009)

Shing, A.W.C., Wong, P.P.L.: Wear of pantograph collector strips. Proc. ImechE J. Rail Rapid Transit. **222**(2), 169–176 (2008)

Suhairy, S.A.: Prediction of Ground Vibration from Railways. SP Report, Boras (2000)

Taylor, G.: A bad wire day. The Rail Engineer (2013)

Vogiatzis, K., Mouzakis, H.: Ground-borne noise and vibration transmitted from subway networks to multi-storey reinforced concrete buildings. Transport (2017). https://doi.org/10.3846/16484142.2017.1347895

Zou, C., Wang, Y., Wang, P., Guo, J.: Measurement of ground and nearby building vibration and noise induced by trains in a metro depot. Sci. Total Environ. **536**, 761–773 (2015)

Zou, C., Wang, Y., Moore, J.A., Sanayei, M.: Train-induced field vibration measurements of ground and over-track buildings. Sci. Total Environ. **575**, 1339–1351 (2017)

Application of Nano-Silica in Concrete Bridges in Vietnam for Sustainable Development

Le Hong Lam[1] and Dao Duy Lam[2(✉)]

[1] Ho Chi Minh City University of Transport, Ho Chi Minh City, Vietnam
[2] University of Transport and Communications, Ho Chi Minh City, Vietnam
daoduylam@utc.edu.vn

Abstract. Concrete bridges are very popular in Vietnam. However, under severe environments of the tropical climate, they can be lead to defects or deterioration with excessive rate. This point demands on materials performance with respect to extremes in not only stress and strain but also the properties of cementitious materials in practical design. In addition, cracking, shrinkage and creep of concrete due to the rapid ambient conditions of temperature and humidity are also a major challenge to researchers in Vietnam. Therefore, improving the sustainability of bridges by applying high performance concrete is very necessary. Recent developments in nanotechnology have given a lot of new challenges to produce the eco-friendly high performance concrete. It has been found out that use of nanotechnology through using nano-silica as a mineral admixture to modify the properties of the cementitious materials has significantly improved the performance of concrete. This paper investigates the effect of nano-silica as an addition on new concrete generation, the advancements in nanotechnology on the properties and performance of cementitious materials; and the application to concrete bridges in Vietnam in near future for sustainable development purpose.

Keywords: Nanotechnology · High performance concrete · Nano-silica
Concrete bridge

1 Introduction

Transport infrastructure system of Vietnam has developed strongly for recent 20 years. For bridge structure, besides common simple spans, continuous spans or concrete filled steel tubes, cable-stayed bridges have been widely used in modern bridge structures. In general, the concrete bridge structure is the most popular. Moreover, the increasing specified requirements for concrete such as: high compressive strength, waterproof, anti-acid corrosion and others requirements such as the Self Consolidating Concrete (SCC) of concrete filled steel tubes are also important and it is necessary to study the high performance concrete. This paper presents the effect of nano-silica on high performance concrete and its possibility for application on concrete bridges in Vietnam.

H. Rodrigues and A. Elnashai (Eds.): GeoMEast 2018, SUCI, pp. 23–31, 2019.
https://doi.org/10.1007/978-3-030-01932-7_3

2 Concrete Bridge Damages in Vietnam

Vietnam is an open sea country in tropical zone with rough conditions such heavy annual rainfall, high humidity and strong sea-wind. These factors contribute to the enhancement of corrosion phenomena occurring strongly in construction and transport projects. In reality, so many concrete bridges are corroded in Vietnam. Not only spans but also abutments are in corrosion.

Over time, the concrete surface gradually converts to calcium carbonate due to the effects of carbon dioxide. This is a swelling phenomenon that breaks the protective concrete and reduces the strength of the concrete.

When water is absorbed, calcium in concrete dissolves into water and exits. This phenomenon is calcium leaching. If carbon dioxide is present, the calcium will lose more and the concrete will decline more rapidly.

In strong inorganic acids such as sulfuric acid, nitric acid, the strength of the concrete is degraded by acid corrosion Fig. 1.

Fig. 1. Damage of Van Kiep bridge – Phu Yen

In addition, during construction and operation of concrete bridges in Vietnam, there are some problems with concrete. According to statistics, most of the bridges generate the crevices in different parts including the components of a pre-stressed concrete beams. It is clearly that the crevices affect the reliability and longevity of this project in the next few decades.

During Super-T guttering, cracking caused by shrinkage of concrete and early-age strength of concrete are not sufficient to withstand the local force of the beam Figs. 2, 3.

Damaged concrete usually encountered during the mining process, may be due to the strength of concrete, manufacturing technology, durability is not guaranteed Figs. 4, 5.

In Vietnam, the unexpected cracks which were observed in most of concrete bridges have created favorable conditions for harmful air and water infiltrating into the structure. It leads to the fact that the structure will be eroded with time resulting in rapid degradation process, declining life expectancy and adversely affecting the quality of exploitation. And, it causes costly maintenance, repairs, reducing the efficiency of

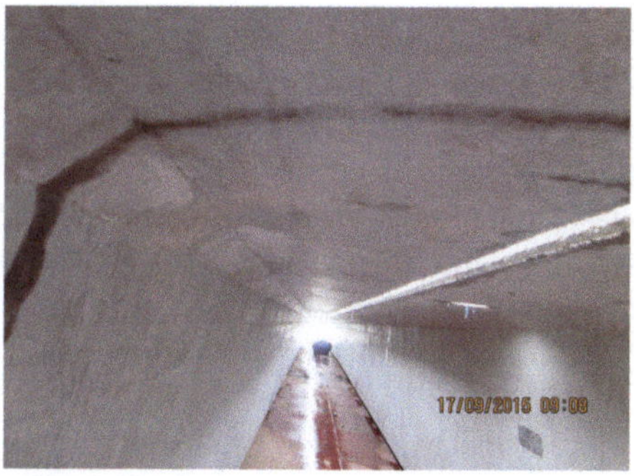

Fig. 2. Shrinkage cracks of concrete beam

Fig. 3. Cracks on Super T beam

Fig. 4. Damaged expansion joints of Lang Um bridge –Thai Nguyen

Fig. 5. Crack in box girder of Nguyen Tri Phuong bridge-Ho Chi Minh City

construction investment. Expenses for the repair, overhaul and maintenance or even replacement of parts of this structure consequently becomes important, accounting for a large proportion of operating costs of repair works after putting into operation.

Ultra - high performance concrete (UHPC) should be the first selection, since with UHPC, we can get high compressive strength, as well as reduce maintenance and repair costs. Among of many kinds of UHPC, nano-silica concrete is currently being studied and used due to its full characteristics such as high strength, early-age strength, waterproofing, anti-acid corrosion, segregation of concrete mix… to overcome the disadvantages of ordinary concrete.

3 Nanotechnology and Nano-Silica Concrete

Ultra-high performance concrete (UHPC) is one of the most promising types of concrete, which has been developed in the last decade. The efficiency of UHPC is particularly dependent on its density. This can be maximized by optimizing the particle packing, thus resulting in ultra-high consolidation of the concrete matrix. The optimized particle packing can be obtained through an almost 'perfect' grain size distribution, by incorporating a homogeneous gradient of fine and coarse particles in the mixture. In this scope, the use of nanomaterials, as pozzolanic additions, is highly effective. In fact, due to their extremely small size, nanomaterials can fill the voids between cement and silica fume particles, leading to higher packing level and also generating a denser binding matrix, with more calcium silicate hydrate (C-S-H). Thus, this causes a significant improvement on both durability and mechanical properties of concrete [1].

Today a wide variety of nanoparticles of different materials, sizes or shapes can be produced in large quantities. Therefore these particles get more and more in focus for the use in construction materials, especially those who have a similar chemical composition as materials that are already common in construction. Nanoparticles made of Al_2O_3, Fe_2O_3, TiO_2 or ZrO_2 were studied [2–6]. Nano-engineering encompasses the techniques of manipulation of the structure at the nanometer scale to develop multifunctional, cementitious composites with superior mechanical performance and durability potentially having a range of novel properties such as: self-sensing capabilities, self-cleaning, self-healing, high ductility, and self-control of cracks [7]. Nano-silica was one of the first nanomaterial that was used in cementitious systems and a lot of methods for the synthesis of nano-silica are known. Vietnam scientists have produced nano-silica by sol-gel or precipitation method, in the scope of the article only refers to the nano-silica concrete to facilitate the application of Vietnam.

Nanoconcrete made with portland cement in which particle sizes ranging from a few nanometer to a maximum of about 100 micrometers and in nano ingredients with at least one dimension of nano meter size. The various constituents of nanocement are as shown in Fig. 6. So the particle size has to be reduced in order to obtain nano-portland cement. Further, if these nano-cement particles are processed with nanotubes and reactive nano-size silica particles, then strong, tough, more flexible, cement can be developed with enhanced properties Fig. 6.

The Nano-silica in concrete or mortar will increase the density, reduces porosity, and improves the bond between cement matrix and aggregates [9–12]. Nano-silica can effectively reduce the amount of portlandite, leading to a denser microstructure of interfacial transition zone (ITZ), between aggregate and paste [7].

Add nano to the concrete, there may be two possible reaction mechanisms during hydration of cement in presence of nano-silica. Hydration of cement can be accelerated by addition of nano-silica. $H_2SiO_4^{2-}$ forms and reacts with the available Ca^{2+} which forms an additional calcium–silicate–hydrate (C–S–H) and these C–S–H particles are spread in the water between the cement particles and serve as seeds for the formation of more compact C–S–H phase.

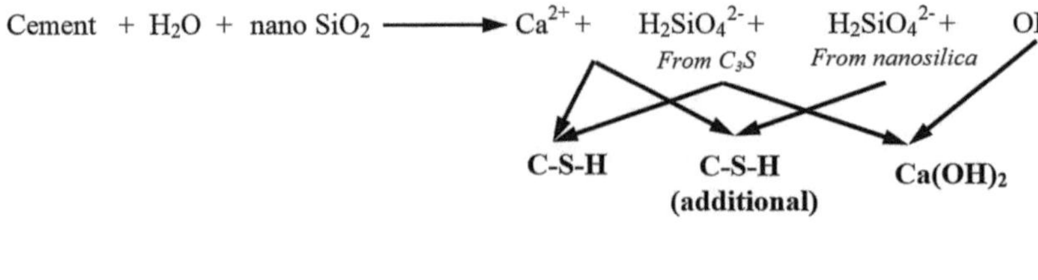

(1)

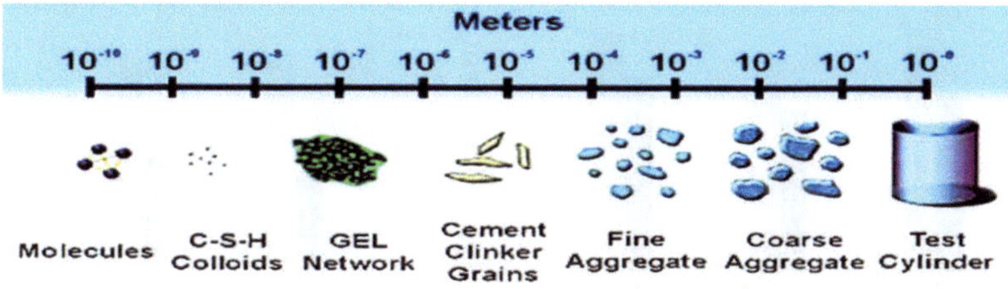

Fig. 6. Scales of various constituents of concrete [8]

The formation of the C–S–H phase is no longer limited on the grain surface alone, as in the pure C3S, but also takes place in the pore space. This process is shown in Eq. (1). The formation of large numbers of seeds thereby causes an acceleration of early cement hydration [13].

Further, the pozzolanic reaction of nano-silica with $Ca(OH)_2$, which is formed during the hydration of cement, produces additional C–S–H which is the main constituent for strength and density in the hardened cementitious system. At the same time $Ca(OH)_2$, which hardly contributes to strength development, is consumed [13]:

$$\begin{aligned} \text{nano-}SiO_2 + H_2O &\rightarrow H_2SiO_4^{2-} \\ Ca(OH)_2 + H_2O &\rightarrow Ca^{2+} + OH^- \\ H_2SiO_4^{2-} + Ca^{2+} &\rightarrow \text{C-S-H(Additional C-S-H)} \end{aligned} \tag{2}$$

Basing on the above nanotechnology, the case studies show high quality concrete according to practical requirements of Vietnam bridge.

Higher compressive strength, flexural strength and early-age compressive strength:

- By adding nano-silica in the corresponding mixtures and accelerates setting time and compressive strength of concrete. Nano silica is used for the preparation of rock-matching grouting, gypsum particle board, additives in tile, workability increaser, anti-bleeder etc. Li et al. [14] demonstrated the effect of addition of nano-silica in high volume fly ash concrete and the results are compared with control concrete.
- Ehsan Ghafari et al. [1] were used to prepare the UHPC mixtures: ordinary Portland cement type I:52.5 R, silica fume (SF), quartz flour (P600), used as a micro filler (particle sizes less than 10 μm), siliceous sand, with maximum aggregate size of 0.6 mm, polycarboxylate ether based superplasticizers (SP), and two types of nano-silica (One type of nano-silica particle was coated with a thin layer of carbon (NSC), which enables the particle to be dispersed easier in the mixture, while the other type was non-coated (NS)). The average values of compressive strength at 7 and 28 days are shown, for all the specimens, in Fig. 7. It can be seen that the addition of nano-silica significantly increased the early-age compressive strength. The higher compressive strength might be due to larger specific surface area of nanoparticles, which can react more quickly with $Ca(OH)_2$, making the microstructure denser. At the age of 28 days, the highest compressive strength was

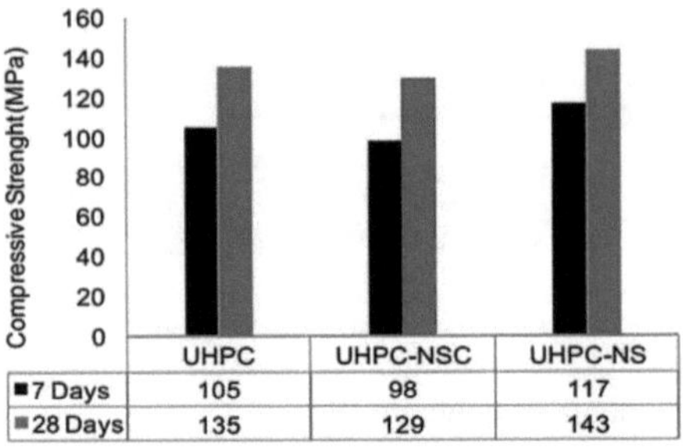

Fig. 7. Compressive strength of UHPC

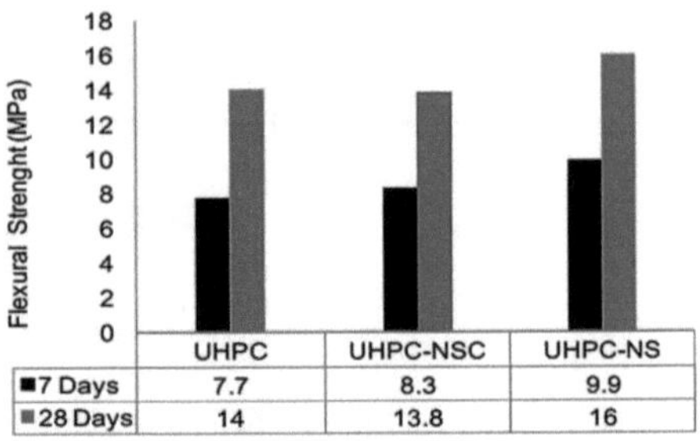

Fig. 8. Flexural strength of UHPC

obtained for the UHPC-NS. The effect of nano-silica coated with carbon was not as effective as not-coated silica, which may result from the fact that this coating reduces the activity of nanoparticle. From Fig. 8. It can be observed that the flexural strength was enhanced by incorporating nano-silica, at the ages of 7 and 28 days.

Lower permeability:

- Shi et al. [15] studied the chloride permeability and microstructure of Portland cement mortar with four different types of nanomaterials. This study confirmed that an addition of nano-silica and nano-clay significantly improved the chloride penetration resistance as well as the general ionic permeability of cementitious mortar. An experimental study performed by Ji [16] showed that the addition of nano-silica to the mixture improved the water permeability resistance of concrete.
- NS and NSC concrete [7] ere found to be effective in reducing the volume of permeable voids as well as water absorption. It can be seen that an addition of nano-silica (NS) reduced VPV in 21% and water absorption in 36%. Table 1 also presents the results concerning the sorptivity coefficient and the porosity measured by vacuum saturation. The UHPC-NS specimen gave the lowest value of porosity (2.12%). This can be due to the discontinuity of capillary pores by formation of more C-S-H gel (Table 1).

Table 1. Sorptivity coefficient and porosity of UHPC mixtures

	Water absorption		Sorptivity Test	Vacuum	MIP Test	
	Absorption*	VPV*	Sorptivity Coef.**	Porosity*	Total porosity*	Capillary Pores*
UHPC	1.212	2.91	0.044	3.11	6.35	2.65
UHPC-NSC	1.11	2.75	0.041	2.94	4.74	2.56
UHPC-NS	0.808	2.28	0.038	2.12	4.3	1.72

*Percent (%); **In unit $Kg/m^2/h^{0.5}$

Reduced calcium leaching: Gaiteroa et al. [17] studied and proved the effect of silica nano particles on the reduction of calcium leaching rate of cement paste and it was concluded that addition of nano-silica can control the C-S-H degradation due to calcium leaching and can increase the overall strength of cement-based materials at every stage.

Anti-acid corrosion: Nanoconcrete is lesser the porosity, lesser will be the chances of acid attack on concrete. In general it is said that all high performance concrete mixtures show a better resistance against acid attack than the reference ordinary type concrete [18].

Self-compacting concrete (SCC), self-consolidating mortars (SCM): The influence of Nano silica and silica fume on the workability and stability of fresh self-consolidating mortars (SCM). Nano-silica and silica fume cause to increase rheological and viscosity properties in mortars. That can have rheological and mechanical synergistic effects and can better increase mechanical performance while maintaining adequate flow properties for self-compacting mortar [19, 20].

4 Recommendation for Nano-Silica Application in Vietnam

In Vietnam, sustainable development is one of the fundamental elements of economic growth model reform and a primary goal of the national economy in which infrastructure should be recognized as the conductor of socio-economic development, the conduit for distributing production forces and accelerating the circulation of goods.

Nano-silica can be a good solution for the new eco-concrete material in infrastructure construction. Vietnam has produced silica nano-particles from rice husk ash by precipitation method [21]. In recent years, together with the experience of application of nano-silica mortar in repairing expansion joint [22] and the results of other studies on nano-silica concrete in the world, there are enough theory and basic practical knowledge for the study and the application of nano-silica technology for concrete bridges in Vietnam.

With the current conditions of Vietnam, more research should be done:

Studying the main characteristics of nano-silica by using modern techniques like electronic microscope techniques (SEM, TEM and others). These characteristics will give recommendations in the manufacture of nano-silica concrete;

Finding the kind of nano-silica suitable to produce nano-silica concrete in Vietnam with high quality and reasonable cost;

Optimal design and testing of nano-silica mortar and concrete structure;

Finally, determining the methods of design and fabrication of nano-silica concrete; optimizing the application of nano-silica for development of new concrete generation with available materials from Vietnam for sustainable development goal.

5 Conclusions

As discussed above, nano-concrete overcomes the disadvantages of ordinary concrete. Moreover, besides the advantages including improvement of compression strength, flexural strength and waterproofing, nano-silica concrete can be considered as eco-concrete. This is due to the face that nano-silica grain made from rice husk ash replaces cement content and reduces CO_2 emissions in the process cement production. Thus, it is possible to propose research and development of nano-silica technology in concrete structures in general, and concrete bridges in particular.

Nano-silica which is produced successful in Vietnam is shown as a good solution for the new eco-concrete material in infrastructure construction for sustainable development however further research shall be done to ensure the possibility of application of nano-silica technology for concrete bridges in Vietnam.

References

1. Ghafari, E., Costa, H., Júlio, E., Portugal, A., Durães, L.: Optimization of UHPC by adding nanomaterials. In: 3rd International Symposium on UHPC and Nanotechnology for High Performance Construction Materials, Kassel, 7–9 March 2012
2. Jayapalan, A., Lee, B., Kurtis, K.: Effect of nanosized titanium dioxide on early age hydration of Portland cement. In: Nanotechnology in Construction 3: Proceedings of the NICOM3; [with CD-ROM], pp. 267–273. Springer, Berlin (2009)
3. Nazari, A., Riahi, S., Riahi, S., Shamekhi, S.F., Khademno, A.: Mechanical properties of cement mortar with Al_2O_3 nanoparticles. J. Am. Sci. **6**(4), 94–97 (2010)
4. Nazari, A., Riahi, S., Riahi, S., Shamekhi, S.F., Khademno, A.: The effects of incorporation Fe_2O_3 nanoparticles on tensile and flexural strength of concrete. J. Am. Sci. **6**(4), 90–93 (2010)
5. Nazari, A., Riahi, S., Riahi, S., Shamekhi, S.F., Khademno, A.: An investigation on the strength and workability of cement based concrete performance by using ZrO_2 nanoparticles. J. Am. Sci. **6**(4), 29–33 (2010)
6. Campillo, I., Guerrero, A., Dolado, J.S., Porro, A., Ibanez, J., Goni, S.: Improvement of initial mechanical strength by nanoalumina in belite cements. Mater. Lett. **61**(8–9), 1889–1892 (2007)
7. Srinivas, K.: Nanomaterials for concrete technology. Nanotechnology Group, Department of Physics, GMR Institute of Technology, Rajam, Andhra Pradesh, India (2014)
8. Chong, K.P.: Smart structures research in the U.S. Keynote paper. In: Proceedings of the NATO Advanced Research Workshop on Smart Structures, held in Pultusk, Poland, Smart Structures, pp. 37–44. Kluwer Academic (1998)
9. Qing, Y., Zenan, Z., Deyu, K., Rongshen, Ch.: Influence of nano-SiO_2 addition on properties of hardened cement paste as compared with silica fume. Constr. Build. Mater. **21**, 539–545 (2007)
10. Lin, K.L., Chang, W.C., Lin, D.F., Luo, H.L., Tsai, M.C.: Effects of nano-SiO_2 and different ash particle sizes on sludge ash–cement mortar. J. Environ. Manage. **88**, 708–714 (2008)
11. Senff, L., Hotza, D., Repette, W.L., Ferreira, V.M., Labrincha, J.A.: Mortars with nano-SiO_2 and micro-SiO_2 investigated by experimental design. Constr. Build. Mater. (2010). https://doi.org/10.1016/j.conbuildmat.2010.01.012

12. Lin, D.F., Lin, K.L., Chang, W.C., Luo, H.L., Cai, M.Q.: Improvements of nano SiO_2 on sludge/fly ash mortar. Waste Manag. **28**, 1081–1087 (2008)
13. Singh, L.P., Karade, S.R., Bhattacharyya, S.K., Yousuf, M.M., Ahalawat, S.: Beneficial role of nano-silica in cement based materials – a review. Constr. Build. Mater. **47**, 1069–1077 (2013)
14. Li, G.: Properties of high-volume fly ash concrete incorporating nano-SiO_2. Cem. Concr. Res. **34**, 1043–1049 (2004)
15. Shi, X.H.A.X.: Chloride permeability and microstructure of Portland cement mortars incorporating nanomaterials. J. Transp. Res. Board, **2070**(1), 13–21 (2008)
16. Ji, T.: Preliminary study on the water permeability and microstructure of concrete incorporating nano-SiO_2. Cem. Concr. Res. **35**(10), 1943–1947 (2005)
17. Gaitero, J.J., Campillo, I., Guerrero, A.: Reduction of the calcium leaching rate of cement paste by addition of silica nanoparticles. Cem. Concr. Res. **38**, 1112–1118 (2008)
18. Khitab, A., Arshad, M.T., Awan, F.M., Khan, I.: Development of an acid resistant concrete: a review. Int. J. Sustain. Constr. Eng. Technol. **4**(2), 33–38 (2013)
19. Ramezanianpour, A.A., Mahdikhani, M., Moghaddam, S.S., Nikravan, M., Mousavi, S.R.: Mechanical properties and durability of self consolidating mortars containing nano SiO_2. In: 3rd International Symposium on UHPC and Nanotechnology for High Performance Construction Materials, Kassel, 7–9 March 2012
20. Hjiaj, M., Dao, D.L., de Saxcé, G.: A family of bi-potentials describing the non-associated flow rule of pressure-dependent plastic models. Acta Mech. **220**, 237–246 (2011)
21. Nguyen, T.T., et al.: Synthesis nano-silica from rice husk ash by using precipitation method. Sci. J. Univ. Can Tho **8**, 58 (2014)
22. Vu, B.T., Thanh, B.T., Lam, N.X., Long, N.N., Anh, L.B.: Using nano and dispersed fiber reinforced cement mortar to repair expansion joints in bridges. Sci. J. Transp. Commun. (2016)

FRP Tubes Filled with Reinforced Concrete Subjected to Cyclic Axial Loading

Asmaa Abdeldaim Ahmed[1(✉)], Mohamed Hassan[1,2], and Radhouane Masmoudi[1]

[1] Department of Civil Engineering, University of Sherbrooke, Sherbrooke, QC, Canada
Asmaa.Ahmed@USherbrooke.ca
[2] Helwan University, Cairo, Egypt

Abstract. Infrastructure safety, durability, and serviceability have been significantly enhanced as a result of improved construction materials over the past two decades. Concrete-filled fibre-reinforced polymer (FRP) tubes (CFFTs) system is one of the most promising technique to protect the reinforced concrete structures from aggressive environmental conditions. Most of the experimental investigations performed on CFFT columns have focused on short, unreinforced, small-scale concrete cylinders, tested under monotonic axial loading. In contrast, only few studies have so far investigated the effects of the internal longitudinal reinforcement type (steel or FRP bars) on the behavior of CFFT long columns. This paper presents preliminary test results of an experimental study on the behavior of concrete-filled FRP tube (CFFT) columns internally reinforced with steel and FRP bars. Six reinforced concrete (RC) and CFFT columns were constructed and tested until failure. The test parameters were: (1) internal reinforcement type (steel, glass FRP (GFRP), and amount, and (2) GFRP tube thicknesses. All columns had 1900-mm in height and 213-mm in diameter. Examination of the test results has led to a number of significant conclusions in regards to the trend and ultimate condition of the axial stress-strain behavior and mode of failure of tested CFFT columns. These results are presented, and a discussion is provided on the influence of the main test parameters in the observed behaviors.

1 Introduction

An important application of fibre-reinforced polymer (FRP) composites is as a confining material for concrete, both in the seismic retrofit of existing reinforced concrete (RC) columns and in the construction of concrete-filled FRP tubes (CFFTs) as earthquake-resistant columns in new construction. The FRP tube acts as a stay-in-place structural formwork, a noncorrosive reinforcement for the concrete for flexure and shear using the multidirectional fiber orientation, provides confinement to the concrete in compression, and the contained concrete is protected from intrusion of moisture with corrosive agents that could otherwise deteriorate the concrete core (ACI 440.2R-08).

H. Rodrigues and A. Elnashai (Eds.): GeoMEast 2018, SUCI, pp. 32–44, 2019.
https://doi.org/10.1007/978-3-030-01932-7_4

To date, most of the experimental investigations performed on FRP confined concrete columns have focused on short, unreinforced, small-scale concrete cylinders, tested under concentric and monotonic axial loading (Mirmiran et al. 2001; Fam et al. 2003; Lam and Teng 2009). In contrast, only few studies have so far investigated the effects of the slenderness ratio and internal longitudinal reinforcement type (steel or FRP bars) on the behavior of FRP confined concrete long columns (Mirmiran et al. 2001; Mohamed et al. 2010; Fitzwilliam and Bisby 2010; Masmoudi 2011; Yuan and Mirmiran 2001). Mirmiran et al. (2001) carried out a comprehensive parametric study on the buckling of over 11 500 CFFT columns. They found that instability of CFFT columns might occur at a lower slenderness ratio than that of ordinary RC columns (without FRP tubes); however, the ultimate capacity of the former might be higher than that of the latter. This attributed to the bilinear stress-strain behavior of the CFFT columns in which the buckling mode of failure initiated at the plastic branch of the curve, which was characterized by a lower Young's modulus. They also recommended that the current slenderness limit of 22 for steel-reinforced concrete columns bent in single curvature be reduced to 11 for CFFT columns. Masmoudi and Mohamed (2011) conducted an experimental investigation on the axial behavior of CFFT columns internally reinforced with steel or carbon FRP (CFRP) bars with different slenderness ratios ranging from 4 to 20. The test results showed that the CFFT columns reinforced with CFRP bars behaved similar to that of CFFT columns reinforced with steel bars. The axial capacity of steel or CFRP-reinforced CFFT decreased as the slenderness ratios increased. This can draw the conclusion that the increase of the slenderness ratio of CFFT columns reinforced internally with steel or CFRP bars might be a critical factor that controls the mode of failure and might prevent such columns from attaining their ultimate load capacity.

To the knowledge of the candidate, no investigations have addressed the behavior of FRP-reinforced CFFT columns under axial cyclic compression loading. To address such knowledge gaps and properly understanding the general behavior of FRP-reinforced CFFT columns under axial cyclic loading more experimental studies are needed. This paper presents the test results of an experimental study aimed at investigating the behavior of CFFT columns reinforced with longitudinal steel or FRP bars tested under axial cyclic compression loading. A total of six RC and CFFT columns were constructed and tested until failure. All columns had 1900-mm in height and 213-mm in diameter. The effect of internal reinforcement type and amount, and GFRP tube thicknesses were addressed.

2 Experimental Work

2.1 Materials

Four materials were used in fabricating the test specimens. These materials are concrete, FRP tubes, steel reinforcing (bars and stirrups), and FRP bars. The following sections provide a description of the different experimental tests conducted to evaluate the mechanical properties of the different materials used herein.

2.1.1 Concrete

All columns were constructed using a ready-mixed normal strength concrete (NSC) with an entrained-air ratio of 5% to 8%. The actual concrete compressive strength was determined from testing six concrete cylinders (150 × 300 mm) on the same day of testing the columns. Figure 1 shows typical axial stress-strain curves for the concrete cylinders. The average concrete compressive strength and tensile strength were 41.5 MPa and 4.2 MPa, respectively.

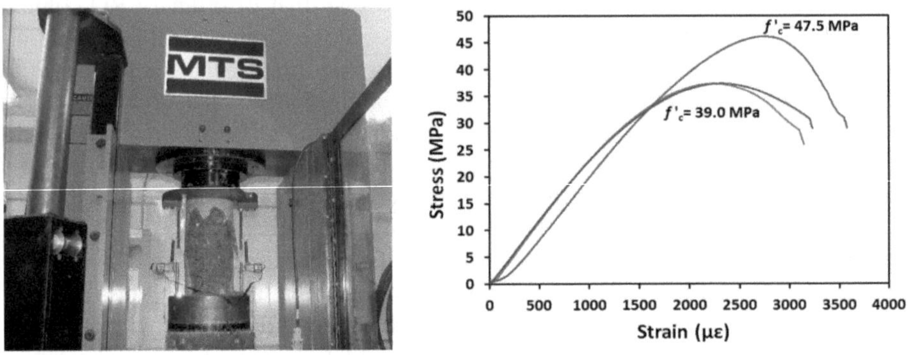

Fig. 1. Typical axial stress-strain relationships for concrete cylinders

2.1.2 Steel and FRP Bars

Two different steel bars were used to reinforce the control and CFFT specimens. Wire mild steel bars 3.4 mm in-diameter were served as transverse spiral reinforcement for the control specimens. Deformed steel bars M15 (16 mm in diameter; 200 mm^2 in cross-sectional area); were used as a longitudinal reinforcement for test specimens. The mechanical properties of the steel bars obtained from standard tests that were carried out according to ASTM 2009, on five specimens for each type of the steel bars. The mechanical properties of the steel bars are presented in Table 1. Two types of sand-coated FRP bars manufactured by a Canadian company (Pultrall, Inc. 2007) were used as longitudinal reinforcement for the tested columns. Sand-coated surface was made to improve the bond between the bars and surrounding concrete. The two types of FRP bars were made from continuous glass fibres with a fibre content of 73% and impregnated in a vinyl ester resin through the pultrusion process. GFRP bars No. 3 and No. 5 (9.5 mm and 15.9 mm in-diameter; 71 mm^2 and 199 mm^2 in cross-sectional area, respectively) were used. Table 1 shows the mechanical properties of the FRP bars as provided by the manufacture. Figure 2 shows the different bars used in this investigation.

2.1.3 FRP Tubes

Two types of GFRP tubes were used as structural stay in-place formwork for the tested specimens herein. The GFRP tubes were fabricated using filament-winding technique; E-glass fiber and Epoxy resin were used for manufacturing these tubes. The two types

Table 1. Tensile properties of the GFRP, and steel bars

Reinforcement type	Nominal diameter (mm)	Nominal area (mm^2)	Tensile modulus of elasticity (GPa)	Yield strength (MPa)	Ultimate strength (MPa)	Ultimate strain (%)
GFRP	9.5	71	45.4	–	856	1.89
	15.9	199	48.2	–	751	1.60
Wire (mild steel)	3.4	9	200	675	850	0.30*
15 M (deformed)	16	200	200	419	686	0.21*

*Yield strain

Fig. 2. Different steel and FRP bars used in this study

of GFRP tubes (types A&B) were used with different thicknesses and having the same internal diameters 213 mm. The thickness of tube (A) equals 2.90 mm, while for tube B equals 6.40 mm. Different fibre angles with respect to the longitudinal axis of the tubes were used (± 60°, ± 65°, ± 45°, and 90°). The fibre orientations of the tubes were mainly in the hoop direction, and no fibres in the longitudinal direction. The glass fibre volume fraction as provided by the manufacture was 68% ± 3%. Figure 3 shows the specimens of FRP tubes for the split-disk test and coupon tensile test. Typical test samples of the coupon tests and split-disk test in the hoop direction for GFRP tubes are shown in Figs. 4 and 5, respectively. Table 2 shows the dimensions and mechanical properties of FRP tubes.

Fig. 3. Specimens of FRP tubes for the split-disk test and coupon tensile test [Masmoudi, R]

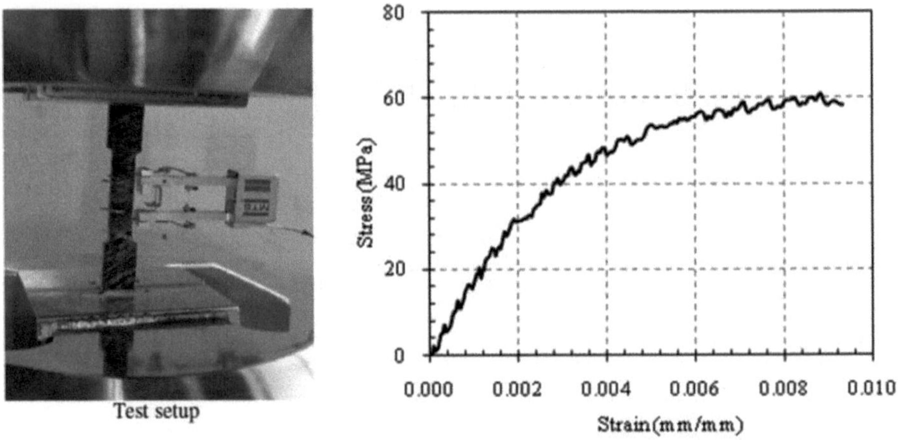

Fig. 4. Test setup and load-strain curve for the FRP tubes for coupon tensile test [Masmoudi, R]

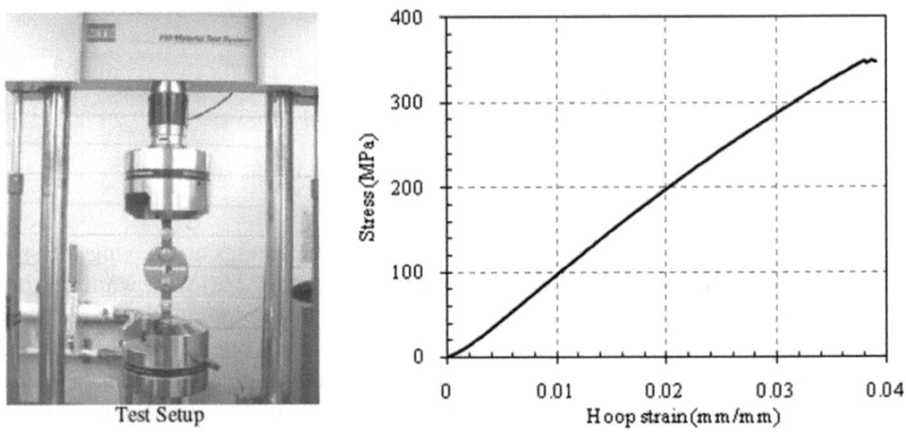

Fig. 5. Test setup and stress-hoop strain behaviour of the FRP tubes for split-disk test [Masmoudi, R]

Table 2. Dimension and mechanical properties of FRP tubes

Tube type	D (mm)	t_{frp} (mm)	No. of layers	Stacking sequence	f_{FRPU} (MPa)	ε_{FRPU} (%)	E_{FRPU} (MPa)	f_X (MPa)	ε_X (%)	E_X (MPa)
A	213	2.90	6	[60°, 90°, 60°]	548	1.70	32260	55.2	0.62	8865
B	213	6.40	12	[±60°, 90°, ±60°, 90°]	510	1.69	30200	59.2	0.75	7897

D and t_{frp} are the internal diameter and thickness of the FRP tubes, respectively. f_{FRPU}, ε_{FRPU}, and E_{FRPU} are, respectively, the ultimate strength, ultimate tensile strain, and Young's modulus in the hoop direction; while f_X, ε_X, and E_x are the ultimate strength, ultimate tensile strain, and Young's modulus in the axial direction, respectively.

2.2 Instrumentation and Testing Procedures

Several strain gauges were mounted on the internal reinforcement bars prior the concrete casting and on concrete or GFRP tube surfaces before testing. Eight strain gauges were located at the column mid-height in both axial and lateral directions to measure the axial and lateral strains, respectively. Two strain gauges were bonded on two longitudinal bars at 180° apart at the mid-height of the column. Figure 6 (a) shows the strain gauges instrumentation GFRP tube surface. Two displacement transducers (DTs) were used to measure the axial deformation of the column over the full height as shown in Fig. 6 (a). Additionally, two in-plane linear variable displacement transducers (LVDTs) were located at the mid-height to record the lateral displacements of each column. All columns were capped with a thin layer of the high strength sulphur to ensure uniform load distribution during testing. Before testing, both ends of the columns were further confined with bolted steel collars made from 10 mm thick steel plates in order to prevent premature failure at their ends. The specimens were loaded under axial compression loading using a 6000-kN capacity-testing machine. Loading and unloading in compression tests were achieved with load control at a rate approximately equal to 2.3 kN/s. During the test, load, axial and lateral displacements, and strain gauges were recorded automatically using a data acquisition system connected to the computer. Figure 6 (b) shows the test specimen inside the testing machine.

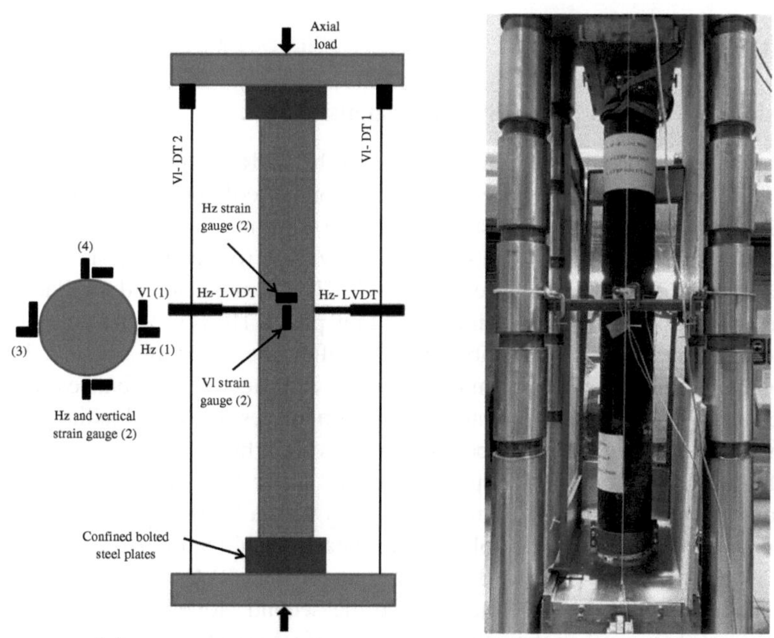

(a) (b)

Fig. 6. Instrumentations and Test setup: (a) schematic of the test setup and placing of specimen instrumentations; (b) test specimen inside the testing machine

2.3 Test Specimens

A total of six RC and CFFT circular columns were fabricated and tested under concentric axial cyclic compression loading. Two RC control columns and four CFFT columns were internally reinforced with steel, or Glass FRP (FRP). The test parameters were: (i) GFRP tubes thicknesses (2.9 and 6.4 mm); and (ii) internal reinforcement type (steel; and GFRP) and amount. All columns had the same height (h = 1900 mm) to diameter (D = 213 mm) ratio of 9.0. The control RC columns were reinforced longitudinally with reinforcement ratio (ρ_L) equal to (3.4%), one specimen reinforced with steel bars and the other specimen reinforced with GFRP bars. Steel spiral stirrups (pitch = 50.6 mm) were used as transverse reinforcement and designed to have approximately similar hoop stiffness as the GFRP tube (Type A). The CFFT columns were laterally confined with GFRP tubes (Type A or B). One specimen was internally reinforced with deformed steel bars (6 M15; ρ_L = 3.4%) and laterally confined with tube type (A). Three specimens were reinforced with 6 GFRP bars No. 3 or No. 5 (ρ_L = 1.2 and 3.4%, respectively) and laterally confined with tubes type (A and B). The test specimens were labeled as follows: the first letter S, A, or B is defining "the type of lateral reinforcement: steel spiral stirrups, GFRP tube type (A), or tube type (B)" then followed by a letter S, or G indicating "the longitudinal reinforcement type: steel, or GFRP bars", respectively. The number between brackets indicates "the longitudinal reinforcement ratio". The final letter (C) refers to unloading/reloading cyclic loading. Table 3 shows the test specimens' details.

3 Test Result

3.1 Axial and Lateral Stress-Strain Responses

As shown in Fig. 7 the stress-strain responses of the GFRP-reinforced control columns behaved similar to that of the steel-reinforced control column up to their peak load. However, the peak axial stress for steel-reinforced column was slightly higher than that of their counterpart reinforced with GFRP bars by 11% (on average). The axial stress-strain curves for GFRP and steel reinforced CFFT columns showed similar shapes of the hysteresis loops for the unloading/reloading paths. However, the steel-reinforced CFFT column hysteresis loop starts to open after the yielding of steel bars. The unloading paths for the CFFT columns reinforced with steel or FRP bars exhibited non-linear behavior. The degree of the non-linearity increases as the unloading axial strain increases. The reloading paths can be resembled as straight lines. The envelop curves of the reinforced CFFT- columns, representing the upper boundary of the axial cyclic stress-strain responses, showed bilinear responses with a transition zone in the vicinity of the unconfined concrete (fc') followed by nearly stabilization of the load carrying capacity as shown in specimens B-G(3.4)-C and B-G(1.2)-C). The initial slope was almost identical for all the specimens while the second slope is highly governed by GFRP tubes stiffness rather than the internal reinforcement type and amount, particularly in thicker tube thickness.

Table 3. Specimen's details and test results

ID	Lateral reinforcement type	Longitudinal bars		P u (kN)	f'_{cc} (MPa)	f'_{cc}/f'_{c}	εcc (με)	$\varepsilon_{cc}/\varepsilon_{o}$	εh, min. (με)	εh, aver. (με)	εh, max. (με)
		Type	Area								
S-S(3.4)-C	ϕ3.4@50.6	Steel	6 M 15	1948	54.60	1.23	−2510	1.04	377	599	836
S-G(3.4)-C	ϕ3.4@50.6	GFRP	6 No. 5	1575	47.20	1.08	−2711	1.12	653	935	1144
A-S(3.4)-C	A	Steel	6 M 15	2402	67.38	1.53	−13749	3.83	2442	4697	9707
A-G(3.4)-C	A	GFRP	6 No. 5	2603	73.06	1.66	−13718	4.63	5172	8087	9610
B-G(3.4)-C	B	GFRP	6 No. 5	3455	96.97	2.20	−15578	5.49	4435	9745	15135
B-G(1.2)-C	B	GFRP	6 No. 3	3272	91.82	2.08	−15563	5.96	11456	13787	16113

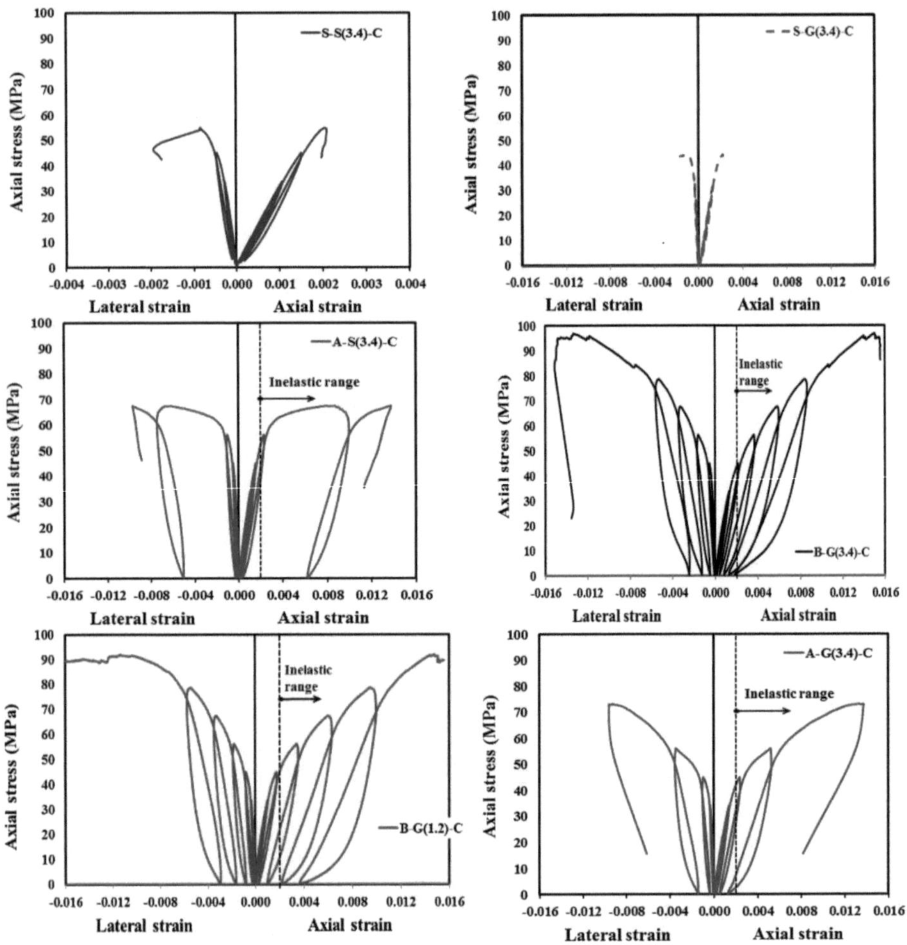

Fig. 7. Axial and lateral stress-strain curves of the tested specimens

3.2 State of Stress in the Fiber Reinforced Polymer Tube

Typical distributions of axial and lateral strains at various loads of selected reinforced CFFT columns over the perimeter of the GFRP tube at the column mid-height are presented in Fig. 8. As shown in this figure, the uniform distribution of the lateral strains in the FRP tubes near loading level of 2000 kN indicates efficient confinement of the tubes. As a result of the instability failure of the reinforced CFFT columns due to buckling produced highly variable lateral confinement and induced significant bending in the column before failure. The maximum, minimum, and average lateral strains in the hoop direction (εh, max, min, aver.) of the FRP tube at the ultimate load are reported in Table 3.

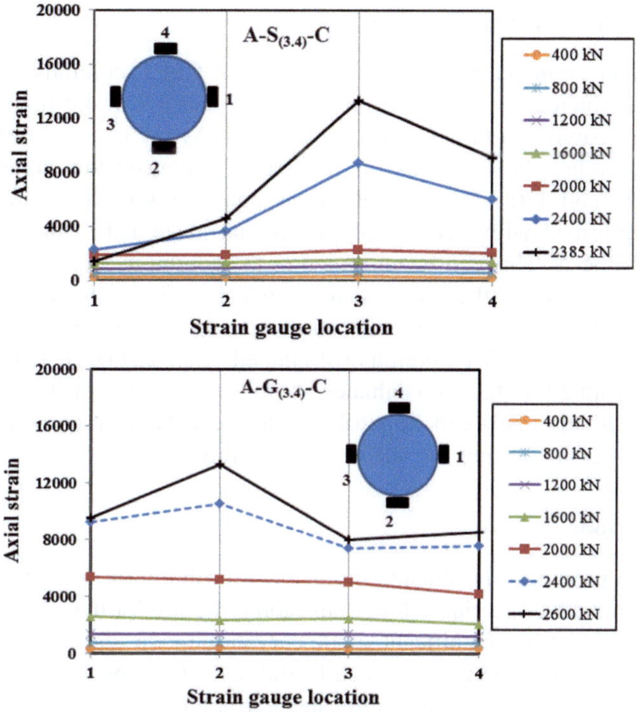

a) Axail Strain

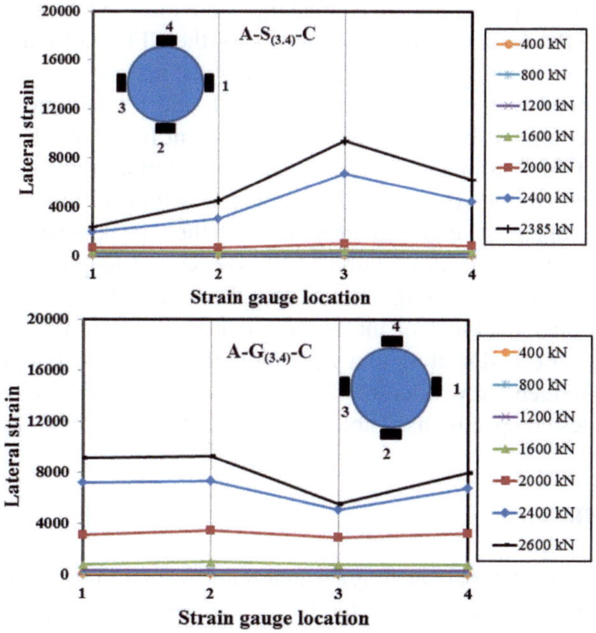

b) Lateral strain

Fig. 8. Strain distribution versus different strain gauges locations surrounding the column perimeter at the mid-height for specimen A-S (3.4)-C and A-G(3.4)-C: (a) Axial strain; (b) lateral strain

3.3 Axial Load Carrying Capacity

The strength and strain enhancement ratios of the CFFT columns (A-$S_{(3.4)}$-C and A-$G_{(3.4)}$-C) were increased ranging from 1.3 to 1.5 and 3.7 to 4.4 times compared to their counterpart control specimens (S-$S_{(3.4)}$-C and S-$G_{(3.4)}$-C), respectively as indicates in Table 3. Providing the FRP tube as in tube A enhanced the strength and strain capacity by 52% and 470%, respectively, in comparison with their control specimens which were reinforced with steel spiral stirrups and designed to have similar lateral stiffness as in Tube A. This can be attributed to the continuity of the FRP tubes rather than the discontinuity of the steel stirrups, which reflects the superior confining behavior of the FRP tubes over the steel stirrups to increase not only the strength but also the ductility of the CFFT columns (Mohamed et al. 2014). Increasing the GFRP tube thickness from 2.9 to 6.4 mm enhanced both the strength and strain ratios by 25% and 12%, respectively. This can be attributed to the enhancement of lateral confinement, as a result of increasing the stiffness of the tube.

3.4 Failure Mode

The GFRP tube provided significant confinement attributing to shift the failure mode from axially dominated material failure to instability failure for the CFFT columns. The instability was evident in a significant single curvature mode shape of the bent column. Despite, the specimens experienced much lateral deflections beyond the ultimate load, the deflected columns were still stable and carried more axial load. Loading the specimens continued until localized failure occurred near the mid height of the column. Finally, GFRP tube rupture, concrete crushing, and local buckling of steel bars or crushing of the FRP bars in the compression side of the CFFT columns were observed. This observation is in agreement with the previous research works conducted on slender FRP-confined columns (Mohamed, H. et al. and Fitzwilliam, J). On the other hand, the control columns showed substantially different failure mode compared to that occurred for the CFFT columns. For both control columns reinforced with steel or GFRP bars showed similar responses. The failure was typically initiated with vertical cracks started to appear at approximately 85% of their peak loads and followed by concrete dilation and lateral deformation of transverse and longitudinal reinforcement leading to concrete cover spalling. Thereafter, the concrete core crushed and spiral stirrups fractured after buckling of the longitudinal bars. Moreover, inclined diagonal shear surface was observed leading to a separation of the concrete core into two column parts causing a sudden drop after reaching the peak load. Figure 9 shows Overall failure modes of tested specimens. Table summarizes the test results for all specimens.

4 Conclusions

The axial behavior of FRP bars as longitudinal reinforcement in compression members has been quite limited, especially for the CFFT columns. This paper reported the test results of six full-scale RC and CFFT columns reinforced longitudinally with steel and GFRP bars tested under cyclic axial compression loads. Examination of the test results

Fig. 9. Overall failure modes of tested specimens

has led to a number of significant conclusions in regards to the general behavior, axial load carrying capacity of steel or FRP-reinforced CFFT columns. On the basis of the experimental test results and discussions of this research, the following conclusions can be drawn:

1. The CFFT columns behaved as long columns. The mode of failure characterized by loss of stability associated with considerable bending of the specimens.
2. Increasing the thickness of the GFRP tubes significantly increased the ultimate axial and strain capacities of the CFFT reinforced tested columns.
3. The CFFT columns reinforced with GFRP bars exhibited similar responses compared to their counterparts reinforced with steel bars at the same longitudinal reinforcement amount. No significant difference was observed in terms of ultimate axial strength and strain capacities.
4. Using FRP bars instead of conventional steel bars in the CFFT columns can provide a step forward to develop a totally corrosion-free new structural system.

References

ASTM A615/A615M-09, ASTM: Standard specification for deformed and plain carbon steel bars for concrete reinforcement, West Conshohocken, PA (2009)

American Concrete Institute (ACI): Guide for the Design and Construction of Externally Bonded FRP Systems for Strengthening Concrete Structures. ACI 440.2R-08, Farmington Hills, Mich (2008)

Fam, A., Greene, R., Rizkalla, S.: Field applications of concrete-filled FRP tubes for marine piles. ACI Special Publication, (Field Application of FRP Reinforcement: Case Studies), vol. SP(215), no. 9, pp. 161–180 (2003)

Fitzwilliam, J., Bisby, L.: Slenderness effects on circular CFRP confined reinforced concrete columns. J. Compos. Constr. **14**(3), 280–288 (2010)

Lam, L., Teng, J.G.: Stress-strain model for FRP-confined concrete under cyclic axial compression. Eng. Struct. **31**(2), 308–321 (2009)

Masmoudi, R., Mohamed, H.: Axial behavior of slender-concrete-filled FRP tube columns reinforced with steel and carbon FRP bars. In: 10th International Symposium on Fiber-Reinforced Polymer Reinforcement for Concrete Structures, Tampa, Florida, April 2011, ACI-SP-275 (2011)

Mirmiran, A., Shahawy, M., Beitleman, T.: Slenderness limit for hybrid FRP concrete columns. J. Compos. Constr. ASCE **5**(1), 26–34 (2001)

Mohamed, H., Abdel Baky, H., Masmoudi, R.: Nonlinear stability analysis of CFFT columns: experimental and theoretical investigations. ACI Struc. J. **107**(6), 699–708 (2010)

Mohamed, H., Afifi, M., Benmokrane, B.: Performance evaluation of concrete columns reinforced longitudinally with FRP bars and confined with FRP hoops and spirals under axial load. J. Bridge Eng., pp. 1084–0702 (2014). ISSN/04014020(12)

Pultrall, Inc.: V-ROD Composite Reinforcing Rods Technical Data Sheet. Thetford Mines, Canada (2007). www.pultrall.com

Yuan, W., Mirmiran, A.: Buckling analysis of concrete-filled FRP tubes. Int. J. Struct. Stab. Dyn. **1**(3), 367–383 (2001)

Ultimate Flexural Capacity Predication of Rectangular FRP Tube Beams Filled with Concrete

Asmaa Abdeldaim Ahmed[1(✉)], Mohamed Hassan[1,2], and Radhouane Masmoudi[1]

[1] Department of Civil Engineering, University of Sherbrooke, Sherbrooke, QC, Canada
Asmaa.Ahmed@USherbrooke.ca
[2] Helwan University, Cairo, Egypt

Abstract. The concept of concrete-filled fiber-reinforced polymer (FRP) tubes (CFFTs) is promising for a variety of structural applications and a good alternative for innovative constructions because of numerous attractive features, including durability and concrete confinement. An extensive research work was conducted on circular (CFFT) with and without internal bars for beams and columns. However, much less attention has been given to rectangular CFFT especially those reinforced with internal reinforcement bars. This paper aims at developing a simplified analytical design method to predicate the ultimate moment capacities of previously tested steel-reinforced rectangular CFFT beams at the University of Sherbrooke. A wide range of test parameters was considered such as the FRP tubes thickness, fibre laminates, and steel reinforcement. The proposed design method was found to be acceptable for predicting the ultimate moment capacities of the beams. The accuracy of the theoretical analysis showed good agreement with the measured values.

1 Introduction

The construction industry is expressing a great demand for innovative and durable structural members. Concrete-filled fiber-reinforced-polymer (FRP) tube (CFFT) provides an innovative and effective system, suitable for several structural applications, including fender piles, girders, and bridge piers in marine and other harsh environments. The FRP tube has several advantages over steel reinforcement as it can offer the flexural, shear, and confinement reinforcement simultaneously, without the labour costs and time involved in building conventional rebar cages. Filling the FRP tube with concrete provides compressive resistance and prevents local buckling of the thin tube. The CFFT system may also include internal longitudinal steel reinforcement, as the tube will protect the steel from water infiltration and corrosion, while the steel reinforcement provides ductility and additional strength and stiffness as well as minimize slip of the tube relative to the concrete core, which can be a major factor in the loss of composite action. Therefore, utilization of FRP tubes filled with plain or reinforced concrete as a hybrid composite system can provide the proposed system many unique

H. Rodrigues and A. Elnashai (Eds.): GeoMEast 2018, SUCI, pp. 45–55, 2019.
https://doi.org/10.1007/978-3-030-01932-7_5

advantages, which impact on civil engineering infrastructures costs, strength and performance, and their sustainability for longer service life (Ahmed and Masmoudi 2016, 2017; Ahmed et al. 2018).

Rectangular sections are more favorable in construction due to erection speeds and easy feasibility to create composite action with other members such as bridge decks and columns. The growing popularity of using circular CFFT members has motivated researchers to expand the investigation of the system feasibility for rectangular sections in both axial and flexural applications. Under flexural loading, Fam et al. (2005) studied the flexural behaviour of rectangular CFFT beams tested under bending loading. The study showed that rectangular CFFT was a feasible system that could offer similar flexural strength to rectangular concrete filled steel tube. The tube laminate structure and its progressive failure contribute to the slightly nonlinear behavior of beams. Amir Fam and Skutezky (2006) studied rectangular and T-beams constructed of rectangular concrete filled and non-filled glass-FRP pultruded tubes without and with concrete slabs attached with shear studs. A comparison of concrete filled and non-filled GFRP tubes showed little additional stiffness but a large increase in strength due to the presence of the concrete core. Conversely, a comparison of concrete filled and non-filled GFRP tubes with the addition of the concrete deck showed a large increase in stiffness but only slight increase in strength. Moreover, adding concrete slabs to concrete-filled GFRP rectangular tubes enhances their flexural strength and stiffness. Failure mode may change from crushing of GFRP compression flange to rupture of tension flange (for long span) or to GFRP web buckling (for short span). Belzer et al. (2013) investigated the composite action and performance of rectangular CFT glass-FRP tubes in four-point bending. The three different configurations include (1) unbonded, (2) bonding of the flanges, and (3) bonding of the flanges and webs. The results showed low composite action for the unbonded beam configuration up to full composite action for the completely bonded beam configuration. They also suggested the feasibility of implement Carbon-FRP into the tension flange of the tube or provides tension reinforcing bars within the tube, could increase the flexural stiffness of rectangular CFFT tubes. An extensive research project has been conducted at the University of Sherbrooke to investigate flexural performance of rectangular CFFT beams totally filled with reinforced concrete (Abouzied and Masmoudi 2015, 2017; Masmoudi and Abouzied 2018). A total of 24 beams were fabricated and tested under a four-point bending moment. The tubes were fabricated filament-wound technique at the University of Shebrooke using glass FRP and Vinyl Ester Resin. The beams have a 3200 mm long and 305 $\times$ 406 mm^2 cross section. The test matrix covered a wide range of test parameters such as tube thickness of 3.4 mm to 14.2 mm, fiber structure laminate, and internal reinforcement ratio. The tested beams were reinforced at the tension side with four steel bars 15 mm diameter (15 M) as flexural reinforcement with a concrete cover of 38 mm. The test results showed that the reinforced-rectangular CFFT beams experience significantly higher ductility, higher stiffness, and superior strength than the RC beams. All beams failed in a flexural failure with no signs of web buckling, or slippage between the concrete core and the tubes, or rupture at tubes corners. Reinforced-rectangular CFFT beams exhibited gradually in a sequential manner (yielding of steel, buckling of compressed tube flange, and finally rupture of the fibers). Even after the ultimate failure, the CFFT beams can maintain a residual strength because of the existence of the steel that withstands high strains and elongation. Also, they concluded that the flexural strength of the reinforced-rectangular CFFT beams increases with increasing the FRP tube thickness until a certain limit of reinforcement ratio (ρf = Af/Ac = 10%, where Af is area of FRP tube and Ac is the area of concrete core) that separates the under-reinforced (flexural tension) and over-reinforced (flexural compression) CFFT section. Furthermore, the study showed that the deflection behaviour can be modelled and predicted. A comparison between fully, partially filled CFFT, and conventional steel-RC beams have also made. The results indicated that, the partially-CFFT beams had an overall strength-to-weight ratio 370% higher than that of the RC beam, while their weight was 30% lighter than the RC beam. Moreover, providing inner hollow GFRP tubes changed the failure from tension to compression failure, which is desirable in FRP composite section design.

In this paper, first a simplified analytical procedure based on force equilibrium, strain compatibility, linear elastic behaviour of FRP and partial stress– strain confinement model for concrete is developed to predict the ultimate moment capacities of previously tested beams at the University of Sherbrooke (Abouzied and Masmoudi 2015, 2017; Masmoudi and Abouzied 2018). Second, a new empirical design expression is proposed to facilitate reasonable predication for the ultimate moment capacities of the tested beams failed under tension failure-controlled. The results from the theoretical analysis as well as the empirical formula are compared against the experimental results to determinate their accuracy. The simplified analytical procedure and empirical design formula are presented in following sections.

2 Simplified Proposed Analytical Method

In this section, a simplified analytical method is developed to predict the ultimate moment capacities corresponding to the failure mode of the tested RCFFT beams. The analytical procedure (as shown in Fig. 1) involves the determination of the position of the neutral axis for a given strain of the extreme compression fibre by using the principles of strain compatibility and cross-sectional forces equilibrium. The assumptions in the analysis include: (1) Euler–Bernoulli beam theory (that plain sections remain plain after deformation), (2) strain distribution throughout the depth of the section is linear, (3) perfect bond between FRP tube section, steel bars and concrete core, (4) GFRP tube section behaves linear elastically until failure, (5) Confinement effect is considered, (6) concrete contribution after cracking below the neutral axial is neglected, (7) $\varepsilon_{bar} = \varepsilon_y = \varepsilon_u = 0.0021$ (yield strain of steel bars), and (8) stress–strain curve for the GFRP tube was based on full section behaviour (based on mechanical properties of coupon test). The flexural analysis procedure is straightforward. The parts of the FRP tube above and below the neutral axis are considered effective in resisting compression and tension forces, respectively. Figure 1 shows the cross section of the reinforced rectangular CFFT beam that was used in the analysis, where the neutral axis is located within the cross section at a depth c.

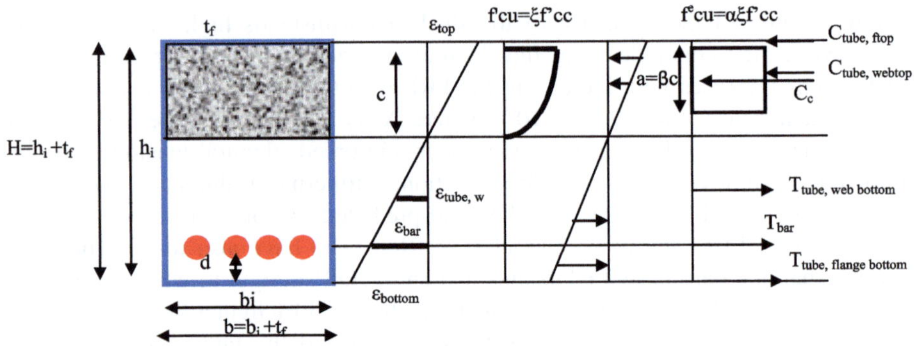

Fig. 1. Strain and stress profile of reinforced rectangular CFFT beam cross section

Referring to the equivalent stress and strain distribution diagram shown in Fig. 1, the internal tensile forces in the FRP tube (bottom flange and web) and steel bars, T, can be expressed respectively, as follows:

$$T_{tube,flangeBottom} = E_{tx}\varepsilon_{bottom}(bt_{FRP}) \tag{1}$$

$$T_{tube,webBottom} = 2E_{tx}\varepsilon_{web,bottom}(H - c)t_{FRP} \tag{2}$$

$$T_{bar} = A_{bar}E_{bar}\varepsilon_{bar} \tag{3}$$

where, $H = H_i + t_{FRP}$, H_i and t_{FRP} are the internal height and thickness of FRP tube, respectively, and E_{tx} tensile modulus in the axial direction. Also, the strains $\varepsilon_{flange,Bottom}$ and $\varepsilon_{web,Bottom}$ can be computed in terms of ε_{top} (the design ultimate axial strain obtained from the axial compression tests) as follows:

$$\varepsilon_{flange,bottom} = \varepsilon_{top}\left(\frac{H - c - t_{FRP}/2}{c + t_{FRP}/2}\right) \tag{4}$$

$$\varepsilon_{web,bottom} = \varepsilon_{top}\left(\frac{(H - c - t_{FRP}/2)/2}{c + t_{FRP}/2}\right) \tag{5}$$

The internal compression forces in the FRP tube (top flange and web), C_{tube}, flange top and C_{tube} web, top, can be expressed as follows, assuming ε_{top} the ultimate compression strain based on coupon tests:

$$C_{tube,flangetop} = E_{tc}\varepsilon_{top}(bt_{FRP}) \tag{6}$$

$$C_{tube,webtop} = 2E_{tc}\varepsilon_{web}(c - t_{FRP}/2)t_{FRP} \tag{7}$$

$$\varepsilon_{web,top} = \varepsilon_{top}\left(\frac{c/2}{c + t_{FRP}/2}\right) \tag{8}$$

Herein for simplicity, the internal compression force in the concrete block at the top is calculated based on an equivalent stress distribution assuming a rectangular stress block with a depth equal to some fraction of the neutral axis depth, where a = βc, and a magnitude equal to some fraction of the concrete compressive strength, fecu = α fcu, [CSA A23.3 (2014)]. α is the ratio of the assumed uniform stress in the rectangular compression block (f ecu) to the maximum partially confined concrete compressive strength (fcu), given by $\alpha = 0.85 - 0.0015\ f'_c \geq 0.6$, β is the ratio of the depth of the rectangular compression block (a) to the depth to the neutral axis (c), where $\beta = 0.97 - 0.0015\ f'_c \geq 0.6$. In fact, these factors need more examination for rectangular CFFT beams with a wide range of concrete strength. Hence, the internal compression force in the concrete block can be expressed as follows:

$$C_c = \alpha f'_{cu}(A) \tag{9a}$$

Where A is the area of concrete in compression zone (=b * a). It is assumed that the ratio of the partially confined concrete compressive strength fcu, to the confined concrete compressive strength f'_{cc} is equal to ξ (0.0 < ξ < 1.0) or can be expressed as follows:

$$f_{cu} = \xi f'_{cc} \tag{9b}$$

Substituting Eq. 9a in Eq. 9b can lead to:

$$C_c = \alpha\, \xi f'_{cc}\, (ba) \tag{9c}$$

The value of ξ presents the level of confinement gained from the interaction between FRP tube and concrete core of CFFT beam under flexural load. Also, this value (ξ) depends on different parameters such as thickness and fiber orientation of FRP tubes, type and reinforcement ratio of internal rebars and concrete compressive strength. Strength of confined concrete f'_{cc} is related to its unconfined concrete strength and the confinement pressure from the FRP tubes. The f'_{cc} values can be calculated from the proposed model by Wu and Wei (2010). The model considered for rectangular column, corner radius and aspect ratio.

$$f'_{cc} = f'_{co}\left[1 + 2.2\left(\frac{2r}{b}\right)^{0.72}\left(\frac{f_l}{f_{co}}\right)^{0.94}\left(\frac{h}{b}\right)^{-1.9}\right] \tag{10a}$$

Where f'_{co} unconfined concrete strength (MPa), r is the corner radius, (2r/b) is the corner radius ratio, h and b is large column length and width, respectively, f_l is the lateral confinement pressure ($f_l = \frac{2f_{frpu,hoop}t_{FRP}}{b} = \frac{2E_{frp}\varepsilon_{frpu,hoop}t_{FRP}}{b}$), f_{frpu} is the ultimate hoop tensile strength of FRP tubes.

Therefore, the nominal moment strength of the beams can be determined by taking the moment of tensile and compression stress resultants of tube and bars about the compression stress resultant in concrete.

$$M_{theo} = T_{bar}(H - d - a/2 - t_{FRP}/2) + T_{tube,flangebottom}(H - a/2 - t_{FRP}/2) + T_{tube,webbottom}(H - a/2 - t_{FRP}/2)/2 + C_{flange,top}(a/2 + t_{FRP}/2) + C_{web,top}(C - a/2) \tag{11}$$

In fact, in Eq. (11), there are three unknowns: strains, neutral axis depth c and ξ. As a result, the design procedure is iterative. Assuming compression failure of CFFT beams is triggered by failure of the tube, which is immediately followed by crushing of concrete as a secondary failure, where the hoop tensile stress (confinement effect is insignificant) (AASHTO 2012). Hence, the analytical procedure is performed by assuming a given strain value (εtop) at the level of the extreme level of the tube based on the compression coupon tests. Flexural analysis of reinforced and unreinforced CFFT beam suggests that the neutral axis is located within a region of 0.25–0.4Hi and 0.2–0.3H (Abouzied and Masmoudi 2015, 2017) and Fam and Skutezky (2006), respectively. Therefore, the analytical procedure for the reinforced rectangular CFFT beams is performed by assuming the neutral axis depth (c) equal to ηHi from the inner height, where η is taken in range between 0.20 and 0.45. The strain values for tube and concrete at any level are determined based on the assumption of a linear strain distribution along the depth of the beam between the extreme compression fibers and the tensile FRP tube reinforcement. The moment capacities of the CFFT beams are determined for all values of η. Then, the experimental bending moments of CFFT beams are compared with the theoretical moments. The optimum value of η is obtained and correlate the theoretical with experimental bending moments for the tested beams. Thereafter, by equating the compressive and tensile forces, the values of ξ (confinement efficiency) could be determined for the different values of η.

3 A Simplified Proposed Design Experssion

In this section, an empirical design formula is proposed. In order to establish an expression for the moment capacity of rectangular CFFTs fully filled with reinforced concrete, a regression analysis for the tested beams were considered in lieu of the rigorous equilibrium and strain compatibility approach. Since the tubes had different tube thickness and different laminate structures, a total reinforcement index parameter ω_t is introduced to represent the different reinforcement ratios and material strengths. The total reinforcement index ω_t is the sum of FRP and steel reinforcement indices ω_f and ω_s and is defined as given in Eq. 1:

$$\omega_t = \rho_f \frac{f_{fut}}{f'_c} + \rho_s \frac{f_y}{f'_c} \tag{12}$$

where the term (ρf) is the FRP reinforcement ratio, defined as the ratio of cross-sectional areas of the tube and concrete core. f_{fut} is the tensile strength of the tube in the longitudinal direction (MPa), which is given in Table 1, and f'_c is the unconfined concrete compressive strength (MPa), (ρs) is the steel reinforcement ratio (=As/bd), b is cross-section width (mm), d is the effective depth (mm), and fy is the yield tensile stress of internal steel bars (MPa). Based on the best fitting of the experimental data (B1 to B4 in given Table 1), the following an empirical formula is proposed to predicate the moment capacity of rectangular CFFTs predominantly governed by tension failure. The new empirical design formula is similar design formula adopted by the AASHTO (2012) for circular CFFT members, however, the tube geometry and internal bars were considered. The FRP tube controls failure in tension. As such, the concrete longitudinal compressive strain at failure may exceed 0.003 (AASHTO 2012).

$$M_u = 0.0054b\,h^2 f_c^{'}\left(100\,.\,\rho_f \frac{f_{fut}}{f_c^{'}} + 100\,.\,\rho_s \frac{f_y}{f_c^{'}}\right)^{0.983} \tag{13}$$

Where h is member total height (mm) (=h_o + $2t_f$).

4 Verification of the Proposed Analytical Procedure and Empirical Model

Table 1 shows the summary of the experimental test results conducted by (Abouzied and Masmoudi 2016). As seen in Table 2 a strong correlation is observed between the theoretical predictions and the test results for the ultimate moment capacity when the values of η are in the range 0.28–0.45. Similar findings have been found in Abouzied and Masmoudi (2016). The corresponding Mu(exp)/Mu(theo) ratios and ξ values are on average in the range 0.92 ± 0.1 to 1.16 ± 0.1, and 0.81 ± 0.4 to 0.45 ± 0.14, respectively. It is interesting to mention that the concrete compressive strengths fcu at ultimate for the reinforced CFFT beams were on average 60% the confined concrete compressive strength f_{cc}. These results confirm that the FRP tubes partially confined the concrete core of the tested CFFT beams. The value (ξ) depends on different parameters such as thickness and fiber orientation of FRP tubes, type and reinforcement ratio of internal rebars and concrete compressive strength. The analytical design procedures presented herein were found to be acceptable for predicting the ultimate moment capacities of reinforced rectangular CFFT beams with different tubes thickness and mode of failure; refer to Fig. 2. The accuracy of the theoretical analysis procedures pertains good agreement with the measured values on average of 0.99 ± 0.1.

On the other hand, comparisons between the theoretical predictions and the test results for the ultimate moment capacity using the empirical design formula are shown in Table 2. As seen in Table 2 a strong correlation is observed between the prediction values and the test results, particularly when the failure predominantly governed by tension failure. The Mu(exp)/Mu(theo) ratio value is on average of 1.0 ± 0.031 for beams B1 to B4. However, the accuracy of the proposed formula slightly overestimates the moment capacity predications when other mode of failure was triggered such as balanced or compression failure. The empirical expression is simple and provide reasonable prediction to use by the designer. However, more experimental tests are needed to cover a wide range of parameters to assess the confinement concrete contribution for rectangular cross-section with different aspect ratio, fiber laminate, steel reinforced ratios.

Table 1. Summary of the experimental test results database (Abouzied and Masmoudi 2016)

Beam	(hi mm × bi mm	t_f (mm)	f′c	Stacking sequence	Mechanical properties	Axial direction			Transverse direction			Mode of failure	Mu kN m
						E_f (GPa)	ff (MPa)	ε_f (mm/m)	E_f (GPa)	F_f (MPa)	ε_f (mm/m)		
B1	(406 × 305)	3.4	49.7	[90°, ±30°, 90°]	Tens. test	14.3	249.0	15.7	16.0	257	21.8	Tension	249
B2					Comp. test	14.0	−92.0	−7.0	17.8	−175	−10.4		267
B3		5.7	48.7	[90°, ±30°, 90°, ±30°, 90°]	Tens.test	14.5	173	15.3	14.4	249	23.9	Tension	404
B4					Comp. test	15.5	−165	−12.5	14.5	−293	−24.0		392
B5		8.7	41.7	[90°, ±3 0°_{2}, 90°, $\pm 30^{\circ}_{2}$, 90°]	Tens.test	16.2	19	18.9	13.7	168	19.2	Balanced	559
B6					Comp.test	17.7	−189	−11.8	13.8	−211	−17.8		560
B7		9.9	48.7	[90°, $\pm 30^{\circ}_{6}$, 90°]	Tens. test	18.6	242	15.3	13.4	125	16.6	Comp.	581
					Comp. test	20.1	−17	−9.	12.2	−217	−24.0		

Table 2. Theoretical results of moment capacities predications (M_{uexp}/M_{utheo}) for the analytical design procedures and empirical design expression and η values

Label	η = c/Hi														Eq. 13
	0.20	0.22	0.24	0.26	0.28	0.3	0.32	0.34	0.36	0.38	0.4	0.42	0.44	0.45	
B1	0.85	0.89	0.93	0.96	0.99	1.02	1.05	1.08	1.10	1.12	1.15	1.17	1.19	1.20	**0.97**
B2	0.91	0.96	1.00	1.03	1.07	1.10	1.13	1.16	1.18	1.21	1.23	1.25	1.27	1.28	**1.04**
B3	0.77	0.82	0.87	0.92	0.97	1.01	1.05	1.09	1.12	1.16	1.18	1.21	1.24	1.25	**1.02**
B4	0.74	0.80	0.85	0.90	0.94	0.98	1.02	1.06	1.09	1.12	1.15	1.18	1.20	1.21	**0.99**
B5	0.66	0.70	0.75	0.79	0.83	0.87	0.90	0.93	0.96	0.99	1.01	1.03	1.06	1.06	0.87
B6	0.66	0.71	0.75	0.79	0.83	0.87	0.90	0.93	0.96	0.99	1.01	1.04	1.06	1.07	0.88
B7	0.65	0.70	0.74	0.78	0.82	0.86	0.89	0.92	0.95	0.98	1.00	1.03	1.05	1.06	0.66
Aver.	0.75	0.80	0.84	**0.88**	**0.92**	**0.96**	**0.99**	**1.02**	**1.05**	**1.08**	**1.10**	**1.13**	**1.15**	**1.16**	**1.00***
SD	0.10	0.10	0.10	0.10	0.10	0.09	0.09	0.10	0.09	0.09	0.10	0.09	0.09	0.10	0.03

*Average is calculated from B1 to B4 (failure dominated by tension failure)

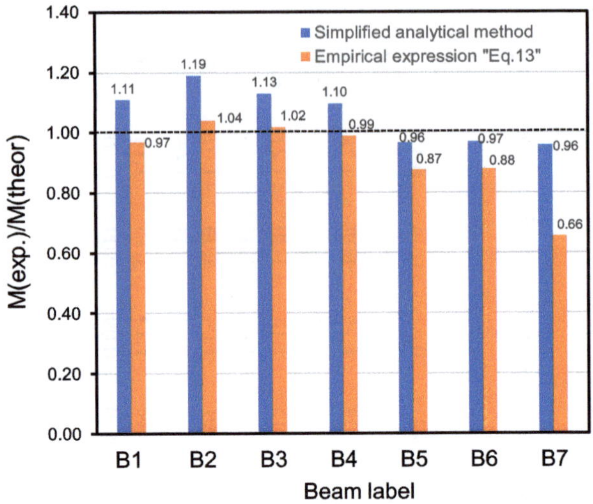

Fig. 2. Experimental to the theoretical moment capacity predications of steel-reinforced rectangular CFFT beams

5 Conclusions

Using conventional beam theory for the analysis and design can be used for the reinforced rectangular CFFT beams. The present theoretical analysis procedures and empirical design formula found to be acceptable and easy to use for predicting the ultimate moment capacities of steel reinforced rectangular CFFT beams with an average of 0.99 ± 0.1 and 1.0 ± 0.03, respectively. A strong correlation observed between the theoretical predictions and the test results for the ultimate moment capacity when the values of η are in the range 0.28–0.45. However, more experimental tests are needed to cover a wide range of parameters to assess the confinement of concrete contribution for rectangular cross-section under flexural with different aspect ratio, fiber laminate, steel reinforced ratios.

Acknowledgements. The research reported in this paper was partially sponsored by the Natural Sciences and Engineering Research Council of Canada (NSERC). The authors also acknowledge the contribution of the Canadian Foundation for Innovation (CFI) for the infrastructure used to conduct testing. Special thanks are due to the technical staff at the University of Sherbrooke.

References

AASHTO: Guide Specifications for Design of Concrete-Filled FRP Tubes for Flexural and Axial Members, 1st edn., Washington, DC, p. 48 (2012)

Abouzied, A., Masmoudi, R.: Flexural behavior of rectangular FRP-tubes filled with reinforced concrete: experimental and theoretical studies. Eng. Struct. J. **133**, 59–73 (2017)

Abouzied, A., Masmoudi, R.: "Structural performance of new fully and partially concrete-filled rectangular FRP-tube beams. Constr. Build. Mater. J. **101**, 652–660 (2015)

Ahmed, A.A., Hassan, M., Mohamed, H., Abouzied, A., Masmoudi, R.: Axial behavior of circular CFFT long columns internally reinforced with steel or carbon and glass FRP longitudinal bars. Eng. Struct. **155**, 267–278 (2018)

Ahmed, A.A., Masmoudi, R.: Concrete filled FRP tube columns longitudinally reinforced with FRP and steel bars. In: CANCOM2017: Canadian international conference on composite materials, Ottawa, Ontario, Canada, 17–20 July 2017 (2017)

Ahmed, A.A., Masmoudi, R.: Behaviour of FRP reinforced CFFT columns under axial compression loading. In: 5th International Structural Specialty Conference, CSCE 2016, London, Ontario, 1 June–4 June 2016, p. 10 (2016)

Amir Fam, A., Skutezky, T.: Composite T-beams using reduced-scale rectangular FRP tubes and concrete slabs. ASCE J. Compos. Constr. **10**(2), 172–181 (2006)

Belzer, B., Robinson, M., Fick, D.: Composite action of concrete-filled rectangular GFRP tubes. ASCE Compos. Constr. J. **17**(5), 722–731 (2013)

Fam, A., Schnerch, D., Rizkalla, S.: Rectangular filament-wound glass fiber reinforced polymer tubes filled with concrete under flexural and axial loading: experimental investigation. ASCE Compos. Constr. J. **9**, 25–33 (2005)

Masmoudi, R., Abouzied, R.: Flexural performance and deflection prediction of rectangular FRP-tube beams fully or partially filled with reinforced concrete. ASCE Struct. J. **5**, 30 (2018)

Wu, Y., Wei, Y.: Effect of cross-sectional aspect ratio on the strength of CFRP-confined rectangular concrete columns. Eng. Struct. **32**(2010), 32–45 (2010)

Reinforced Concrete Multi-Rib Arch Bridge Strengthened by Changing Structural System

Yiqiang Xiang(✉) and Bishnu Gupt Gautam

Department of Civil Engineering, Zhejiang University, Hangzhou, China
{Xiangyiq, bishnu}@zju.edu.cn

Abstract. Before and after the 1980s, the China built many reinforced concrete multi-rib arch bridges. With the increase of traffic volume and load, the carrying capacity of the multi-rib arch bridges are obviously lower, it is necessary to take various measures to strengthen and enhance bearing capacity of these bridges. This paper proposes a reinforcement method by adding cable towers in two ends abutment of arch bridges and the inclined cables to improve behavior of RC multi-rib arch bridge. The main idea is that additional bridge towers are set on two side abutments of the RC rib arch bridge with single span to hang stayed cables. Newly-casting RC longitudinal beams are set in two sides of the bridge and the transverse prestressed steel tendons with $4\varphi^{j}15.24$ steel strands on the both sides of every cross beam were stretched, which are wrapped by the steel tubes and connected with longitudinal beams, to enhance the integrality and stiffness of the multi-rib arch bridge. Then, the longitudinal beams of each side of arch bridge are supported by using stayed cables through the cable tower. Comparison of the values gained by space finite element and field test confirmed that the proposed method for strengthening the rib arch bridge and enhancing the bearing capacity is effective.

1 Introduction

Many of the reinforced concrete arch bridges were constructed in the 1970s to 1980s in China, in which the design loads level of quite a bit bridges was Truck-15 and Trailer-80 (1989a, b). When comes into the 21st century, with the fast development of Chinese economy and increase of traffic volume and truck standard, China's Ministry of Transport revised relative codes for design of highway bridges and culverts and enhance load levels of bridge design (2004a, b, 2008a, b). Hence, some of bridges need to strengthen to improve their behavior and enhance their load carry capacity. The strengthening technologies for bridge superstructures mainly include shotcreting reinforcement, pasting steel reinforcement plates on surface of beam and box girder, pasting carbon fiber or other fiber materials, increasing the cross section of beam and box girder, adding external prestressed tendons, changing the structural system, adding main girders or strengthening lateral connection among main girders and other strengthening methods etc. (Xiang et al. 2014a, b). Author et al. have made a series of explorations and practices in bridge strengthening and testing from 2000 (Xiang et al. 2002, 2004, 2007). In this paper, aiming at the existing lower load carrying capacity of the early reinforced concrete multi-rib arch bridges, a reinforcement method for

H. Rodrigues and A. Elnashai (Eds.): GeoMEast 2018, SUCI, pp. 56–64, 2019.
https://doi.org/10.1007/978-3-030-01932-7_6

changing multi-rib arch bridges to arch-cable system bridges is proposed. The main idea is that adding cable tower in two ends abutment of multi-rib arch bridges and the inclined cables improve behavior of the RC multi-rib arch bridges. In order to confirm effectiveness of the proposed strengthening method, the bridge is analyzed by spatial finite element method according to requirement of strengthening and design load standard. The results are compared with values gained by field test.

2 Changing Multi-Rib Arch Bridge to Arch-Cable System Bridge

Figure 1 is elevation of an early typical reinforced concrete multi-rib arch bridge in China-Baibu Bridge. The bridge was a two-way two lane bridge, which after reinforcement it still maintains two-way two lane. The bridge width arrangement is 0.5 m (Handrail + longitudinal beam) + 7.0 m (non-motor vehicles and motor vehicles shared lane) + 0.5 m (Handrail + longitudinal beam). The cross slope of bridge deck was 1.5% (Fig. 2).

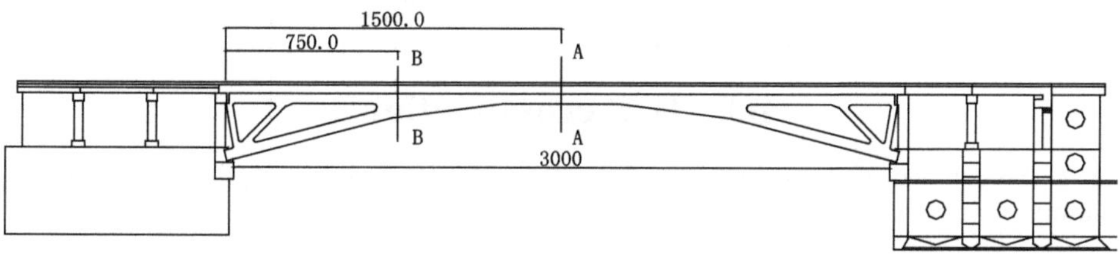

Fig. 1. Elevation of the typical RC multi-rib arch bridge- Baibu Bridge before strengthening (Dimension unit: cm)

Due to the bigger deflection problem in the arch crown of the bridge, in order to guarantee the traffic safety of the bridges and enhance the load bearing capacity of the bridges from Truck-15 and Trailer-80 to Truck-20 and Trailer-100 (2004). This paper proposes the reinforcement method for changing multi-rib arch bridges to arch-cable system bridges. That is changing the original reinforced concrete rib arch bridge into the arch-cable composite structure by adding cable tower and cables. One end point of cables is fixed in the end reconstructed cross beam of the rear load bearing wall on the open caisson foundation and the other end point is connected with two side newly-casting longitudinal beams near quarter span locations in the multi-rib arch bridge. The cross beams in the rear load bearing walls can resist the up-pulling forces of stay cable in side span, the rear wall is also connected with open caisson foundation by implanting steel bars. The two towers were newly built in the front of the open caisson foundation. The towers fixed with adding longitudinal beams in two external sides of bridges. The objective of doing this is that, on the one hand, it can partly restore rib arch crown altitude in mid-span of the bridge. On the other hand, it can significantly reduce moment in the mid-span of the bridge by adding cables, and enhance the load bearing

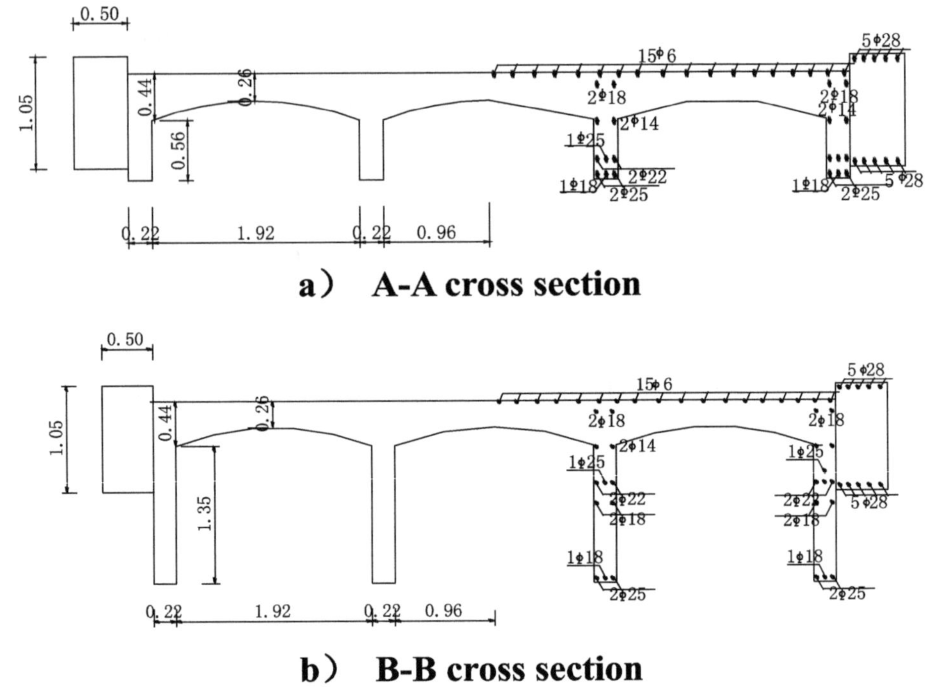

Fig. 2. Section and reinforcement drawing of Baibu Bridge before strengthening (Dimension unit: m)

capacity of the whole structure. Figure 3 shows strengthening scheme of Baibu Bridge and the elevation and cross section layout of reinforcement design. The related analysis and testing research have been conducted.

To make the arch ribs of reinforced concrete rib arch bridge and cables to work together and distribute the load, the 0.5 m railings on two external sides of the old rib arch bridge were removed and the old concrete surface was roughened to pour reinforced concrete rectangular longitudinal beams along two external sides, whose height is 1.05 m and width is 0.5 m. At the same time, the Φ16 steel bars are implanted into the cross beams, where cross beams connected with the longitudinal beam. On the basis of the original cross beam (clear distance 7.0 m) of the bridge, two steel pipes with φ 100 × 3 mm are set in the both sides of every cross beam of the old bridge. The prestressed steel tendons with $4\varphi^{j}15.24$ go through the steel tubes and are drawn. The main role of adding and setting the longitudinal beams in two sides of bridge is to bear horizontal forces caused by the side cables anchored in the longitudinal beams and strengthen the integrity and stiffness of the bridge (Fig. 4).

The newly built towers are reinforced concrete column structure. The dimension of its section is a 0.5 m (lateral) × 1.0 m (longitudinal) rectangle. There is no transverse brace connection between the top of towers due to low tower. A cap at the top of every tower column was designed by considering beauty, which is 0.7 m in width and 0.96 m in

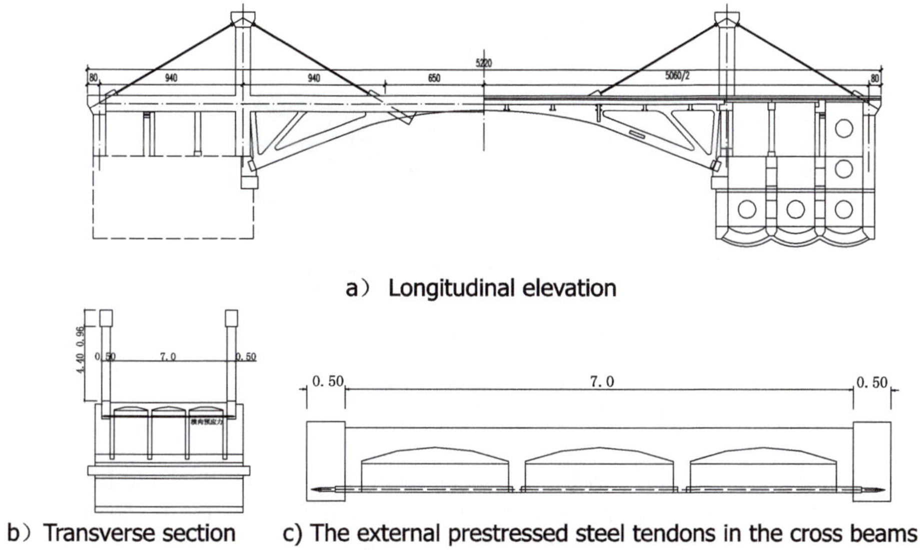

Fig. 3. Strengthening project of Baibu Bridge (Dimension unit: cm)

Fig. 4. Layout of the external prestressed steel tendons on the cross beams

height. The tower body below beam is also reinforced concrete rectangular structure, in which the transverse width varies from 0.5 m to 0.8 m and longitudinal width is 1.0 m.

Stayed cable uses the cables of OVM-LZM7-37 with outsourcing PE. The wire strength is 1670 MPa. For the bearing capacity of abutment foundation, due to the only increment of the self-weight of the 5 m-high concrete towers, two pairs of cables and strengthening concrete beams, the total increment of dead load at one side abutment foundation is about 200 kN, the stresses increase of foundation bottom caused by increased loads is about 2.5 kPa. The bearing capacity of soil foundation has been largely improved enough to withstand the load increment by considering fully consolidated action of the original bridge foundation after twenty years of operation. Figure 5 shows view of the near completion of the bridge reinforcement engineering.

Fig. 5. View of near completion of the bridge reinforcement project

3 Analysis of Finite Element Method

For the multi-rib arch cable bridge, the calculation of static and dynamic behavior is complex. No simplified and practical method is available. Here, the modeling of the arch-cable structure (the new structure after reinforcement) is established by the general FEM. The web dividing of finite element calculation model showed in Fig. 6. Concrete elastic modulus is 3.0×10^4 MPa, density of concrete is 25 kN/m^3, Poisson's ratio is 0.167. Top end beam on the open caisson foundation is constraint vertically and horizontally. Because the original arch bridge is laid on the soft soil foundation, in order to analyze its mechanical characteristics correctly, a model considering foundation's action is established. In the arch feet, boundary condition is considered as level spring, in-plane rotational spring and vertical spring support to simulate the elastic support of soil on abutment. The rest of the degrees of freedom are constraint. For the Baibu Bridge engineering background, whose foundation is shallow foundation with a buried depth of 5 m, the bearing stratum of foundation soil is silt and clay. Here, foundation coefficient adopted the "m" method to calculate the foundation soil proportion coefficient according to Xiang et al. (2002).

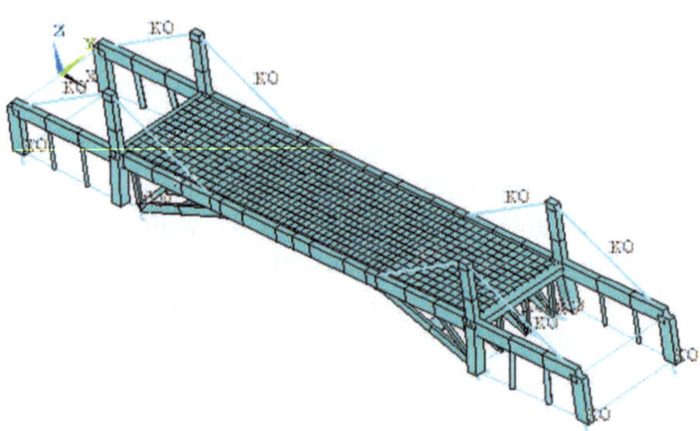

Fig. 6. Spatial finite element model after strengthening

4 Field Static Load Test

After the completion of the reinforcement engineering and maintenance of the bridge, field static load test of the bridge are carried by the equivalent truck load of the reinforcement design load standard Truck-20 and Trailer-20. The test work mainly includes deflections of the arch ribs in the vault(I-I), quartile(II-II) section, the normal strain of the arch ribs in the vault(I-I), quartile(II-II) section in longitudinal direction of the bridge and possible crack and the width and length of crack development. During construction of the bridge, the cable forces and varying was also tested and the controlled, the control values of cable forces are 800 kN in terms of design and calculation. The strain testing use the static resistance strain gauge to measure the strain of concrete, and the temperature compensation is paid attention to during test. The deflection and displacement of arch ribs in the control section of the bridge, due to restriction of the navigation under bridge are observed by the deflection point crosshairs installed on the cross section and the total station, which· is erected in the location of sighting all deflection points. The four deflection points of II-II section are measured directly with dial indicators. The layout of the bridge test sections is shown in the Fig. 7.

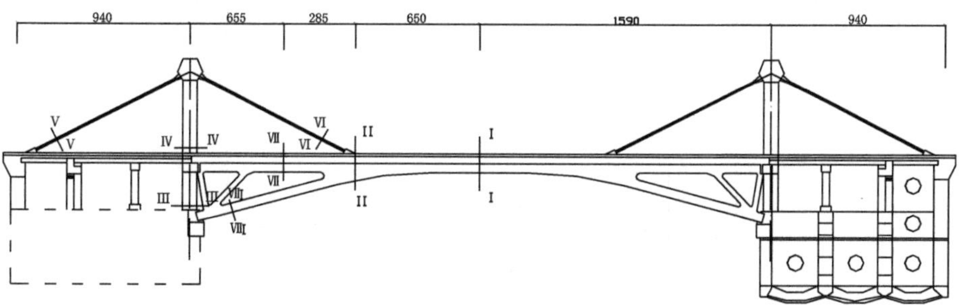

Fig. 7. Layout of bridge test sections (Dimension unit: cm)

5 Comparison of Analysis Values and Testing Results

In terms of spatial finite element analysis of typical reinforced concrete multi-rib arch bridge after reinforcement, we can get the rib's deflections of each crown mid-span section in each loading conditions. To validate the applicability and validity of the reinforcement method, this paper compared spatial calculation results obtained by FEM with the test values gained by the field testing of the background bridge. The detailed process of field test is omitted.

Table 1 lists the comparison of measured deflections and theoretical values of arch ribs of the control section in different loading cases. Figure 8 further shows the distribution along bridge width of the arch rib deflections of the control section in the typical loading cases.

From Table 1 and Fig. 8, it is found that the spatial calculated results agree well with the measured values. The general distribution trend of the rib deflections of control sections predicted by two methods is almost the same. It indicates that spatial

Table 1. Comparison of arch rib deflections of the control sections in different loading cases

Load cases	Control section	No. of arch ribs	Measured deflection Se (mm)	Theoretical values Ss (mm)	Se/Ss
Maximum bending moment at arch crown	I-I	No. 1	1.95	1.87	1.04
		No. 2	2.3	2.37	0.97
		No. 3	2.2	2.37	0.93
		No. 4	1.9	1.87	1.02
Minimum bending moment at L/4 section	II-II l/4	No. 1	0.49	0.469	1.04
		No. 2	0.45	0.498	0.90
		No. 3	0.47	0.498	0.94
		No. 4	0.44	0.469	0.94
Maximum bending moment of upper chord	I-I	No. 1	1.9	2.09	0.91
		No. 2	2.3	2.44	0.94
		No. 3	2.6	2.44	1.07
		No. 4	2.2	2.09	1.05
Minimum bending moment at base of main tower III–III	II-II l/4	No. 1	1.42	1.4	1.01
		No. 2	1.55	1.56	0.99
		No. 3	1.51	1.56	0.97
		No. 4	1.47	1.4	1.05

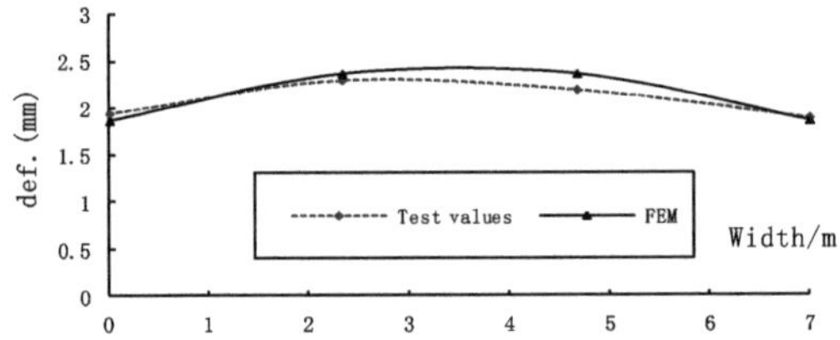

a) Deflection distribution of mid-span section under loading case of maximum bending moment at arch crown

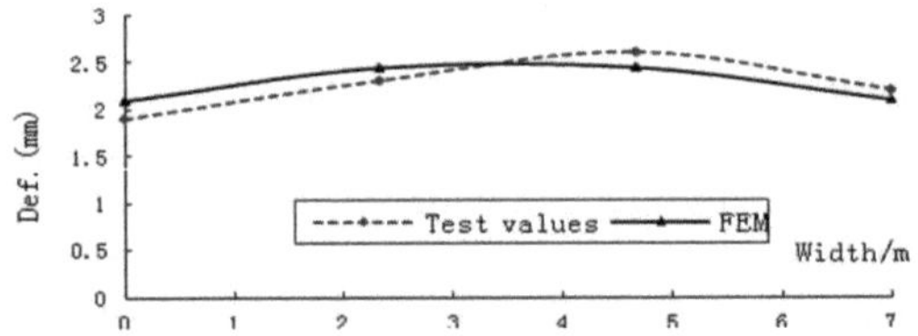

b) Deflection distribution of mid-span section under loading case of minimum bending moment at L/4 section

Fig. 8. Comparison of theoretical deflection values and test values of arch ribs under typical loading cases

finite element calculation can predict and give the actual stress state of reinforced concrete multi-rib arch bridge after reinforcement. At the same time, the proposed reinforcement method for changing multi-rib arch bridge to the **arch-cable system** bridge is feasible and effective.

6 Conclusions

Aiming at the existing lower load carrying capacity of the early reinforced concrete multi-rib arch bridges, this paper proposes a reinforcement method for changing the multi-rib arch bridge to the arch-cable system bridge. The details of reinforcement and the reinforcement plan are described. The reinforcement bridge is analyzed by spatial finite element method according to requirement of strengthening and design load standard. The field load test is completed. The results shows the spatial finite element analysis values agree well with the measured values of the field bridge test. It meets the expectation of the bridge reinforcement design. It can effectively enhance the load level and ultimate bearing capacity by changing the multi-rib arch bridge to the arch-cable system bridge. It is a kind of simple, practical, safe reinforcement design method. It not only has theoretical significance, but also has a certain practical value to bridge engineering.

Acknowledgments. The authors gratefully acknowledge the support of Department of Zhejiang Provincial Transportation.

References

China's Ministry of Transport: General Code for Design of Highway Bridges and Culverts. People Communication Press House, Beijing (1989a). JTJ021-85

China's Ministry of Transport: Code for Design of Ground Base and Foundation of Highway Bridge. People Communication Press, Beijing (1989b). JTJ024-85

China's Ministry of Transport: General Code for Design of Highway Bridges and Culverts. People Communication Press, Beijing (2004a). JTG D60-2004

China's Ministry of Transport: Code for Design of Highway Reinforced Concrete and Prestressed Concrete Bridges and Culverts. People Communication Press, Beijing (2004b). JTG D62-2004

China's Ministry of Transport: Specifications for strengthening design of highway bridges. People's Communications Press, Beijing (2008a). JTG/TJ22-2008

China's Ministry of Transport: Technical specifications for strengthening construction of highway bridges. People's Communications Press, Beijing (2008b). JTG/TJ23-2008

Xiang, Y.Q., Li, X.S., Shen, Y.G. et al.: Theory analysis and experimental study of spatial girder and arch combination bridge. China J. Highw. Transp., Xi'an Highw. Univ. (2002). 1001-7372(2002) 01-0067-05

Xiang, Y.Q., Yao, Y.D., Hu, F.Q. et al.: Study on static and dynamic behavior of t-shape rigid frame bridge reinforced by external tendons. China J. Highw. Transp., Xi'an Highw. Univ. (2004). 1001-7372(2004)01-0039-06

Xiang, Y.Q., Yang, W.L., Pan, R.Q.: Load Lateral distribution of rigid frame arch bridge strengthened with stayed cable and its experimental study. China J. Highw, Transp., Xi'an Highw. Univ. (2007). 1001-7372(2007)04-0091-05

Xiang, Y.Q., Zhang, X., Xing, Y.: Analysis and strengthening of superstructure diseases of concrete bridges in China. In: Bridge Maintenance, Safety, Management and Life Extension-Proceedings of the 7th International Conference of Bridge Maintenance, Safety and Management, Taylor and Francis-Balkema, Routledge (2014a). 20143618129434

Xiang, Y.Q., He, X.Y., Xing, C. et al.: Disease analysis and strengthening of concrete bridge substructure in China. In: Safety, management and life extension - proceedings of the 7th international conference of bridge maintenance, safety and management, Taylor and Francis–Balkema, Routledge (2014b). 20143618129435

The Influence of Spectral Responses on the Structures Heights

Mohamed Ahatri[1(✉)], Khadija Baba[2], Soukaina Touijrate[1], and Lahcen Bahi[1]

[1] 3GIE Laboratory, Mohammadia Engineering School, Mohammed V University, Rabat, Morocco
mohamed.ahatri@gmail.com
[2] GCE Laboratory, High School of Technology-Salé, Mohammed V University, Rabat, Morocco

Abstract. The earthquakes are abrupt and sudden, and, in few seconds, they can lead to huge casualties. Nowadays, several approaches can be used to better know the seismic movement than we can expect when and where will have an earthquake. These approaches kind, always begins with a full study of the seismic action: the seismic birth, the wave formation and propagation, accelerometers and associated spectrum representation, etc.

The aim of this study, on the one hand, is to construct and examine the spectral response of the ground motion for many earthquakes, on the other hand, is to study their influence on the structures height.

We studied the acceleration files for the loma prieta (6,9 Mw) and the landers earthquakes (7,3 Mw) which had an important magnitude and caused significant damage to human life and property.

1 Introduction

The seismic action can be described as a transitory disturbance due to the aleatory propagation of seismic waves in the soil. The motion in the given site is generally governed by three main factors: the source, the path travelled by the seismic waves and the local geotechnical conditions.

The damage assessment caused to structures during an earthquake is very complex, for that, the different tools have been developed to quantify the damaging potential of an excitation. We focus, in this paper, on the response spectra corresponding to the maximum response of a simple oscillator for different fundamental frequencies.

Among the seismic motion characteristics, we have the released energy which propagates in the elastic waves form. These waves are formed near of the focus and propagate towards the ground surface. There are two elastic wave types which are the compression and the shear waves, each one has its own velocity.

The elastic waves propagation and the environment properties crossed by these waves have been studied in recent years by researchers in the seismology field using the data from the response spectra.

H. Rodrigues and A. Elnashai (Eds.): GeoMEast 2018, SUCI, pp. 65–76, 2019.
https://doi.org/10.1007/978-3-030-01932-7_7

The purpose of this paper is to define the influence of the earthquake characteristics on the structures height and subsequently define the height limit of the structures in each zone or the recommendations to be made in case of exceeding this height.

A MATLAB code is developed, in the first part, to read the time versus acceleration data of the two earthquakes, which obtained from PEER Strong Motion Database. In the second place, to solve the ordinary differential equation and give various results like accelograms, displacement response spectrum, velocity response spectrum and acceleration response spectrum.

In the second part, we studied the structures stability by the spectral method in order to analyze the influence of the earthquakes characteristics on the structures knowing that they have different heights.

2 Earthquakes Data

Firstly, the 1989 Loma Prieta earthquake occurred in Northern California on October 17 at 5:04 p.m. local time (1989-10-18 00:04 UTC). The shock was centered in The Forest of Nisene Marks State Park approximately 10 mi (16 km) northeast of Santa Cruz on a section of the San Andreas Fault System and was named for the nearby Loma Prieta Peak in the Santa Cruz Mountains. With a moment magnitude of 6.9 and a maximum Mercalli intensity of IX (Violent), the shock was responsible for 63 deaths and 3,757 injuries [USGS].

Secondly, the Landers Earthquake, at 4:57 a.m. local time (11:57 UTC) on June 28, 1992, a magnitude 7.3 earthquake awoke much of Southern California. Though it turned out it was not the so-called “Big One” as many people would think, it was still a very strong earthquake. The shaking lasted for two to three minutes. The damage and loss of life were minimized by its location in the sparsely-populated Mojave Desert [USGS].

3 The Seismic Motion Parameters

3.1 Acceleration

From an accelerogram, it is generally possible to specify three phases: the initial phase when the accelerations value is low, the second phase with a high level of acceleration and the final phase with a progressive return to rest with the accelerations value very low.

3.2 Duration

The earthquake duration is difficult to define directly from the accelerogram. Some seismologists proposed to define it from the Husid diagram which is a convenient way to quantify the excitation level variation in time. By definition, if a (t) is the acceleration as a time function, T is the total earthquake duration, the integral variation is (Pecker 1984):

$$h(t) = \frac{\int_0^t \{a(t)\}^2 dt}{\int_0^T \{a(t)\}^2 dt} \quad (1)$$

3.3 Response Spectra by the Duhamel Integration Method

The acceleration response spectrum is the curve giving the maximum acceleration of a simple oscillator. A simple oscillator is a hardware system with a single degree of freedom of translation formed by a mass M and connected to the ground by means of a rigidity rod K. This system is also characterized by its damping C.

The motion equation of forced oscillations after the earthquake application in the accelerogram $a_g(t)$ form will be the following (Kramer 1996):

$$ma(t) + Cv(t) + kD(t) = -ma_g(t) \quad (2)$$

The differential equation resolution makes it possible to know the maximum acceleration undergone by the mass M.

Taking into account the damping definition ζ, the equation can be written in this form (Kramer 1996):

$$a(t) + 2\omega v\zeta(t) + \omega^2 d(t) = -a_g(t) \quad (3)$$

Where:

$$T = 2\pi\sqrt{\frac{m}{k}} \quad (4)$$

$$f = \frac{1}{T} \quad (5)$$

$$\omega = \frac{2\pi}{T} = 2\pi f \quad (6)$$

k: The system stiffness;
ω: The non-damped system pulsation (in radian/ second);
T: Period (in seconds); the motion being periodic, the cycle duration is called the motion period;
f: Frequency (in hertz).

The solution is given by Duhamel's integral. The relative displacement is obtained (Kramer 1996):

$$d(t) = -\frac{1}{\omega_D}\int_0^t a_g(\tau)e^{-\zeta\omega(1-\tau)} \sin\omega_D(t-\tau)d\tau \quad (7)$$

$\omega_D = \omega\sqrt{1-\zeta^2}$: Pseudo-pulsation of the free oscillations damped;
τ: Integration variable.

It notes that the displacement value depends only on the pulsation ω, on the damping coefficient ζ and on the ground acceleration a_g.

The structures having always a very low damping, we can consider that $\omega_D = \omega$, which corresponds to a very little damped oscillator; in neglecting the secondary terms, the seismic motion components become:

The relative displacement (Kramer 1996):

$$d(t) = -\frac{1}{\omega}\int_0^t a_g(\tau)e^{-\zeta\omega(1-\tau)}\sin\omega(t-\tau)d\tau \tag{8}$$

The relative velocity (Kramer 1996):

$$v(t) = -\int_0^t a_g(\tau)e^{-\zeta\omega(1-\tau)}\cos\omega(t-\tau)d\tau \tag{9}$$

The pseudo-acceleration (Kramer 1996):

$$a(t) = +\omega\int_0^t a_g(\tau)e^{-\zeta\omega(1-\tau)}\sin\omega(t-\tau)d\tau \tag{10}$$

The formulas below with the earthquake registration (accelerogram) allow to calculate systematically for all simple oscillators possible (i.e. for the whole periods range and possible damping) the maximum response values in the displacements terms [d(t)] max and draw the corresponding graphics, called responses spectra travel. Similarly, we can draw the maximum responses in the velocity terms [v(t)] max and the acceleration [a(t)] max.

The displacement response spectra (Kramer 1996):

$$S_d = [d(t)]_{max} = S_v/\omega = D_{max} \tag{11}$$

The velocity response spectra (Kramer 1996):

$$S_v = [v(t)]_{max} = \omega S_d \tag{12}$$

The acceleration response spectra (Kramer 1996):

$$S_a = [a(t)]_{max} = -\omega^2 d(t) = \omega S_v = \omega^2 S_d \tag{13}$$

4 Case Study

In this paper, we studied different structures with different heights (4 floors, 5 floors, 7 floors, 9 floors and 11 floors) by spectral analysis using the earthquake response spectra (Earthquakes Data). We performed the analyze using the Robot Analysis software which works by the finite Element method.

5 Results and Discussion

Using a MATLAB program which developed to solve the ordinary differential equation with the time history records of the two earthquakes. The acceleration, the time duration and the response spectra are obtained.

5.1 Acceleration

Based on the seismic inputs of the Loma Prieta 1989 and The Landers 1992 earthquakes, the acceleration response of a freedom single degree system is obtained.

For the Loma Prieta, the maximum acceleration is 0.3674 g at 6.35 s and the minimum acceleration is −0.2827 g at 4.66 s.

For the Landers, the maximum acceleration is 0.6418 g at 10.16 s and the minimum acceleration is −0.7803 g at 14.83 s.

The Figs. 1 and 2 show the acceleration as a time function of the two earthquakes.

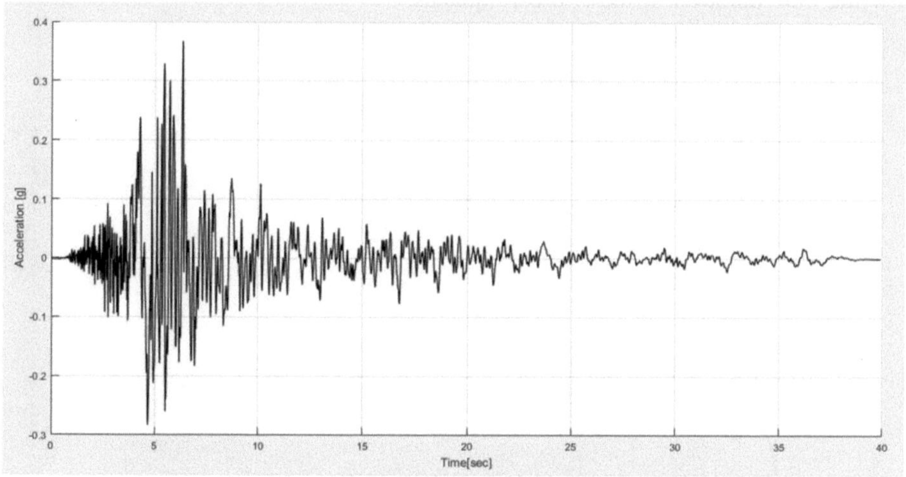

Fig. 1. Acceleration curve – The Loma Prieta 1989

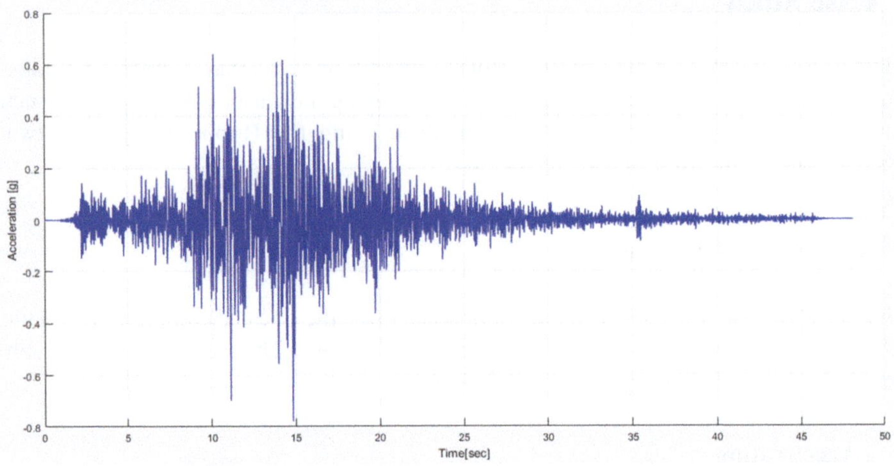

Fig. 2. Acceleration curve – The Landers 1992

5.2 Time Duration

The significant earthquake duration defined as the time interval between 5% and 95% of the total acceleration energy. For the Loma Prieta, time duration is 11,37 s and for The Landers Earthquake, is 13,73 s.

The earthquake duration is calculated from the Husid diagram, the Figs. 3 and 4 show the time duration of the two earthquakes.

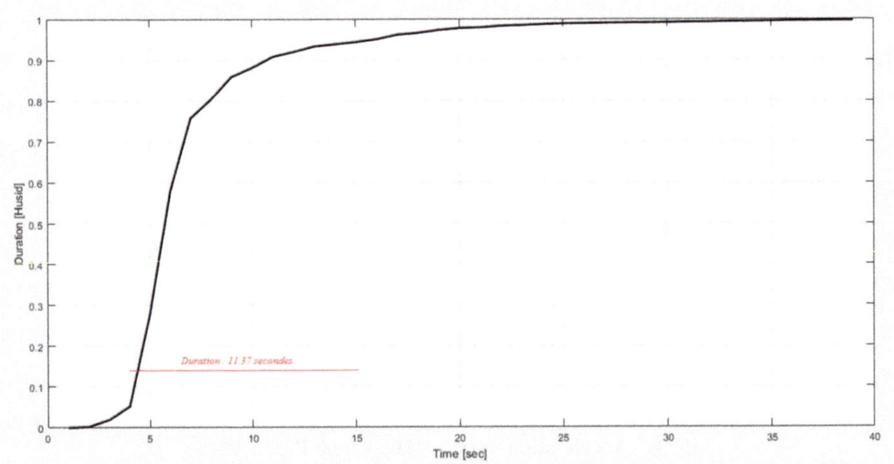

Fig. 3. Husid's diagram - The Loma Prieta 1989

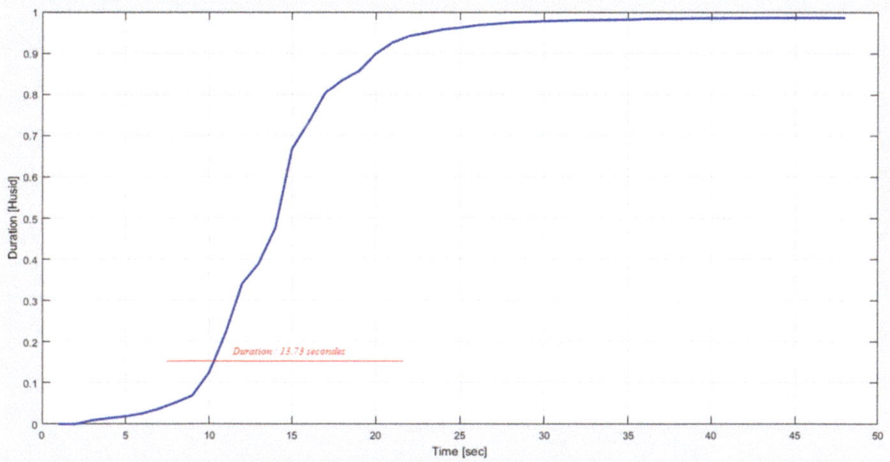

Fig. 4. Husid's diagram - The Landers 1992

5.3 The Response Spectra of the Two Earthquakes

The Figs. 5, 6 and 7 show the response spectra of 5% damping.

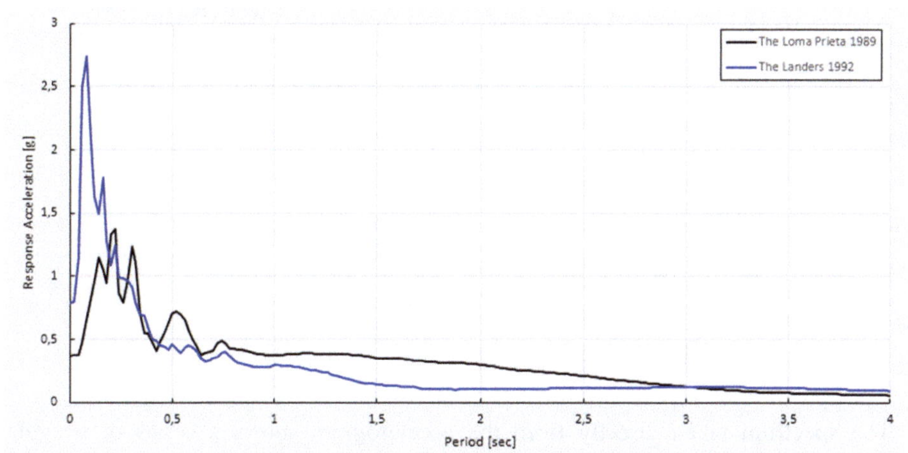

Fig. 5. Acceleration response spectrum (Sa)

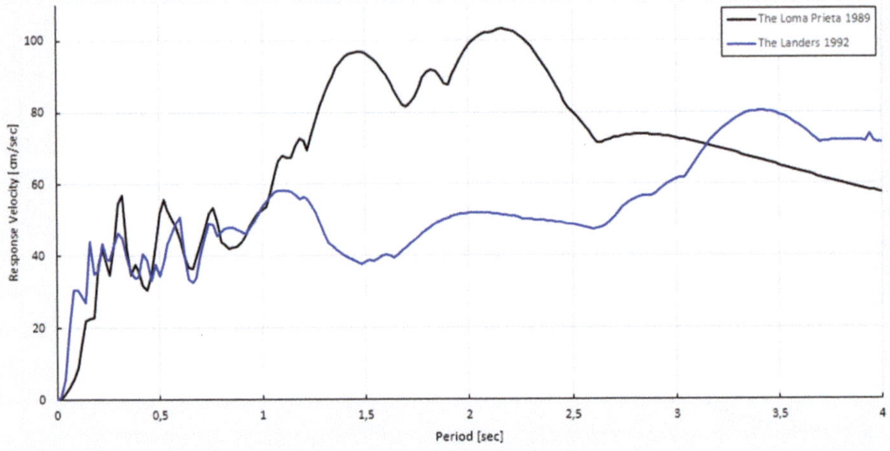

Fig. 6. Velocity response spectrum (Sv)

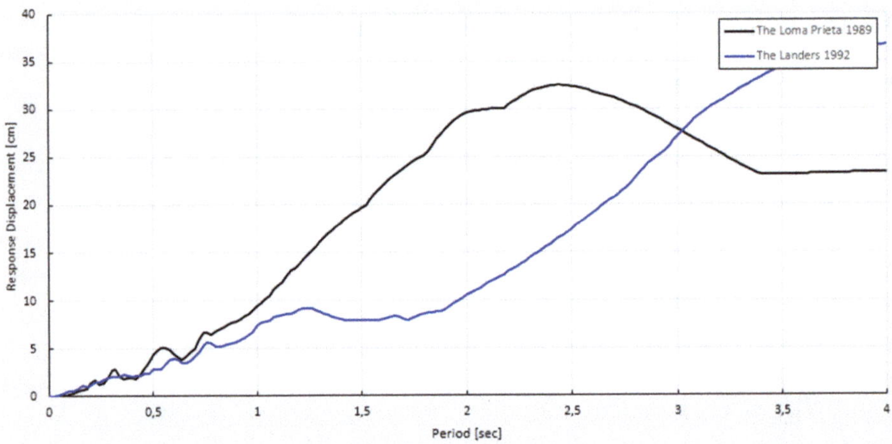

Fig. 7. Displacement response spectrum (Sd)

The spectrum taken directly from the accelerogram shows a series of irregular peaks. However, the spectral displacements increase with period, thus, the spectral velocities and accelerations, as expected, show a variation in their tendency with a decrease of the period.

Finally, depending on the derived Eqs. (11, 12, 13) (Kramer 1996), the differential motion response spectra can be obtained for any period interval.

5.4 The Response Spectra Influence on Structures in Several Heights

Using a Robot Structural Analysis software, the reactions, the displacements and the forces are obtained, for five structures with a different floors number: 4 floors, 5 floors, 7 floors, 9 floors and 11 floors.

Using the structure study based on the spectral method by the Duhamel integration for the structures design which uses the finite elements method, we had been able to see the response spectra influence on structures in several heights (Table 1).

Table 1. The Loma Prieta 1989 and the Landers 1992 earthquakes data

Earthquake	The Loma Prieta 1989	The Landers 1992
Time	00:04 UTC on October 18	11:57 UTC on June 28
Emplacement	North California	South California
Moment magnitude	6.9 Mw (Clough et al. 1994)	7.3 Mw (Landers Earthquake 2014)
Maximum perceived intensity	IX (Violent) (Stover and Coffman 1993)	IX (Violent) (Landers Earthquake 2014)
Depth	19 km (Stover and Coffman 1993)	1,09 km [USGS]
Epicenter	37.04°N 121.88°W	34.217°N 116.433°W
Type	Oblique-slip	Strike-slip

As shown in the Tables 2, 3 and 4 for the Loma Prieta earthquake, we observe that reactions and the global displacements increase from a structure to another according to the levels number, as soon as we go beyond the 8 levels, we no longer have a significant increase and these values almost stagnate. By against, the efforts continue their evolution with the height increase.

Table 2. Structure reactions values according to the the Loma Prieta earthquake 1989

Floors	FX [T]	FY [T]	FZ [T]	MX [kNm]	MY [kNm]	MZ [kNm]
4	4,75	8,37	89,9	236,72	103,8	2,83
5	5,39	9,63	109,66	276,3	121,05	3,94
7	6,01	10,95	152,15	318,09	137,96	5,17
9	6,2	11,03	196,18	345,59	143,29	5,66
11	6,3	11,32	234,4	368,37	138,04	6,13

Table 3. Displacements global values according to the the Loma Prieta earthquake 1989

Floors	UX [cm]	UY [cm]	UZ [cm]	RX [Rad]	RY [Rad]	RZ [Rad]
4	1,5	1,8	0,1	0,003	0,003	0,001
5	2,4	2,5	0,1	0,005	0,003	0,001
7	4	3,7	0,2	0,005	0,003	0,002
9	5,2	4,5	0,2	0,005	0,003	0,003
11	5,2	4,5	0,2	0,005	0,003	0,003

Table 4. Structure forces values according to the the Loma Prieta earthquake 1989

Floors	FX [T]	FY [T]	FZ [T]	MX [kNm]	MY [kNm]	MZ [kNm]
4	89,9	8,1	15,13	44,55	95,52	122,19
5	109,66	8,93	15,47	56,3	95,46	134,96
7	152,15	9,81	16,12	70,17	95,68	148,35
9	196,18	10,14	16,4	73,59	95,78	153,52
11	234,4	10,43	16,58	76,23	95,85	158,77

Table 5. Structure reactions values according to the the Landers earthquakes 1992

Floors	FX [T]	FY [T]	FZ [T]	MX [kNm]	MY [kNm]	MZ [kNm]
4	3,99	5,72	89,61	157,78	83,63	2,12
5	4,21	4,71	110,19	128,75	89,65	2,56
7	3,42	4,19	149,29	113,79	69,23	1,96
9	3,48	5,43	188,14	152,6	71,08	2,2
11	3,93	6,71	234,67	192,53	83,22	2,82

Table 6. Displacements global values according to the the Landers earthquake 1992

Floors	UX [cm]	UY [cm]	UZ [cm]	RX [Rad]	RY [Rad]	RZ [Rad]
4	1,2	1,2	0,1	0,003	0,003	0,001
5	1,4	1,1	0,1	0,003	0,003	0,001
7	1,5	1,2	0,1	0,003	0,003	0,001
9	2,4	2,1	0,1	0,003	0,003	0,001
11	3,8	3,2	0,1	0,003	0,003	0,002

As shown in the Tables 4, 5 and 6 for the Landers earthquake, we observe that the global displacements and the forces according to X increase according to the structure height, according to Y, the efforts almost stagnate. Concerning the reactions, we have another figure case, the values increase and decrease in a very slight way (Table 7).

Table 7. Structure forces values according to the the Landers earthquake 1992

Floors	FX [T]	FY [T]	FZ [T]	MX [kNm]	MY [kNm]	MZ [kNm]
4	89,61	7,31	14,79	37,69	95,52	110,59
5	110,19	7,61	14,76	39,21	95,58	115,43
7	149,29	6,98	14,39	35,86	95,87	104,23
9	188,14	7	14,59	45,31	96,02	106,4
11	234,67	7,53	14,85	56,06	96,14	114,76

Comparing the two earthquakes results, we noticed that the Loma Prieta earthquake has greater values of displacements, efforts and reactions than the Landers earthquake, knowing that it is stronger than the first one.

6 Conclusions

The Robot structural analysis software, based on the finite element method, has been developed and improved for structural calculation, in which 5% damping has been considered for soil and structures.

The results show that the response spectra influence on the structures increases according to the latter height, knowing that this evolution remains less important when a certain number of levels are exceeded.

The comparison of this influence between the two studied earthquakes shows that the Loma Prieta earthquake is more destructive than the Landers earthquake, which shows that the response spectra influence on the structures at different heights does not depend only on the magnitude, but also on the acceleration allure, duration, depth and slip type of the earthquake.

References

Pecker, A.: livre de Dynamique des sols, Presse de l'école nationale des Ponts et Chaussées (1984)

Clough, G.W., Martin II, J.R., Chameau, J.L.: The Geotechnical Aspects. Practical lessons from the Loma Prieta earthquake, National Academies Press, pp. 29–46. ISBN 978-0309050302

Landers Earthquake: Southern California Earthquake Data Center. Accessed 15 Aug 2014
Kramer, S.L.: Geotechnical Earthquake Engineering. Prentice Hall, Upper Saddle River (1996)
Stover, C.W., Coffman, J.L.: Seismicity of the United States, 1568–1989 (Revised). U.S. Geological Survey Professional Paper 1527, United States Government Printing Office, pp. 98, 99, 180–186 (1993)
USGS: US Geological Survey
USGS: M7.3—Southern California. United States Geological Survey

Effect of Steel Slag on Cold Bituminous Emulsion Mix

Amol Rakhunde(✉) and N. A. Hedaoo

Department of Civil Engineering, College of Engineering, Pune, India
amolsr92@gmail.com, nah.civil@coep.ac.in

Abstract. Due to limited modes of practice of utilization, huge amount of iron and steel slag dumped in yards of each production unit and engaging of important agricultural land and gave pollution to whole environment. The effective use of steel slag, as by product of steel industry which is sustainable construction material, in cold bituminous emulsion mix (CBEM) has ultimate benefits such waste disposal, saving natural resources and lowering cost of roadways. Effect of mixes prepared with steel slag were compared with the mixes prepared with replacement of cement and control mix. Experimental work was carried out on the sample of control mix, mix with cement replacement and modified mix. For modified mix, steel slag were replaced by 10%, 20%, 30% and 40% of aggregate having same size as of steel slag in gradation. Filler were replaced by 1%, 2% and 3% of weight of total aggregate by cement. Optimum emulsion content of each mix obtained by using Marshall stability test and comparison of stability values were carried out. The comparison of results of indirect tensile strength test and retained stability test on CBEM prepared with steel slag, cement and control mix. Significant improvement in Marshall stability, retained stability and indirect tensile strength of mixes prepared with the replacement of steel slag as compared to control mix and mixes prepared with replacement of cement.

1 Introduction

Sustainability is a primary focus of 21st century engineering, and therefore, the use of sustainable materials has been investigated for their economic, environmental, and social benefits. Different types of fillers can be used for enhancement of performance of cold bituminous emulsion mixes. Hydrated lime and Ordinary Portland Cement (Al Nageim et al. 2012) are most commonly used as filler to improve the performance of cold bituminous emulsion mixes. Some waste materials and by product from industries such as fly ash, GGBS, rice husk ash, silica fume (Nassar et al. 2016) can also be used as fillers in the cold mixes.

Steel slag is a co-product of the steel industry and is, with proper design, a sustainable construction material. In an integrated steel plant, 2–4 tonnes of wastes (including solid, liquid and gas) are generated for every tonne of steel produced. Steel slag is an industrial by-product obtained from the steel manufacturing industry.Current total production of steel slag in India, are around 12 million tonnes per annum (Indian Minerals Yearbook 2017), which is far behind the developed countries. Presently in

H. Rodrigues and A. Elnashai (Eds.): GeoMEast 2018, SUCI, pp. 77–86, 2019.
https://doi.org/10.1007/978-3-030-01932-7_8

India, due to limited modes of practice of utilization, huge amount of iron and steel slag dumped in yards of each production unit and engaging of important agricultural land and gave pollution to whole environment. An efficient approach to overcome these problems is the slag utilization. Physical and chemical characterization of steel slag is a deciding factor of steel slag utilization as recycled raw material as road aggregate, cement and concrete admixture, soil stabilizer and construction material, etc.

Therefore, it is imperative to carry out a comprehensive study, based on local materials and conditions, of the performance of CBEM that contains steel slag and compare that performance with CBEM that contains conventional aggregates. In a present study, medium setting emulsion was used for preparation of cold bituminous emulsion mixes. Marshall mix design was carried out to determine the optimum emulsion content. The retained stability and indirect tensile strength tests were performed on the cold bituminous emulsion mixes prepared at optimum emulsion content. Details of different mixes prepared are shown Table 1.

Table 1. Details of mixes

Type of mix	OPC %	Steel slag %
Control Mix (CM)	0	0
CM + 1% OPC	1	0
CM + 2% OPC	2	0
CM + 3% OPC	3	0
CM + 10% steel slag	0	10
CM + 20% steel slag	0	20
CM + 30% steel slag	0	30
CM + 40% steel slag	0	40

2 Materials

2.1 Aggregate

The physical properties of the coarse aggregates used in this study are given in Table 2. Basalt-type aggregates bought from Samarth crusher site located Jambhulawadi, Pune district in Maharashtra, India were used in this investigation. Aggregate gradation adopted as per Ministry of Road Transport and Highway (MoRTH 2013) for the preparation of CBEMs shown in Table 3.

2.2 Emulsion

The results of different tests carried out on the emulsion are given in Table 4. The results of viscosity and residue by evaporation tests indicate that the grade of emulsion used in this study was a medium setting. The results of a particle charge test indicate that the emulsion was cationic.

Table 2. Physical properties of aggregates

Property of aggregate	Type of test	Test method	Maximum value (%)	Results (%)
Hardness	Los Angeles abrasion test	IS: 2386 (Part IV)	40	15.52
Toughness	Aggregate impact test	IS: 2386 (Part IV)	30	7.56
Crushing resistance	Crushing value test	IS: 2386 (Part IV)	30	9.35
Flakiness and elongation Index	Shape test	IS: 2386 (Part I)	30	22
Water absorption	Water absorption	IS: 2386 (Part III)	2	0.4

Table 3. Aggregate gradation

IS sieve size (mm)	Cumulative % weight passing by weight of total aggregate	
	Prescribed range	Adopted value
26.5	100	100
19	90–100	95
13.2	-	-
9.5	60–80	70
4.75	35–65	50
2.36	20–50	35
0.30	3–20	11
0.075	2–8	5

2.3 OPC

53 grade of OPC was used in this investigation. Filler was replaced with the same mass of OPC. The OPC was added in three different proportions, 1, 2 and 3% by weight total of aggregates. The specific gravity of the OPC was 3·10.

2.4 Steel Slag

Steel slag used in this study were brought from Rambo Industry, Coimbatore, India. Specific gravity of steel slag is 2.41. 96% of steel slag is retained on the 2.36 mm and 300μ sieve size. So, same size of aggregate from aggregate gradation are replaced by steel slag from 10 to 40% (of weight aggregate of same size). Chemical composition of steel slag is obtained by Scanning Electron Microscopy (SEM) analysis carried out in Metallurgy department, College of Engineering, Pune are given in Table 5.

Table 4. Emulsion test result

S. No.	Characteristics	Standards	Reference	Results
1	Viscosity by saybolt furol viscometer, sec, at 50 °C	50–300 s	IS 3117: (2004) (A)	74 s
2	Coagulation of emulsion at low temperature	Nil	IS 8887: (2004) (E)	Nil
3	Particle charge	Positive	IS 8887: (2004) (E)	Positive
4	Coating ability and water resistance:			
	1. Coating, dry aggregate	Good	IS 8887: (2004) (F)	Good
	2. Coating, after spraying	Fair	IS 8887: (2004) (F)	Fair
	3. Coating, wet aggregate	Fair	IS 8887: (2004) (F)	Fair
	4. Coating, after spraying	Fair	IS 8887: (2004) (F)	Fair
5	Residue by evaporation, Min	65%	IS 8887: (2004) (J)	68%

Table 5. Chemical properties of steel slag

Elements	% Weight
Oxygen (O)	48.33
Silicon (Si)	24.63
Iron (Fe)	14.83
Aluminium (Al)	4.15
Manganese (Mn)	3.87
Calcium (Ca)	2.45
Magnesium (Mg)	1.75
Total	**100**

3 Specimen Preparation

The procedures adopted for the preparation of specimens for different tests were as follows.

3.1 Determination of Initial Emulsion and Water Content

The CBEMs were prepared by following IRC:SP:100-2014 (IRC 2014). The following steps were involved.

Determination of initial emulsion content by using, $P = 0.05A + 0.1B + 0.5C$, where P is the quantity of initial emulsion content (%), A is the percentage of aggregates retained on a 2·36 mm sieve, B is the percentage of aggregates passing the 2·36 mm sieve and retained on a 90 μm sieve, and C is the percentage of aggregates passing a 90 μm sieve.

Determination of premixed water content by preparing different samples of aggregates, initial emulsion content and varying water content. Maximum coating of the aggregate were visually observed and taken as premixed water content.

Determination of optimum emulsion content by using Marshall stability test.

3.2 Sample Preparation for Different Tests

As per IRC:SP:100-2014 (IRC 2014), the specimens were prepared with different sizes of aggregates as per MoRTH (2013) were blended with premixed water content and emulsion content varied from 7.5 to 9% with increment 0.5% to determine the optimum emulsion content. Three specimens were tested for each percentage. After thorough mixing, the mixture was dried for about 1–2 h under a fan and then oven dried at 40 °C for 2 h. The cold mix was then transferred to a Marshall mould with a filter paper on the base plate and compacted with 50 blows by a Marshall hammer on both faces. The Marshall specimens were then extracted from the moulds after 24 h and cured in an oven at 40 °C for 72 h before subjecting them to the Marshall stability test. Tests for retained stability and indirect tensile strength were performed on specimens prepared at optimum emulsion content.

4 Experimental Details

4.1 Marshall Stability Test

The specimens for the Marshall stability test were prepared by following the procedure given above. The test was conducted at 25 °C, following IRC:SP:100-2014 (IRC 2014) and the emulsion content corresponding to 4% air voids in a compacted specimen was selected as the optimum emulsion content.

4.2 Retained Stability Test

The loss of adhesion of aggregates with bitumen is studied by using Retained Stability test to examine the effect of additive on resistance to moisture induced damage. It is the ratio of the stability test performed in conditioned and unconditioned states. The test was performed by following the Asphalt Institute Manual Series No. 14 (MS-14) (AI 1989).

4.3 Indirect Tensile Strength

The indirect tensile strength test was carried out on specimens prepared at optimum emulsion content. Three specimens were tested for each mix. Samples of 100 mm diameter and 63 ± 2·5 mm height were used. The test was conducted at 25 °C, following ASTM D 6931 (ASTM 2012).

5 Results and Discussion

5.1 Marshall Stability Test

The optimum emulsion content and different Marshall properties obtained for different mixes are given in Table 6. From Table 6 it can be observed that the optimum emulsion content increased with an increase in the proportion of steel slag and OPC. From Fig. 1, the stability values of mixes prepared with the addition of 30% steel slag and 2% OPC were higher than other mixes prepared with steel slag and OPC respectively. However, the stability values of mixes prepared with steel slag were marginally higher than those of mixes prepared with OPC. From Fig. 2, it can be observed that air voids decreases as the emulsion content increases. As per MoRTH, minimum 17% of Voids in Mineral Aggregate (VMA) should be present. From Fig. 3, it can observed that all type of mix has higher VMA than 17%. From Fig. 4, it can be observed that Voids Filled Bitumen (VFB) increases as the emulsion content increases. As per IRC SP 100, there must be at least 2 mm of flow. From Fig. 5, all the type of mix has higher flow value than 2 mm.

Table 6. Design emulsion content and Marshall test results

Mix	Optimum emulsion content (%)	Stability (kN)	Flow (mm)	VMA (%)	VFB (%)-
Control mix (CM)	7.70	14.00	3.25	20.22	80.00
CM + 1% cement	8.05	14.86	3.75	20.89	80.16
CM + 2% cement	8.50	18.30	4.25	21.72	81.09
CM + 3% cement	8.90	15.90	5.00	22.17	82.66
CM + 10% steel slag	8.10	16.00	3.80	20.80	79.72
CM + 20% steel slag	8.25	18.00	4.00	20.25	82.57
CM + 30% steel slag	8.60	19.90	4.40	21.45	81.10
CM + 40% steel slag	8.90	16.60	5.10	21.90	82.63

5.2 Retained Stability Test

The stability values of mixes in conditioned and unconditioned states are shown in Fig. 6. The values of retained stability ratio for different mixes are shown in Fig. 7. From Fig. 7, it can be observed that all mixtures, other than the control mix, passed the minimum required retained stability ratio of 50% as per MoRTH (2013). The mixes prepared with 40% steel slag and 3% OPC showed a significant increase in retained stability ratio.

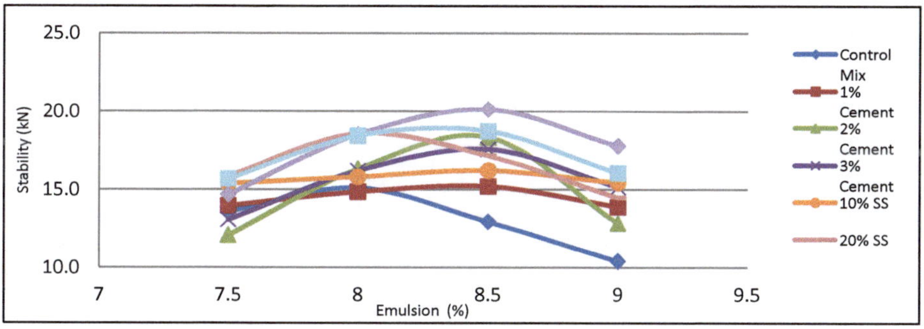

Fig. 1. Emulsion (%) vs. stability (kN)

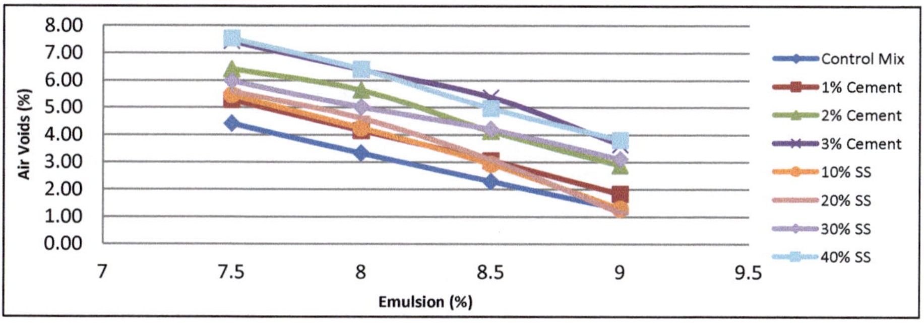

Fig. 2. Emulsion (%) vs. air voids (%)

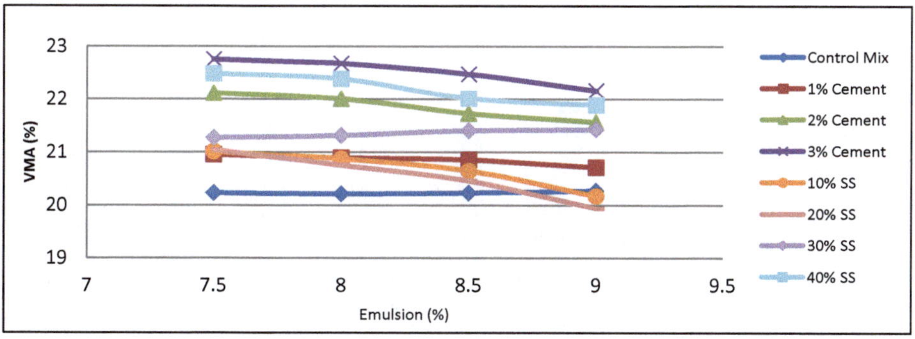

Fig. 3. Emulsion (%) vs. VMA (%)

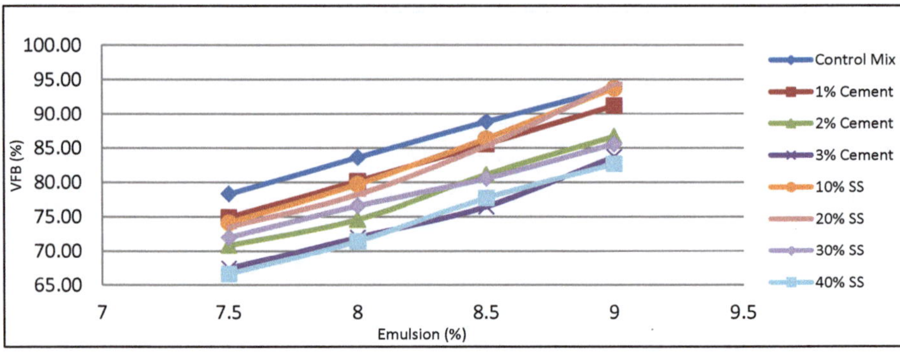

Fig. 4. Emulsion (%) vs. VFB (%)

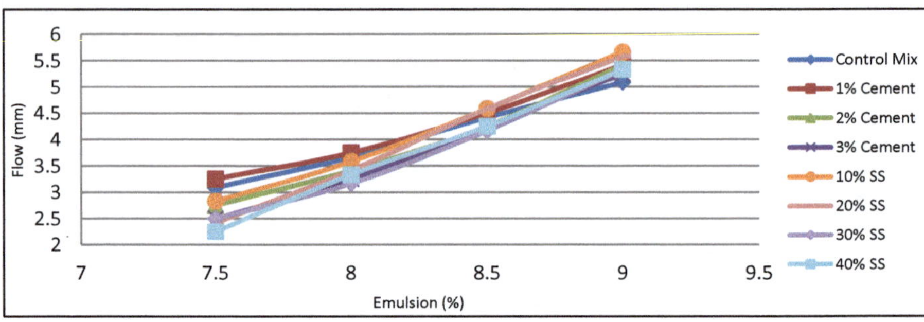

Fig. 5. Emulsion (%) vs. Flow (mm)

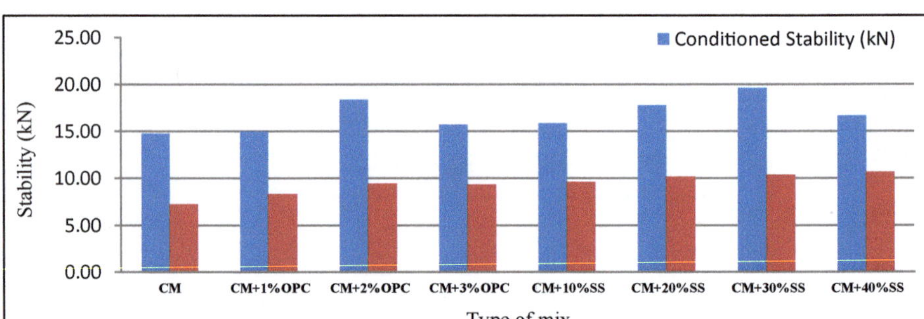

Fig. 6. Retained stability test results

5.3 Tensile Strength

The indirect tensile strength test was performed as per ASTM D 6931 (ASTM 2012), the results are shown in Fig. 8. From Fig. 8, it can be observed that the tensile strengths of mixes prepared with steel slag and OPC were higher than that of the control mix. The tensile strength of the mixes increased with increasing proportions of steel slag and OPC.

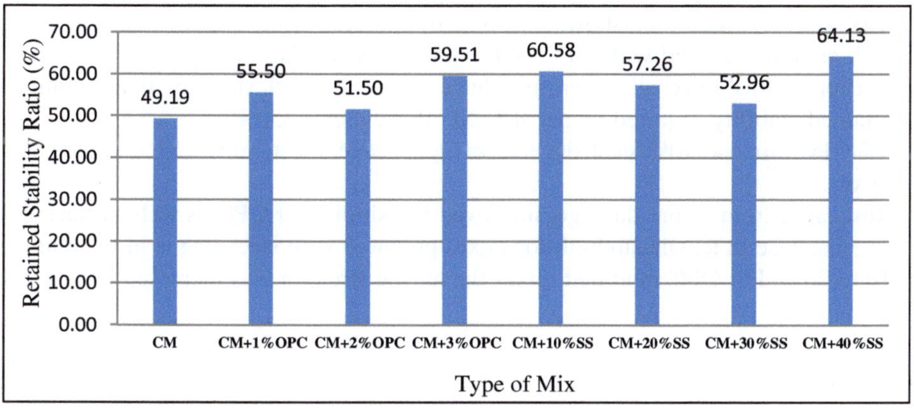

Fig. 7. Retained stability ratios

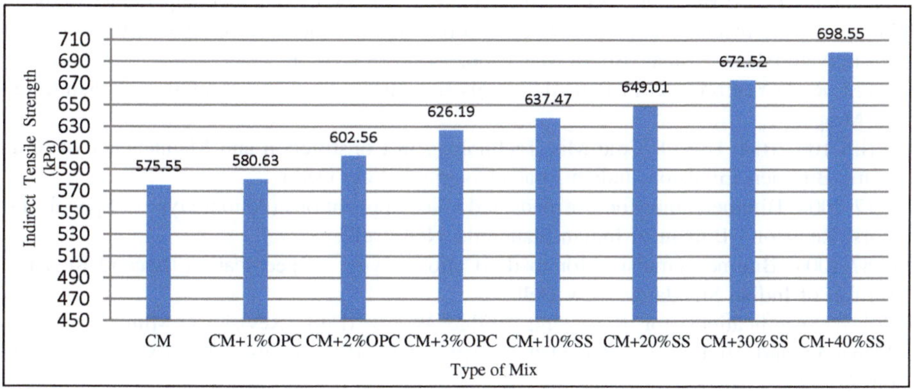

Fig. 8. Indirect tensile strength values of different mixes

6 Conclusions

Initially the optimum emulsion content were determined by using Marshall stability test and results of Marshall stability, Retained stability and Indirect tensile strength obtained at respected optimum emulsion content. Then by comparing all the results of Marshall stability, Retained stability and Indirect tensile strength for control mix, cement replacement mix (1%, 2% and 3% OPC) and modified mix (10%, 20%, 30% and 40% steel slag), following conclusion can be drawn.

- Mix with OPC (1 to 3%) and steel slag (10 to 40%) have more stability value than control mix. Hence by modifying mixes with OPC and steel slag there is an increment of stability value compared to control mix.
- Mix with 2% OPC shows higher stability values as compared to control mix, 1% and 3% OPC.

- Mix with 30% steel slag shows higher stability value as compared to control mix as well as mix prepared with OPC and remaining steel slag mixes.
- Except control mix all other mix have more retained stability than minimum retained stability required as per IRC SP 100 which is 50%.
- 3% OPC mix and 40% steel slag shows higher retained stability ratio compare other mixes.
- Resistance to moisture damage, improved by addition of OPC as well as steel slag.
- The increase in tensile strength for mixes prepared with steel slag is in the range of 10 to 20%. For OPC modified mixes the tensile strength was increased by 0 to 5%.

References

Nassar, A.I., Thom, N., Parry, T.: Optimizing the mix design of cold bitumen emulsion mixtures using response surface methodology. J. Constr. Build. Mater. **104**, 216–229 (2016)

Al Nageim, H., Al-Busaltan, S.F., Atherton, W., Sharples, G.: A comparative study for improving the mechanical properties of cold bituminous emulsion mixtures with cement and waste materials. J. Constr. Build. Mater. **36**, 743–748 (2012)

Indian Minerals Yearbook-2017 Part- II : Metals and Alloys, Slag- Iron And Steel, Indian Bureau Of Mines

IRC: SP: 100-2014: Use Of Cold Mix Technology in Construction and Maintenance of Roads Using Bitumen Emulsion. Indian Roads Congress, New Delhi

IS 3117-2004: Bitumen emulsion for roads and allied applications (Anionic type) - Specification (First Revision), Bureau of Indian Standards, New Delhi

IS 8887-2004: Bitumen emulsion for roads (Cationic type) - Specification (Second Revision), Bureau of Indian Standards, New Delhi

MORTH: Specifications for Road and Bridge Works (Fifth Revision). Ministry of Road Transport and Highways, New Delhi, Section 500, Bituminous cold mix, Clause 519.1, pp. 227–232 (2013)

Koranne, S.S., Valunjkar, S.S.: Utilisation of steel slag in roads of Marathwada region. Int. J. Innov. Eng. Res. Technol. **2**(7) (2015). https://www.ijiert.org/paper-details?paper_id=140251

Post-consumer Cullet and Potential Engineering Applications in North America

Ifeanyi Ndukwe and Qiuyan Yuan(✉)

Department of Civil Engineering, University of Manitoba, Winnipeg, MB R3V5V6, Canada
qiuyan.yuan@umanitoba.ca

Abstract. Finding viable markets for post-consumer cullet generated in North America has been very challenging due to the predominant collection practice of glass commingled with other recyclables such as paper, ceramics, food particles, and glass of fluctuating colors and compositions. Thus, more than 60% cullet still ends up in the landfill. In this paper, potential end markets that could utilize this type of cullet with minimal level of processing have been identified. Based on several field trials conducted as well as the evaluation of the engineering properties, environmental impact, and safety issues, mixed cullet aggregates could be successfully blended with natural aggregates at different proportions for road-based applications, asphalt pavement, and concrete, utility and other construction projects. The debris level of the cullet aggregates for most of the applications should be maintained at 5%. Despite the vast potential end markets being identified, market demands still remain low. One of the major obstacles is the relatively high processing cost for cullet in comparison to competing natural aggregates. Therefore, to enhance recovery and develop strong market demands for cullet, there is a need for government interventions through the provision of recycling incentives, landfill ban, raising of tipping fees, and enforcing laws on the use of cullet as a construction aggregate.

Keywords: Cullet · Engineering application · Glass recycling Aggregate

1 Introduction

Post-consumer cullet comprises of vast glass products recovered from the consumer market after usage. This post-consumer cullet typically comes from a variety of sources including the municipal solid waste (MSW), industrial, construction and demolition (C&D), end-of-life vehicles, and electric/electronic streams. However, the majority of this type of cullet (~87%) comes from the MSW stream. This cullet occurs predominantly in the form of container glass including wine and liquor bottles, beer and soft drink bottles, jars of food and other consumer products such as flat glass, and bulb glass. According to the United States Environmental Protection Agency (USEPA) reports, post-consumer cullet constituted about 4.6% (11.6 million tons) of all MSW generated in the United States in 2012 with a recovery rate of less than 30% (USEPA 2014). In Canada, it has been estimated that out of the approximately 850,000 tons of

H. Rodrigues and A. Elnashai (Eds.): GeoMEast 2018, SUCI, pp. 87–101, 2019.
https://doi.org/10.1007/978-3-030-01932-7_9

post-consumer cullet that enters the Canadian marketplace, only slightly less than half was recovered through various recycling programs (Rutledge and Gagnon 2002). The outstanding amount of the unrecovered cullet ends up in the landfill. Glass cullet is considered a valuable resource due to its recyclability. Therefore, its disposal in the landfill is not an acceptable practice. Theoretically, glass cullet is a 100% recyclable material and therefore it could be recycled indefinitely without the loss of its quality. Another problem associated with glass cullet disposal in the landfill is the non-biodegradability of this material. It takes about a thousand years or more for glass cullet to completely biodegrade in the landfill. Hence, this practice subsequently leads to the consumption of a valuable landfill space. Therefore, the implementation of a rigorous recycling program for glass cullet is vital to optimize the utilization of this valuable resource as well as the conservation of a valuable landfill space.

2 Post-consumer Cullet Recycling Process

Post-consumer cullet from the MSW stream majorly consist of glass container products ($\sim$89%) of food and drink bottles as well as jars of different colors including clear, green, and amber. Once collected, depending on the existing collection program in place, these cullet are transported along with other recyclables to the material recovery facility (MRF) where they are processed and transported to the local end markets (Fig. 1). As shown in Fig. 1, the first step of processing involves the sorting and separation of glass cullet from other recyclables through manual (hand picking) or mechanical means (magnetic separation). They are then conveyed to the pulverising unit for size reduction. Depending on the technological advancement of the facility and the predominant local market, an automated sorting system (optical sorting) could be employed for improved color sorting and separation of opaque non-glass materials.

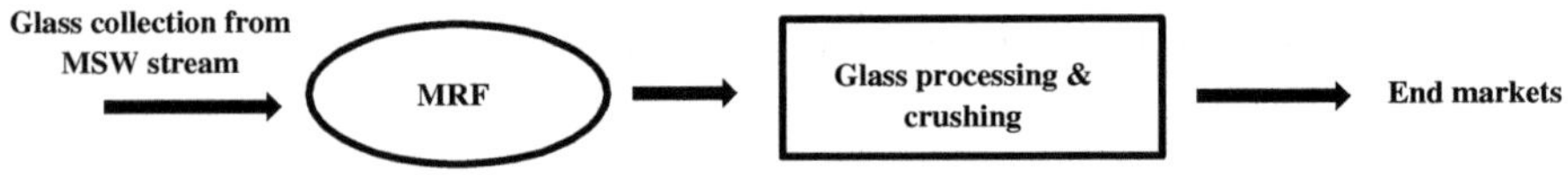

Fig. 1. Simplified block diagram for glass cullet recycling process

3 Established Markets for Processed Glass Cullet in North America

Numerous consumer markets have been identified for processed glass cullet in North America. However, its utilization for new glass containers and fiberglass still remain the most established and predominant markets. According to the Glass Packaging Institute, the average rate of cullet incorporation into new glass containers in 2003 was 34% with the goal of achieving 50% incorporation of glass cullet before 2020 (GPI 2014). The demand for cullet by fiberglass manufacturers continues to grow each year. In 2014, about 51% of processed glass cullet was incorporated into the fiberglass mixture by the manufacturers in the state of California with the goal of increasing cullet utilization in the near future (Vallette 2015). A company called Nexcycle and Strategic Materials Inc. is well known for post-consumer cullet processing and distribution to the dominant markets in North America. Another company named Owens Corning Company is a leader in processed glass utilization for fiberglass production.

4 Major Factors Affecting Cullet Recovery and Usage

Despite the continuous growth in demand for processed cullet, its recovery and utilization rate in North America continues to underperform which results in higher deposition of post-consumer cullet in the landfill. Certain factors including contamination, collection practises, and local market demands are major concerns to cullet recovery and usage.

4.1 Contamination

Cullet contamination is a major concern for cullet recovery and usage. These contaminants constitute glass of different colors and composition, ceramics, metals, cookware, light bulbs, and organics including paper, food remains, and plastics. A significant amount of cullet breakage during handling and truck compaction makes it much tougher for the recyclers to carefully separate the glass cullet from contaminants which diminish the market demand for the processed cullet. For example, the glass container and fiberglass markets have a very strict requirement for processed cullet. For the glass container industry, only glass containers of similar composition and color can be utilized. The cullet must be color sorted and completely free from any other contaminants. This is because these contaminants could possibly damage the glass melting surface and cause undesirable tint to the glass containers (SNC-Lavalin Inc. 2006). The presence of contaminants even at a relatively small concentration has been reported to cause problems in furnace and production line of fiberglass making process. Sensitivity is thus a limiting factor in the utilization of processed cullet for this application. In Canada, Owen Corning has stopped accepting post-consumer cullet as a result of contamination issues (SNC-Lavalin Inc. 2006).

4.2 Collection Practice

The collection practice of glass cullet in a particular region is a major determinant of the level of contaminants present in the processed cullet. In North America, two major collection practices exist for cullet; the deposit return program (DRP) and the multi-material curbside collection. According to USEPA (2001), DRP is defined as a "combination of a product charge (the deposit) and a subsidy for recycling or proper disposal (the refund)". The handling fees incurred by manufacturers and vendors of products are partly balanced by the interest earned on deposits, unclaimed deposits, and sales of collected used products. A refundable deposit is made available for the return of beverage containers including wine, spirit, and beer glass containers to the designated centres. A 2012 report on beverage container collection rate in Canada showed that refillable beer collection rate has increased, averaging 97% nationally

(CM Consulting 2014). In the United States, only 10 states till date have passed the beverage container deposit legislation commonly referred to as the bottle bills. It has been reported that the states that passed the bill law have achieved an average beverage container recycling rate greater than 70% (USEPA 2001; Walls 2011). Cullet collected from this program is usually free from contaminants; which are recycled and sold to container glass and fiberglass markets.

The multi-material curbside collection practice in different municipalities across North America involves glass container (food and drink bottles, Jars) collection commingled with other recyclables including glass of different compositions and color, plastics, paper, metals, cookware, and ceramics. This practice has contributed to increased public and household participation in valuable resource recovery. However, of all recyclables collected through the curbside program, post-consumer cullet is the most affected as it is almost impossible to prevent glass breakage during transportation to the Material Recovery Facilities (MRFs). These contaminants have resulted in recovering low-quality cullet which is rejected by the glass container and fiberglass markets. Thus, the low market demand for this type of cullet has significantly limited its recovery and recycling rate in North America (Rutledge and Gagnon 2002; USEPA 2014). According to the Container Recycling Institute (CRI), an average 40% of post-consumer cullet collected through the single stream/commingled option are processed and returned for new glass and fiberglass production while the remaining 40-60% ends up in the landfill as a result of contamination. However, for the glass cullet collected through the container deposit system, approximately 98% are utilized in the fiberglass markets and for producing glass containers (CRI 2009).

On the contrary, the recovery and recycling rate for post-consumer cullet have been particularly impressive in most European countries. A recent report has shown that the European member states on an average collect and recycle about 73% of all glass containers in the marketplace. Sweden, Belgium, Luxembourg, Austria, and Germany have the highest glass cullet recycling rates. Close to 80% of all the glass containers collected by the member states are processed locally for glass container production (FEVE 2015). Several collection systems including the bring point/drop-off centres and to a lesser extent, the door-to-door separate collection systems for glass cullet have played a key role in achieving the high recovery rate. Glass cullet recovered through these methods is usually not susceptible to cross-contamination and therefore could be easily recycled and returned back to the market. For example, in the city of Brussels, Belgium, majority of post-consumer cullet within the region are collected through the bring points/drop-off centers. However, depending on demands, they could also be collected through the curbside door-to-door separate collection system. In this case, glass is collected separately in a single bin without other recyclables. The recovery rate of cullet within the city is estimated to be higher than 80% (European Commission 2015).

5 Improving Cullet Recovery and Usage

Post-consumer cullet recovery and utilization have not been particularly impressive in North America (less than 30%) due to the use of predominant single stream collection scheme. This collection system results in cross contamination and inconsistency in the cullet supply to the final consumers. Therefore, to improve cullet recovery and usage, certain policies have to be introduced including; the creation of dual-stream (separate collection for glass) recycling programs and drop-off centers, the creation of local markets for mixed cullet by the government, and educating the public about the importance of active involvement in the careful disposal of recyclables.

5.1 Dual-Stream Recycling Programs and Drop-Off Centers

Though the single stream recycling program in North America has led to the high participation rate and large amounts of recyclables turnout, contamination still remains a major issue. Therefore, introducing a separate collection system for glass and/or drop-off centers will help to process high quality and contaminant-free cullet (Morawki 2010). A report has shown that mixed cullet collected from the dual-stream program resulted in an average of about 90% cullet being recycled and reused in the glass container and fiberglass markets (CRI 2009).

The transition from a dual stream to single stream (commingled) recycling program in a community has been reported to stir a significant increase in the collection of recyclables. Recycling rate increment has been observed to vary from 10%–100% accompanied by low collection costs (Abramowitz and Timpane 2010; Fitzgerald et al. 2012). However, single-stream MRFs rely heavily on advanced/sophisticated recycling process to sort the commingled materials due to contamination issues. Very high processing and maintenance costs are typically incurred (Lakhan 2015). On the other hand, dual-stream recycling system/drop-off centers are generally reported to requiring fewer investment capitals at the MRF to sort the materials due to less contamination/debris. However, as mentioned earlier, higher collection costs and reduced level of community participation are the major problems facing this program.

Collection savings from single-stream systems have been reported to vary from 5-25% according to the University of Wisconsin Solid and Hazardous Waste Education Center (SWHEC 2005). On the contrary, other reports have shown that the processing cost for the single-stream system is about 7–50% higher (Lantz 2008; Stantec Ltd. 2009), which will help offset the cost of collection from the dual system. A case study analysis comparing the overall collection and processing costs for recyclables through the single-stream and dual-stream systems in Ontario, Canada, concluded that the single-stream system was about 29% more expensive than the dual-stream recycling program (Lakhan 2015).

5.2 Creation of Local Markets for Mixed Cullet

Mixed cullet as the name implies, consists of glass of different compositions and color with some level of other contaminants present. Mixed cullet utilization as an aggregate in various engineering applications is a potentially emerging market. Creation of a strong market demand for this type of cullet will improve recovery and usage rate. Therefore, using mixed cullet for engineering applications in abundant supply and not suited for container glass and fiberglass markets could potentially result in the increased recovery rate. This is also an inexpensive utilization option for glass cullet in comparison to either container glass or fiberglass markets due to the low processing costs required. Depending on the engineering application, up to 10% contaminants/debris level (by weight) could be allowed in the mix (Reddy 1999).

For cullet incorporation into engineering applications, certain parameters have to be evaluated. They include the engineering performance, environmental impact, safety and handling, and the economic analysis of the aggregate material (Nash et al. 1995).

6 Engineering Performance of Cullet Aggregate

The engineering performance parameters of interest include the specific gravity, gradation, permeability, leachability, durability, unit weight, hydraulic conductivity, shear strength, thermal conductivity, compactability, and chemical resistance. Table 1 summarises the engineering properties of cullet in comparison to sand tested according to the ASTM standards.

Table 1. Engineering properties of cullet and sand (Reddy 1999)

Property	Test method	Cullet	Sand
Specific gravity	ASTM D 854	1.96–2.52	2.6–2.8
Unit weight (pcf)	ASTM D 4254	76–109	80–130
Hydraulic conductivity (cm/s)	ASTM D 2434	10^{-1}–1	10^{-3}–10^{-2}
Durability	ASTM C 131	30–42	NA
Compactability	ASTM D 698		
Optimum moist content (%)		5	8–10
Maximum dry weight (pcf)		105–124	116–133
Shear strength	ASTM D 3080		
Cohesion (c′)		0	0
Frictional angle (ϕ)		49–53	32–38
Thermal conductivity (w/mk)	ASTM C 518	0.26–0.64	2.9–7.7
Chemical resistance (%)	ASTM D 3042	99.8	70.1

The debris level can have a significant impact on the engineering properties of glass cullet. The debris found in the cullet may include paper, foil and plastic labels, paper bags, wood debris, food residue, etc. Thus, for engineering applications, the allowable debris level is 10%. The pieces of glass cullet are predominantly angular in shape and therefore may have the potential to cut, puncture, or wedge into a material (geomembrane material) or moving parts of construction equipment. It is therefore recommended that direct contact should be avoided or cullet should be crushed to a fine size aggregate if direct contact to a material is applied (HDR Engineering Inc. 1997).

Specific gravity is the measure of a material's density. It affects the dry, partially saturated, and saturated unit weights of a porous media. The specific gravity and unit weight of cullet are less than that of sand/natural aggregate (Table 1).

Gradation (Grain size distribution), obtained through sieve analysis, is defined as proportions by mass of soil or fragmented rock distributed in specified particle-size ranges. It directly influences properties such as compaction, permeability, filtration, and shear strength. The gradation of cullet is controlled by the processing method and has been reported to be similar to crushed rock and gravely sand (HDR Engineering Inc. 1997).

Durability assesses the hardness and toughness of the material. Glass cullet's resistance to abrasion is less than that of natural aggregate (Table 1) which results in much greater wear in glass cullet than the natural aggregate. Durability is not affected by debris level.

Shear strength, as expressed by the angle of internal friction (ϕ) and measured in degrees, is defined as the maximum resistance of a soil or rock to shearing stresses. It is a design consideration that affects bearing capacity. From Table 1, the shear strength of cullet is slightly higher than the natural aggregate. Debris level has no effect on the shear strength of glass cullet aggregates.

Compaction is the densification of a soil by means of mechanical manipulation. It is a design consideration that affects density control. Owing to the compactness of the bulk material, the high shear strength of the individual particles and the high inter-particle frictional resistance, cullet compacted to a dense state is rigid and strong (Reddy 1999). The insensitivity of glass cullet to moisture content means that it can be compacted in wet weather conditions.

Hydraulic conductivity relates to the permeability of a material. The permeability of glass cullet is higher than that of natural aggregate (Table 1). Low permeability leads to the build-up of hydrostatic pressure.

Thermal conductivity is the ability of a material to conduct or resist heat flow. Table 1 shows that glass cullet has lower conductivity than the natural aggregate. Therefore, it conducts heat more slowly. Cullet has an excellent chemical resistance even under highly acidic conditions.

Based on several studies, engineering properties of glass cullet have been evaluated analysed in comparison to the natural aggregate for potential engineering applications, the maximum allowable level of cullet aggregate with debris for each application has been established which is shown in Table 2.

7 Engineering Applications of Cullet

7.1 General Construction Backfill

Glass cullet aggregate could be employed as a filling material for general construction including backfilling beside foundations, landscaping, filling under fluctuating loads such as reciprocating pumps and compressors. Major design considerations on the use of an aggregate material as a construction backfill include gradation, compaction, shear strength, and leachability (HDR Engineering Inc. 1997). Based on the Clean

Table 2. Engineering applications of glass cullet (Reddy 1999; CWC 1997)

Application	Cullet content (%)	Debris level (%)
General construction backfill	15–30	5
Non-structural backfill	100	10
Roadway construction	15–30	5
Drainage	100	5
Landfill construction	100	10
Utility construction	100	5–10
Asphalt aggregate/concrete	5–20	5

Washington Center (CWC) study on the best practices of glass recycling, cullet aggregate blending not exceeding 15% and 30% are permitted for fluctuating loads and stationary loads, respectively. However, for the non-loaded general fill, cullet aggregate blend up to a 100% could be used. The debris level is not expected to exceed 10% (CWC 1997). The Washington state department of transportation (WSDOT) specifications allow blending of up to 15% glass cullet aggregate to natural aggregates for generation construction backfill with no more than 10% of the aggregate material passing through the 4.75 mm sieve (WSDOT 1997). The state of Hawaii by law requires the use of 100% crushed cullet aggregate when available, for non-structural backfill in all state and county projects (Yamane 2014).

7.2 Roadway Construction

Glass cullet aggregate for roadway construction includes utilization in the base course, sub-base, sub-grade, and embankments. Certain aggregate characteristic tests including specific gravity, shear strength, gradation, workability, durability, compaction, and permeability have been conducted for cullet aggregate utilization in roadway construction. Based on the preliminary results obtained by the CWC, several state departments of transportation including Connecticut, New Hampshire, New York, and Washington have approved and written specifications for the use of cullet as a roadway construction aggregate in base, sub-base, and embankments (CWC 1993). Cullet aggregate not exceeding 15% have been recommended for base and sub-base applications while a higher cullet content, up to 30%, is recommended for the construction of embankments.

The cullet aggregate particle sizes are expected not to exceed 19 mm with debris level less than 5% (Nash et al. 1995). Glass cullet with a maximum particle size of 4.75 mm, similar to crushed natural and quarried aggregates, has been proven satisfactory for road work applications (Disfani et al. 2009). Blending about 5% glass cullet aggregate for base course application have been approved in New Zealand with aggregate sizes not exceeding 9.5 mm (Arnold et al. 2008).

7.3 Asphalt (Glassphalt)

Asphalt containing glass cullet as an aggregate is commonly referred to as "glassphalt". It is similar to regular hot-mix asphalt with the exception that a certain proportion of the rock and/or sand aggregate is replaced by cullet. The advantages of glassphalt include high heat retention capacity and rapid drying rate in comparison to the conventional asphalt (USEPA 1992). Several field evaluations involving the use of glassphalt for road pavement have been tested in several communities across the United States. Field results showed that blending more than 20% of cullet aggregate into the mix might lead to low skid resistance at a speed above 40 mph. Rutting is also common when studded tires are driven over glassphalt pavement (USEPA 1992; CWC 1996).

Several state departments of transportation in the United States have already approved the use of glass cullet as an aggregate for asphalt pavement structures including; California, Florida, New Hampshire, New York, Pennsylvania, and Virginia. The majority of the state specifications recommend the use of cullet as a blend material not exceeding 5% of the total weight with debris level not more than 1% (by weight). It is expected that the cullet should meet the gradation of 100% passing the 9.5 mm sieve and not more than 20% passing through the 0.075 mm sieve (Nash et al. 1995). Florida State department of transportation recommends blending higher cullet content (up to 15%) to the total asphalt aggregate. An anti-stripping agent such as hydrated lime has been used with satisfactory results when blending glass and asphalt. The anti-stripping agent helps to overcome the problem of stripping caused by loss of adhesion between asphalt and glass in the presence of water.

Field trial analyses in several communities across New Zealand have successfully incorporated between 10–15% of glass cullet into asphalt aggregate mix for pavement construction. However, the projects were discontinued due to the expensive cost of using glassphalt rather than using asphalt-based natural aggregates (Thomas 2005).

7.4 Utility Construction

This involves the use of glass cullet aggregates for trench bedding and backfill. Certain design considerations including compaction, permeability, thermal conductivity, and filtration are very important for the utilization of aggregates in utility construction. CWC recommends blending up to 100% cullet (by weight) as a backfill aggregate material not exceeding five feet below the final grade. However, for backfill loading within five feet of the loading area, cullet aggregates blend might vary from 15-100% (CWC 1997). The debris level should not exceed 5%. For pipe bedding and backfill, the glass cullet shall meet the particle size distribution of 100% passing through the 9.5 mm sieve and a maximum of 10% passing through the 0.075 mm sieve (ARRB group 2010). The states of Washington and Pennsylvania have recommended the use of glass cullet for pipe bedding and backfill.

Field trials employing 100% glass cullet aggregate as pipe bedding materials have been conducted in different counties in the state of Pennsylvania including Lackawanna County. In 2003, Berks County conducted a similar field project by utilizing 700 tons of crushed glass. Though the project was successful, certain problems were encountered during compaction attempts. The cullet exhibited the tendency to "slide" rather than compact due to the non-uniform gradation ranges of the aggregate material from the on-site material testing. Therefore, one of the ways suggested to overcoming this problem in the future, is through compulsory full gradation testing of supplier stockpile prior to usage (PennDOT 2005). Similar field trials in pipe bedding have also been carried out in several water main sites in southern and western Sydney, Australia. Blends of cullet aggregates used for the trials were 25%, 50%, and 100%. Glass cullet passing through 4 mm sieve was employed. According to New South Wales department of environment and climate change, the field trials result showed that, just like the natural aggregate, the glass cullet were easy to handle and compacted naturally with no difficulty (NSW DECC 2007).

7.5 Landfill Cover

Glass cullet could be used as a suitable media aggregate for daily landfill cover to control odour, limit moisture infiltration from the surface, and minimize leachate generation in the waste cell. Cullet required for daily landfill cover would normally undergo very minimal processing to tolerate a high level of contaminants. This is a low value, high volume application for glass cullet. Several states in the United States including; Florida, Washington, New York, Wyoming, Arizona, and Texas have already been using glass as a daily landfill cover (USEPA 2003).

7.6 Concrete (Glasscrete)

Concrete is a major component in building constructions and other civil works such as footpaths. Glass cullet can be also used as a blend aggregate in concrete production. This concrete containing glass cullet as an aggregate is referred to as glasscrete. However, the alkali-silica reaction (ASR) between cement aggregate and glass has proved to be a major drawback for its utilization. The first reported use of coarse cullet aggregate in concrete was published by Johnson (1974). This and a later report by Figg (1981) concluded that the use coarse cullet aggregates in most cases resulted in considerable ASR expansion and cracking. The concrete of this mixture is typically characterised by poor workability resulting in inefficient compaction and low strengths (Arnold et al. 2008). However, a study by Polley et al. (1998) demonstrated that ASR could be avoided by using fine-sized glass aggregate (limited to less than 1.5 mm) and blending a maximum 20% of the cullet to the concrete's aggregate. The fine-sized glass aggregate is known to exhibit pozzolanic reactive activity in the blended mix to form a higher quality hydrated cement matrix with improved compressive strengths. Incorporating about 25% fly ash to the concrete mixture is also recommended to help mitigate ASR (Polley et al. 1998). In the city of Montreal, Québec, designers of the center of sustainable development have already incorporated fine-size cullet aggregated to the concrete that have been used some of their buildings (Fily-Paré and Jolin 2013).

8 Environmental Impacts of Using Cullet Aggregate

As earlier discussed, post-consumer cullet aggregate typically contain contaminants/debris such as plastics, metals, ceramics, food residue, and paper. Although glass is a relatively inert material, contaminants present can influence its chemical constituents. Thus, the aggregate material such as…might be potentially deleterious to the environment when it comes in contact with soil, surface water, or groundwater during several construction projects. Therefore, to evaluate the environmental impact of using glass cullet as an engineering aggregate, CWC sponsored the analysis of cullet aggregates obtained from different sources based on an analytical method for toxicity characteristic leaching procedure (TCLP) of metal testing and harmful contaminants. The result of the analysis showed that only the lead concentration exceeded the maximum acceptable limit in some of the cullet aggregates obtained from different sources. This lead present has been associated with the lead in the foil wrappers typically used in wine bottles. The semi-volatile organic analysis also showed some presence of phthalate compounds, a biodegradation product of plastics. However, it was concluded that neither the organic contaminants nor the lead concentrations observed from some of the cullet aggregates will pose any significant risk to the environment (CWC 1993; HDR Engineering Inc. 1997).

Similar analytical studies of chemical and physical contaminants as well as asbestos, pH, and electrical conductivity of cullet aggregates from different sources were also conducted in New South Wales, Australia for both total and leachable concentrations. Based on the analysis, the metal concentrations and organic pollutants were observed to be well below the specified maximum limits. No asbestos was detected from the glass cullet samples. Therefore, it was concluded that the glass samples will pose no risk to the environment (NSW DECC 2007).

9 Safety and Handling of Glass Cullet

A study conducted by CWC to assess potential airborne risks of cullet aggregates have shown that it composed primarily of amorphous silica and less than one percent crystalline silica (CWC 1993). Crystalline silica is a known carcinogen and the cause of silicosis. Given the low concentrations of crystalline silica in glass cullet, it is thus, neither considered hazardous nor a significant contributor to cancer or silicosis. Cullet is also an abrasive material and can cause skin irritations and cuts during handling. This hazard can be prevented by the use of protective clothing similar to the one used for natural aggregates. Cullet particle sizes less than 19 mm pose similar skin cuts hazard as crushed natural aggregate while cullet of smaller particle sizes (less than 6 mm) has been shown to be non-hazardous during handling (NSW DECC 2007).

10 Economics of Using Glass Cullet as an Aggregate

With the current collection rate of glass cullet less than 40% in North America, a significant amount of this valuable material still ends up in the landfill. Several end markets have already been identified and discussed for mixed glass cullet. However, in order to be sustainable for large scale utilization, the end markets must cope with the economics of cullet collection, processing/recycling, and local demands. Competing materials play vital roles in determining the net market value of a product to the end users. For example, for cullet aggregate utilization in pipe bedding, the consumer is not expected to pay more than the natural aggregate it replaces. Depending on the end markets, the glass cullet requires a level of aggregate size processing which might differ considerably.

Table 3 showcases the overall processing cost and market value of cullet aggregates for different end markets in comparison to the natural aggregates. This is based on the information gathered from recyclers in the state of Hawaii, USA. The glass cullet used for these applications were processed to a 9.5 mm aggregate size. It could be seen from the table that the net market value of cullet aggregate usage for the various end markets might not be competitive with the natural aggregates. Therefore, the use of glass cullet might not save cost and recyclers may find it difficult to recycle it for this purpose due to the potential financial loss which might be incurred.

Table 3. Comparison of the processing costs of cullet aggregate and natural aggregate in the state of Hawaii (Yamane 2014)

Applications	Cullet processing cost/ton ($)	Cullet market value/ton ($)	Natural aggregate cost/ton ($)
Non-structural backfill	140	20–30	5
Glassphalt	180	100	10
Landfill uses	140	15–30	5

For glass cullet usage as a construction aggregate to be considered an economically viable option, life cycle analysis approach for post-consumer cullet has to be examined. An economic model developed by the Clean Washington Center has been used to compare the economic viability of mixed cullet usage as a construction aggregate in comparison to other available alternatives such as landfill disposal and glass container markets. The model inputs costs associated with transportation, landfill prices, labor, collection, and processing which might vary from region to region (CWC 1993). It has been used to compare the economics of glass cullet for these three options in the city of Houston, Texas. Based on the analysis, the use of glass cullet as a construction aggregate will result in an annual savings of $1,375 and $46,375 in comparison to glass container markets and landfill disposal respectively (Nash et al. 1995; Woolverton 1996). Further analysis showed that the rate of return (ROR) on using cullet as a construction aggregate over a 7-year period in comparison to glass container markets and landfill disposal were 4.8% and 309% respectively (Woolverton 1996).

11 Future Prospects and Conclusion

Even though there are potential markets in existence for mixed cullet aggregates, more than 60% still ends up in the landfill in North America. The current practice wastes a valuable resource and the consumption of a valuable landfill space. Low market demands, inadequate knowledge of cullet utilization as a construction aggregate by engineers, and relatively low cost of competing natural aggregates have significantly hampered its recovery and utilization rate. Therefore, there is an urgent need to facilitate its recovery and usage rate.

The creation of high-volume and stable markets for cullet aggregates will improve its economic sustainability in the market for both recyclers and end users. Developing these local markets would require intervention by the government through the provision of recycling incentives to offset the huge processing costs, landfill disposal ban and raising tipping fees as well as enforcing the local use of cullet as a construction aggregate when available. The co-operation of government as well as public and private sectors to set up seminars educating construction engineers on the use and acceptability of cullet as an alternative material for construction projects is also very vital to the improvement in its utilization rate. Switching to the dual-stream collection option for post-consumer cullet might also help to lower the overall processing cost and making it more competitive to the natural aggregates.

In conclusion, post-consumer cullet recovery rate is still a major cause of concern in North America owing to its non-biodegradability in the landfill and waste of a valuable resource. Several potential markets including road applications, asphalt pavement, landfill cover, concrete, and backfill have been identified. Based on the engineering properties, environmental impacts, and safety concerns analysed for crushed cullet aggregate, it could be successfully blended with crushed natural aggregates at varying proportions for different construction projects. Therefore, the development of the aforementioned end markets will be crucial in achieving a near complete diversion of cullet away from the landfill.

Acknowledgements. This research was financially supported by Green Manitoba (WRAPP 14-028) and Natural Sciences and Engineering Research Council of Canada (NSERC RGPIN-2014-05510).

References

Abramowitz, R., Timpane, M.: Consolidation question Answers are provided from a recent study that addressed a present hot-button industry issue: Does switching from dual-stream to single-stream collection and processing of recyclables reduce greenhouse gas impacts? Resour. Recycl. **14** (2010)

Arnold, G., Werkmeister, S., Alabaster, D.: The effect of adding recycled glass on the performance of base course aggregate. New Zealand transport agency research report 351 (2008)

Arrb Group Ltd.: Specifications for recycled crushed glass as an engineering material. The packaging stewardship forum of the Australian food and grocery council (2010)

CM Consulting: Who pays what. An analysis of beverage container collection and costs in Canada (2014)
CRI: Understanding economic and environmental impacts of single-stream collection systems (2009)
Clean Washington Centre, CWC: Glass feedstock evaluation project, testing program design. Dames and Moore Inc. Report No. GL-93-2 (1993)
Clean Washington Centre, CWC: Preparation and placement of glassphalt. Best practises in Glass recycling (1996)
Clean Washington Centre, CWC: Model specifications for glass aggregate. Best practises in Glass recycling. BP-GL4-01-04 (1997)
Disfani, M.M., Arulrajah, A., Suthagaran, V., Bo, M.W.: Geotechnical characteristics of recycled glass-biosolids mixtures. In: Proceedings of the 17th International Conference on Soil Mechanics and Geotechnical Engineering, Alexandria, Egypt, 5–9 October 2009, pp. 201–204 (2009)
European Commission: Assessment of separate collection schemes in the 28 capitals of the EU. Final report, European Commission-DG ENV, Brussels (2015)
FEVE: Glass recycling: years of improvement (2015)
Figg, J.W.: Reaction between cement and artificial glass in concrete. In: 5th International Conference on Concrete Alkali Aggregate Reactions, Cape Town, South Africa, pp. 252–257 (1981)
Fily-Paré, I., Jolin, M.: The use of recycled glass in shotcrete, pp. 14–16 (2013)
Fitzgerald, G.C., Krones, J.S., Themelis, N.J.: Greenhouse gas impact of dual stream and single stream collection and separation of recyclables. Resour. Conserv. Recycl. **69**, 50–56 (2012)
GPI: Achieving 50% recycled content for glass containers. Efforts, challenges, and opportunities ahead for the North American Glass Container Industry (2014)
HDR Engineering Inc.: Glass cullet utilization in civil engineering applications. Nebraska State Recycling Association (1997)
Johnson, C.D.: Waste glass as coarse aggregate for concrete. J. Test. Eval. **2**(5), 344–350 (1974)
Lakhan, C.: A comparison of single and multi-stream recycling systems in Ontario, Canada. Resources. **4**, 384–397 (2015)
Lantz, D.: Metrowaste paper recovery study. Resour. Recycl. **12**, 24–30 (2008)
Morawki, C.: Understanding economic and environmental impacts of single-stream collection systems. Container recycling institute reports (2010)
Nash, P.T., Jayawickrama, P., Tock, R.W., Senadheera, S., Viswanathan, K., Woolverton, B.: Use of glass cullet in roadway construction: laboratory testing and specification development. Research report number 0-1331-2F (1995)
NSW DECC: Trial of recycled glass as pipe embedment material. New South Wales department of Environment and Climate change. ISBN: 978 1 74122 416 0 (2007)
PennDOT: Strategic recycling program fact sheet. PennDOT district 5-0 SR 222 highway project (2005)
Polley, C., Cramer, S.M., de la Cruz, R.V.: Potential for using waste glass in Portland cement concrete. ASCE J. Mater. Civ. Eng. **10**(4), 210–219 (1998)
Reddy, K.R.: Use of glass cullet as backfill material for retaining structures. In: International Conference on Solid Waste Technology and Management, Philadelphia, pp. 1–8 (1999)
Rutledge, S.O., Gagnon, G.A.: Comparing crushed recycled glass to silica sand for dual media filtration. J. Environ. Eng. Sci. **1**, 349–358 (2002)
SNC-Lavalin Inc.: Nova Scotia Glass Study, Resource Recovery Fund Board Inc. SNC-Lavalin Project. No. 017075-0001 (2006)
Stantec Ltd.: Recycling collection operations review, CIF Project 176 (2009)

SWHEC: Single stream collection. University of Wisconsin-Extension. Solid Waste and Hazardous Waste Education Center (2005)

Thomas, C.: Market study for recycled glass in the South Island. Zero Waste New Zealand Trust (2005)

USEPA: Markets for recovered glass. Office of solid waste (0S-301). EPA/530-SW-90-071A (1992)

USEPA: Deposit Refund Systems (Chap. 5), pp. 57–66 (2001)

USEPA: Reuse/recycling of glass cullet for non-container uses. USEPA archive document (2003)

USEPA: Municipal solid waste generation, recycling, and disposal in the United States: facts and figures for 2012 (2014)

Vallette, J.: Post-consumer cullet in California. A healthy building network evaluation for Stop Waste and the optimizing recycling collaboration (2015)

Walls, M.: Deposit-refund systems in practise and theory. Discussion paper. RFF DP 11-47 (2011)

Woolverton, A.B.: Use of glass cullet in roadway construction. M.Sc. thesis, Texas Technical University (1996)

WSDOT: Washington state department of transportation specifications for glass aggregate. Technology brief. CWC fact sheet (1997)

Yamane, J.K.: A study to identify local alternatives to shipping non-deposit glass out of the state of Hawaii. Report No. 14-17 (2014)

Production of Building Bricks Using Cement Kiln Dust CKD Waste

Ahmed H. Abdulkareem[1(✉)] and Saadoon O. Eyada[2]

[1] Civil Engineering Department, Engineering College, University of Anbar, Ramadi, Iraq
ahml973ed@gmail.com
[2] Civil Engineering Department, Selcuk University, Konya, Turkey
saadooneyada@gmail.com

Abstract. Cement production is one of the most important industry and produce a lot quantity of by-product (waste) materials which called cement kiln dust, CKD. In Iraq about 1600 tons of CKD produced each day. Bricks can be produced using different methods, most of them uses the clay as row material and consists of firing process. Another method consists of using waste materials such as Fly Ash, FA. Most of researches focused on the use of CKD in road pavement applications and soil stabilization, etc., but rare researches used the CKD to produce bricks. Two types of CKD, two types of sand and one type of cement have been used in this study. The experimental work consists of three stages. In stage I, one type of CKD and one type of sand has been chosen to complete the stage II due to enhancement in mechanical properties (compressive strength, CS and absorption, Ab) of the produced brick. In stage II, the selected CKD and sand have been mixed with water and cement and pressed to form a molds using a pressure of 6 Mpa. The stage III aimed to enhance the compressive strength, CS and absorption of the produced brick by increasing compressive strength to 11 Mpa without increasing cement percentage to decrease the production cost. All of the properties of the bricks that were tested found to be satisfactory according to the ASTM C62-04(SW, MW and NW) and Iraqi Specification Standard ISS 25/1988, for type A, B, and C. The results of this study has the nobility of producing new type of brick using waste materials with low cost and without adverse effect on the environment and agriculture.

Keywords: CKD · Pressed brick · Waste · Recycling · Novelty

1 Introduction

Recycling is the process by which many industrial waste can be reused to produce the same or other beneficial materials. It's the concerns of nowadays researches since it will reduce the costs, increase the revenues and enhance the environment. Cement production is one of the most important industry and produce a lot quantity of by-product (waste) materials which called cement kiln dust (CKD). Environmental concerns related to the production of Portland cement, specifically atmospheric emissions and the elimination of CKD have become progressively large. In Iraq about 1600 tons of CKD produced each day (Adil and Saadoon 2010). These large quantities are

H. Rodrigues and A. Elnashai (Eds.): GeoMEast 2018, SUCI, pp. 102–113, 2019.
https://doi.org/10.1007/978-3-030-01932-7_10

disposed in an open area without any attempt for recycling. The beneficial uses of CKD include serving as filler material in highway construction and maintenance and as a component of cement mortar/concrete (Mahrous and Yang 2011).

2 Previous Studies

Cement kiln dust (CKD) is a by-product of cement manufacturing. Cement kiln dust is a fine powdery material similar in appearance to Portland cement. It is composed of micron- sized particles collected from electrostatic precipitators during the production of cement clinker (Al-Jabri et al. 2006).

Bricks can be produced using different methods, most of them uses the clay as row material and consists of firing process. Another method consists of using waste materials such as fly ash FA which mixed with sand and small amount of cement with water and then forced into a molds using high pressure to produce a Non-Fired brick or pressed brick.

Most of researches focused on the use of CKD in road pavement applications, soil stabilization and agricultural purposes, but very few researches used the CKD to produce bricks. In this field of application, most of researches concerned with the use of CKD to produce cement blocks which consists of aggregate and cement mixed with water.

Rahman et al. (2011), used the CKD to make cement blocks as a partial replacement with cement using two types of aggregates lime stone origin which designated (Type-N) and light-weight pozzollanic aggregates (Type-P), they indicated that optimum CKD/cement ratio by 20% results in significant gain in strength of the blocks by 1.5 times.

Ghazaly et al. (2012), at the same manner of partial replacement of cement with CKD, they produced blocks that meet the Saudi Arabia standards of Non-Bearing wall block by partial replacement of cement with 20% of CKD. They concluded that the increase in CKD percentage will decrease strength and density and increase the water absorption.

CKD used at percentages of 0, 5, 10, 15, 20, 25, 30, 35 and 40% of the weight of the cement gave promising results for use in the bricks for building walls (El-Didamony et al. 2001; Younoussa et al. 2008; Kung-Yuh et al. 2008).

A patent No. WO2012138399 A1, owned to Loutfy et al. in (2012), demonstrated the beneficial effect of CKD as a row material to produce bricks by using up to 97% of CKD with some additives and cured for 16 h in a furnace at 65 °C, to produce bricks and blocks. Also, the authors used FA bricks as a guide method to produce CKD bricks, by which a blend of CKD, sand and cement are mixed with water at specified percentages and then molded using pressing machine.

The molded bricks are cured at different procedures and tested for mechanical properties and then compared with local and international standards.

The advantages of this research are:

(a) This work is the one in Iraq that concern the use of CKD in brick production and its positive effect on the cost of bricks production.

(b) There is a good chance to reuse the large quantities of materials that exposed on landfill which have an adverse effect on environment for the production of bricks which is the most important building material in Iraq.

3 Materials and Methods

The materials used in this work are discussed in the following paragraphs.

3.1 Sand

Two types of quartz origin sands are used, Namely, Zangora sand from Zangora quarry which designated as (Z), and Kutnia washed river sand from the west bank of Euphrates river which designated as (K). The particle size analysis for each type is conducted according to ASTM D422-02. The gradations of two types are shown in Figs. 1 and 2.

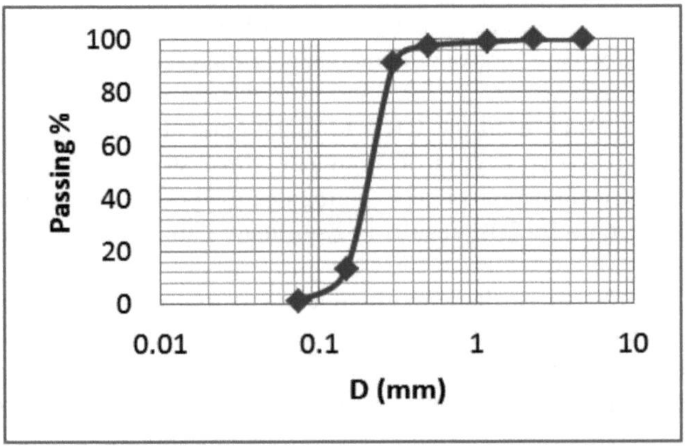

Fig. 1. Particle size analysis of Zangora sand.

Both Z and K sands are exhibited the same physical properties that related to particles size distribution. The Coefficient of curvature (Cc) of Z and K sands are 1.01 and 0.91 respectively to be classified as Well graded. Also they show a Uniform grain size distribution with Coefficient of Uniformity (Cu) of 1.64 and 1.61 respectively. About 50% of Z sand of 210 micron in diameter, while a 50% of K sand grains have a diameter of 240 micron.

3.1.1 Cement Kiln Dust CKD

Two types of CKD are used, Kubaisa KCKD, the by-pass of Kubaisa Cement Factory and Falluja FCKD, the by-pass of Falluja White Cement Factory. Each type materials are passing the sieve No. 30 (0.6 mm).

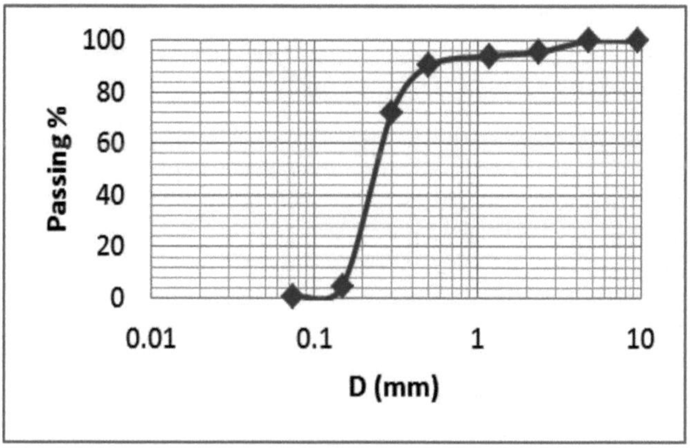

Fig. 2. Particle size analysis of Kutnia sand.

The physical properties consist of particle size gradation with hydrometer analysis which illustrated in Figs. 3 and 4 and the specific gravity which was tested according to ASTM D854-02. The results for two types are tabulated in Table 1.

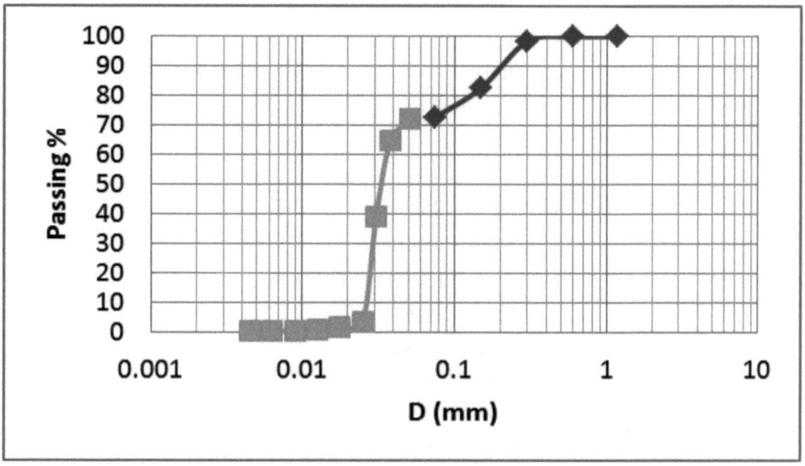

Fig. 3. Particle size analysis for KCKD.

The particle size analysis for both types of CKDs are very similar. So that it's expected that the physical effect of CKD type will depend on specific gravity. The chemical compositions of each type of CKD are presented in Table 2.

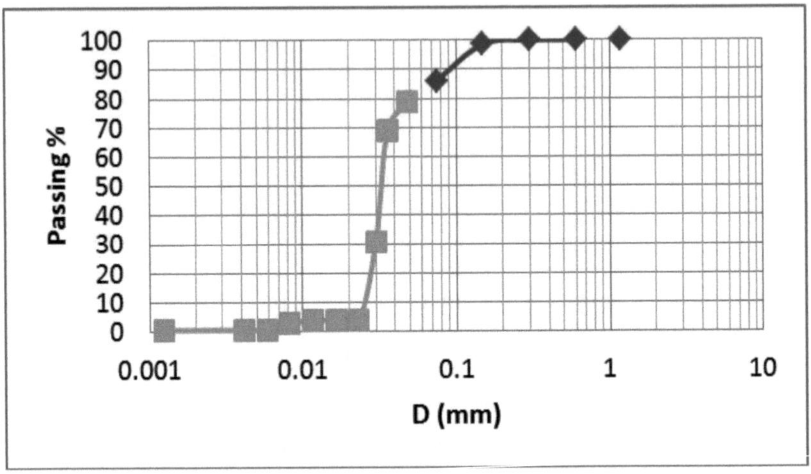

Fig. 4. Particle size analysis for FCKD.

Table 1. Physical properties of KCKD and FCKD.

CKD type	Cc	Cu	D50 (50% diameter), micron	Specific gravity
KCKD	1.23, (well graded)	1.00, (well graded)	34	2.527
FCKD	0.98, (uniform)	1.26, (uniform)	33	2.830

Table 2. Chemical compositions of each type of CKD.

Oxide	% Oxide	
	KCKD	FCKD
CaO	49.65	46
SiO2	14.82	13.3
Al2O3	5.25	4.6
Fe2O3	1.98	0.4
SO3	6.33	3.63
MgO	3.35	1.26
L.O.I	9.77	29.2

3.2 Cement

One type of ordinary cement type I is used to be mixed with CKD and sand, from Kubaisa Cement Factory.

4 Experimental Study and Results

The experimental program consists of the following stages:

4.1 Stage I

In this stage KCKD is mixed with FCKD at different percentages keeping the percentage of the sand Z fixed at 50%. Each percentage is mixed with water at different percentages following to Standard Proctor procedure according to ASTM D698, to determine the Optimum Moisture Content OMC.

Using the specified OMC, the water is added to a mixture of Sand, KCKD and FCKD at specified percentage and thoroughly mixed. Then, a specified weight of the mixture that produce the same density achieved from Standard Proctor method is poured into a cylindrical mold (4 in, in diameter) and is pressed using pressing machine with a pressure of approximately 6 Mpa to produce a specimen of 80 mm (± 3 mm) in thickness. The specimens are cured for 7 days using sealing bags to keep the mixing moisture during curing time. The results of this stage for duplicate specimens are summarized in Table 3.

Table 3. The results of Stage I.

Mixture type	OMC, %	Dry density, g/cm^3	Compressive strength, MPa
50%S+20%K+30%F*	19.45	1.609	2.36
50%S+30%K+20%F	19.10	1.612	3.17
50%S+40%K+10%F	20.09	1.566	4.82
50%S+50%K+0%F	19.11	1.559	4.68

*This mixture consists of 50% Sand, 20% KCKD and 30% FCKD.

The aim of this stage is to choose the effective CKD which has a positive effect on compressive strength. Figure 5, shows the positive effect of KCKD on the compressive strength of the specimens, Meanwhile, declare the negative effect of FCKD on the compressive strength. This effect is associated with the presence of calcium silicates in the CKD (Marku et al. 2012). Furthermore, the low specific gravity of KCKD, has a reduced effect on the dry density of the mixture at different percentages, which can help in reducing the dead loads of the whole building. According to this assumption the increase in KCKD and the decrease in FCKD percentage, increases the compressive strength of specimens, so the KCKD has been chosen for the Stage II.

Accordingly, the mixture type (50%S+50%K+0%F) will be modified to increase the compressive strength and other physical properties. This mixture has been chosen in spite that the mixture (50%S+40%K+10%F) has a higher compressive strength, because the difference is negligible for practical purpose, since the use of one material (KCKD) is more practical than the use of two materials (KCKD and FCKD).

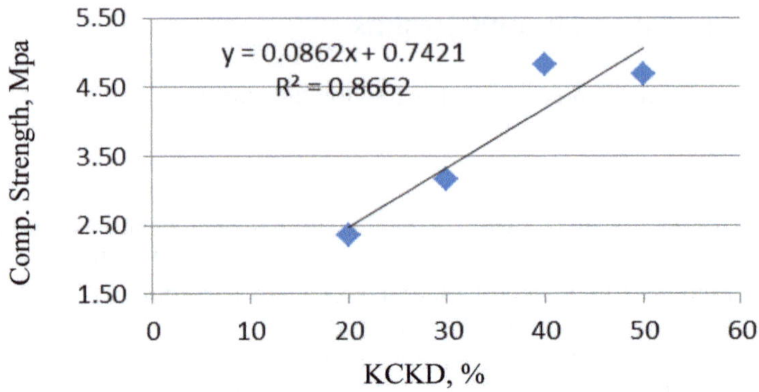

Fig. 5. The effect of KCKD% on compressive strength.

4.2 Stage II

In this stage, the selected mixture (50%S+50% KCKD), is mixed with Cement at different percentages using the same OMC and Dry density achieved from Stage I. In addition to, the sand type K is used instead of Z due to its availability and low cost.

In stage I, part of mixing water was consumed in the hydration of KCKD, due to a 2% of water will be added in this stage to the OMC of (Sand+KCKD) mixture to be 21.11%. At the same manner, an additional water of 45% of cement weight will be added to the blend of Sand+KCKD+Cement to substitute the demand of water for hydration process.

The blend of (50%S+50%K) and the specified percentage of cement are mixed with water at specified water content as described early. A specified weight of the final mix that will produce the desired dry density is poured into the mold to achieve a total thickness of 80 mm (±2 mm), using a pressure about 6 MPa.

The molded specimens are then cured using two procedures, immersing in water (SC) and using closed bags (CC), at different periods (7,14 and 28 days). Figure 6(a, b and c), shows the effect of cement percentage and curing type at different curing periods for duplicate specimens for each test.

It's natural to see that the increase in cement percentage leads to increase in compressive strength due to formation of complex C-S-H. Also the increase in curing period make the specimens gain more compressive strength and its related to more hydration period and more interaction between the free silica and free lime to form a new combinations of C-S-H.

The curing type has a significant effect on the compressive gain of the specimen as shown in Fig. 7. As mentioned previously, two types of curing are used in this research. The curing with closed bags of poly ethylene CC, which means that humidity of 100% will be available for the specimen to be cured and this is the exact moisture needed for the completion of the interactions (hydration). The other type of curing is by immersing the specimen in the water SC, that means that excessive water will be exist, part of the water will be consumed in the interactions while the other great part which surrounding the specimen will play the role in the dissolving of the salts exists in the

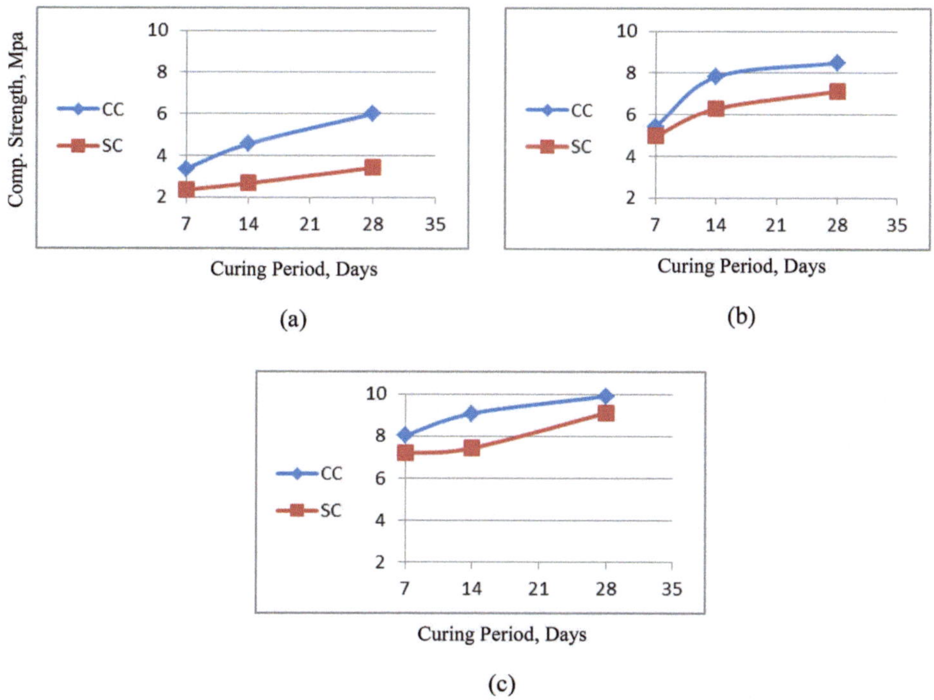

Fig. 6. The effect of cement percentage and curing type on compressive strength for different periods of curing. (a) cement = 3%, (b) cement = 6% and (c) cement = 10%

specimens specially (SO3). This salts are precipitated in the bottom of the curing basin after the super saturation state is occurred. The crystallization of the dissolved salts which exist in KCKD leads to swelling of the specimen and increase in the volume. That explain the low density of specimens cured with immersing in water. The low density of the specimen has an adverse effect on the final compressive strength comparing with closed bag curing and that explain the reduction in compressive strength gain of the specimens with SC. Figure 8 shows that effect by comparing the density of the specimens at the end of the curing period (28 days) for two types of curing. It's obviously, the low density of the specimens cured by SC compared to that cured by CC.

Finally, the compressive strength and absorption after 28 days of curing for each cement percentage are compared with local and international standards. The final compressive strength for each cement percentage after 28 days of curing are presented in Table 4.

Internationally, ASTM C62-04, classifies bricks due to the compressive strength and absorption to different types namely, (SW, MW and NW). For NW type, the compressive strength required is 8.6 MPa for individual and 10.3 MPa for the average. So that, the 9.91 MPa compressive strength achieved using cement percentage 10% is very close to that for type NW and need a very small modification. No limits for absorption are required for NW type.

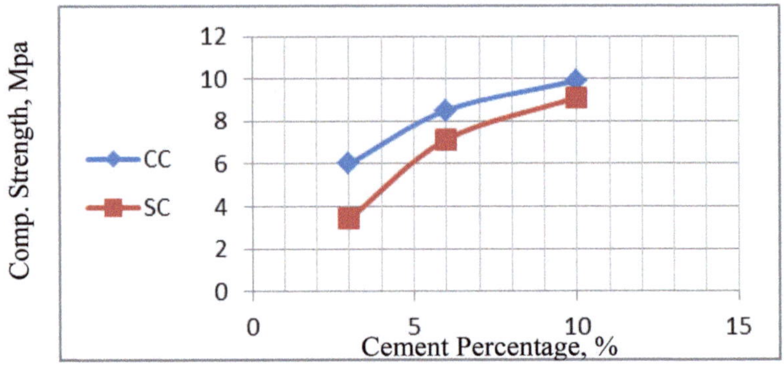

Fig. 7. The effect of curing type on compressive strength of the specimens at different cement %, for a period of 28 days.

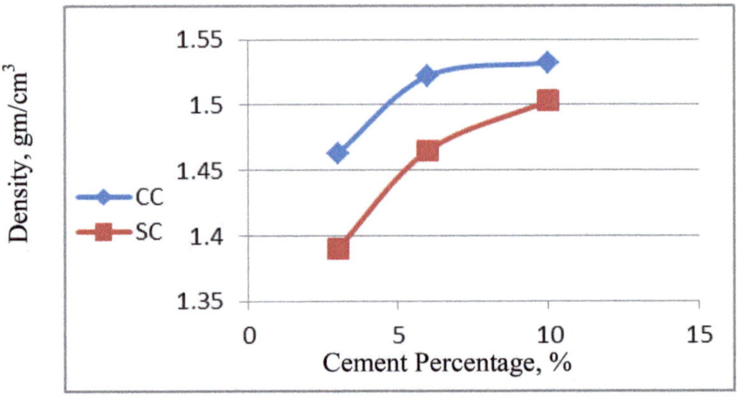

Fig. 8. The effect of curing type on density of the specimens at different cement %, for a curing period of 28 days

Table 4. The results of Stage II at different percentages of cement.

Cement percentage, %	Compressive strength, MPa	Absorption, %
3	6.00	24.2
6	8.50	21.2
10	9.91	18.8

Another standard can be achieved with this compressive strength is the Turkish standard TS 705 (UDK 691.421 1985), which required a minimum of 7.8 Mpa for Brick type 1.8/100. With respect to absorption, a threshold of 18% is the max allowable absorption percentage which is very close to 18.8% achieved using 10% cement.

Also the achieved compressive strength is within the Indian standard IS 4139:1989 (Indian Standard 1989), which required between (7.5–10) Mpa for the bricks type 7.5. Nothing mentioned concerning the absorption.

Locally, here in Iraq the Iraqi Specification Standard ISS 25/1988 (ISS 1987), limits the compressive strength for type C, Non-Bearing Wall Brick by 7 Mpa for individual and 9 Mpa for average. The maximum absorption is 26% for average and 28% for individual, so that the absorption percentage for the mix with 10% cement is within the Iraqi specification. That means that the produced brick by this paper with 9.91 Mpa and 18.8% absorption can be used as Non-Bearing Wall Bricks.

4.3 Stage III

The goal of this stage is to enhance the properties of the specimens produced in the previous stage to achieve the ISS as well the ASTM C62-04 (Annual Book of ASTM Standards) type MW. To do so, attention should be taken to the importance of cost effectiveness. Accordingly, No further cement will be added at this stage. To increase the compressive strength and reduce the absorption, there is a need to increase the density of the specimen to about 1.7 gm/cm^3 as well as increase the pressing pressure to about 11 Mpa. The specimens of this stage will be made using a new mold made of steel to produce a Brick specimen of dimensions (24 × 12 × 7.5) cm which is resembled to most available brick. Figure 9 illustrates the set up used for the preparation of new brick.

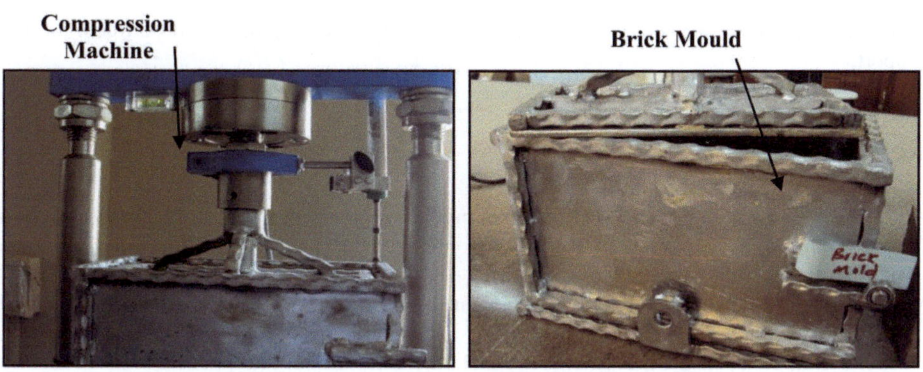

Fig. 9. Set up of brick preparation and pressing.

Five samples have been prepared using the fabricated set up and cured using the sealing bags. After 28 days the samples have been tested for the compressive strength and absorption. The average compressive strength of samples is 21.9 and the absorption is 15.5%. These results are very successful and refer to the enhancement achieved by using the CKD. The results indicate that the bricks produced using this procedure can satisfy the following specifications:

(a) Iraqi Specification Standard ISS 25/1988 Type (A, B and C). That means that the bricks can be used for Non-Bearing and Bearing wall purposes,
(b) ASTM C62-04 type (NW and MW),
(c) Indian standard IS 4139:1989 and
(d) Turkish standard TS 705.

5 Conclisions

- The results indicate that the CKD can be used for the production of bricks.
- The bricks can satisfy many specifications which means that the results are of universal application.
- The produced bricks can be used for many purposes as a Non-Bearing walls and Bearing walls.
- The results will encourage the using of pressed bricks as a new method for producing bricks in Iraq.
- The reuse of waste materials such as CKD will produce new economical building bricks.
- The use of CKD to produce bricks will decrease the use of clay for bricks production which can enhance the agriculture.
- The use of CKD will decrease the pollution caused by disposal of thousands of tons of this materials.

Acknowledgments. The authors thank Anbar University, Civil engineering department for their support.

References

Adil, N.A., Saadoon, O.E.: The use of cement production waste to improve the properties of hot mix asphalt concrete. In: 2nd Regional Conference for Engineering Science, Al-Nahrain University-Baghdad, pp. 622–634 (2010)

Mahrous, A., Yang, H.-S.: Utilization of cement kiln dust in industry cement bricks. Geosyst. Eng. **14**(1), 29–34 (2011)

Al-Jabri, K.S., Taha, R.A., Al-Hashmi, A., Al-Harthy, A.S.: Effect of copper slag and cement by-pass dust addition on mechanical properties of concrete. Constr. Build. Mater. **20**, 322–331 (2006)

Rahman, M.K., Rehman, S., Al-Amoudi, O.S.B.: Literature review on cement kiln dust usage in soil and waste stabilization and experimental investigation. IJRRAS J. **7**(1), 77–87 (2011)

Ghazaly, M.H., Almaghrabi, M.N., Ebrahiem, E.E.: A study on the reuse of cement Kin Dust in the production of cement concrete. Minia J. Eng. Technol. **31**(2), 63–68 (2012)

El-Didamony, H., Elwan, M., El-Alfife, A.: Effect of sand and bypass cement dust as non-plastic materials on clay bricks. In: 7th International Conference on Mining, Petroleum and Metallurgical Engineering (MPM7), Asyut–Egypt, 10–12 February 2001, pp. 333–342 (2001)

Loutfy, R.O., Magdi, M.N.: Method of fabrication of construction material from industrial solid waste. A patent No. WO2012138399 A1 (2012)

Marku, J., Dumi, I., Lico, E., Dilo, T., Çakaj, O.: The characterization and the utilization of cement Kiln Dust (CKD) as partial replacement of portland cement in mortar and concrete production. ZAŠTITA MATERIJALA **53**, 334–344 (2012)

TS 705 for solid and vertically perforated clay bricks, Turkish Standard Institute, UDK 691.421; March (1985)

Calcium Silicate Bricks Specification, Second Revision, Indian Standard (1989)

Iraqi Specification Standards for Building Materials (ISS) (1987)

Younoussa, M., Mohamed, H., Raguilnaba, O.: Cement-lateritic gravels mixtures: Microstructure and strength characteristics original. Res. Art. Constr. Build. Mater. **22**(10), 2078–2086 (2008)

Carbon Fiber Reinforced Polymers (CFRP) for Strengthening and Seismic Retrofitting of Historic Circular Masonry Stone Columns

Sayed Hemeda[1,2](✉)

[1] Conservation Department, Faculty of Archaeology, Cairo University, Giza 12613, Egypt
sayed.hemeda@cu.edu.eg
[2] Civil Engineering Department, Aristotle University of Thessaloniki, Thessaloniki, Greece

Abstract. Background: Many historical and archaeological masonry structures located in seismically moderate and active areas are not capable of sustaining seismic and dynamic actions. Furthermore, recent earthquakes in urban areas like Cairo, Egypt have clearly demonstrated an urgency to upgrade and strengthen these deficient historic masonry structures. Retrofitting with CFRP is now extensively being used as a seismic retrofitting method all over the world. It increases the load bearing capacity of structural elements like beams, slabs, walls and columns; Improve the seismic ductility of masonry stone and brick columns and piers; Improve the seismic response of concrete beam-column connections, shear walls and collector elements; Improve the seismic performance of masonry shear walls and in-fill walls; Restore structural capacity to damaged or deteriorated historic and modern concrete structures.

Objective: The aim of the present research is to investigate the application and durability of historical circular masonry stone columns completely strengthened by carbon fiber reinforced polymer (CFRP) laminates. This paper presents a comprehensive review with complete project experimental and implementation of CFRP in strengthening and seismic retrofitting of eight internal circular masonry stone supporting columns in the Saint George church in old Cairo area in Egypt and other important monuments in Cairo.

Results and Conclusion: Strengthening and seismic retrofit with CFRP materials has gained notable acceptance from the archaeological structural restoration communities in recent years. CFRP is characterized with high strength to weight ratio, excellent resistance to creep and fatigue, extremely durable and low aesthetic impact. Its benefits include the add strength without adding dead load, Withstands sustained and cyclic load conditions, extremely resistant to wide range of environmental conditions, and easy to conceal.

Keywords: CFRP · Historic structures · Multiple stone columns
Preservation of architectural heritage · Seismic retrofitting

H. Rodrigues and A. Elnashai (Eds.): GeoMEast 2018, SUCI, pp. 114–137, 2019.
https://doi.org/10.1007/978-3-030-01932-7_11

1 Introduction

While sustaining an historic buildings and archaeological sites, one must focus first on perceiving how it is working, assessing its execution for static and dynamic stresses. By then, the issues and furthermore evaluation pathologies must be experienced and at precisely that point one can start prescribing the essential geotechnical and fundamental responses for the intervention retrofitting strengthening and rehabilitation.

Old stone work structures have normal pathologies that are mostly identified with their age. For example the wood can be harmed because of the progressions and presentation to mugginess. It can have growth assault or bug assault. Then again the workmanship can be additionally harmed. If there should arise an occurrence of the workmanship the most essential pathologies have basic cause and can be converted into integration, pulverizing, break, clasping and splitting. The most widely recognized causes are connected with soil and establishment settlements, even pushes transmitted by slanted rooftops or curves, the presence of concentrated and high loads (for the most part connected with adjustment of the working to new utilizations) and Earthquakes which have an extensive impact in a material that is both overwhelming and fragile.

For quite a long time, numerous analysts have been completed to discover contrasting options to steels, combinations and solid components to fortifying, seismic retrofitting and battle the high expenses of repair and upkeep of harmed memorable structures incited by mechanical (static and dynamic) and physiochemical activities. Since the 1940s, composite materials like Fiber Reinforced Polymers, framed by the blend of at least two particular materials in an infinitesimal scale, have increased expanding prevalence in the building and auxiliary reclamation of landmarks fields.

FRPs have been around for a long time, offering us better approach to fortify or strengthen solid, brick work and different structures. Glass strands were proposed for use in fortifying bars and prestressing ligaments in the 1960s. In the 1980s a couple of groups started working autonomously on the utilization of FPRs in solid structures: in Switzerland on the utilization of Carbon strands for reinforcing; in Germany on glass filaments for fortifying and prestressing; in the UK and Netherlands on utilization of aramid strands for prestressing; in the USA on carbon and glass strands for the most part to strengthen and seismic retrofitting; this work extended quickly, e.g. in Belgium, Canada, Greece and Italy in mid 1990s.

CFRP are commonly sorted out in a cover structure; to such an extent that every lamina (or level layer) contains a course of action of unidirectional Carbon strands or woven filaments textures implanted inside a thin layer of light polymer lattice material. The strands, ordinarily made out of carbon give the quality and solidness. The network, generally made of polyester, Epoxy or Nylon tar, ties and shields the Carbon filaments from harm, and exchanges the anxieties and loads between Carbon strands.

A considerable lot of existing unreinforced masonry (URM) structures are naturally, geotechnical and seismically powerless and should be retrofitted. CFRP application is generally constrained to RC structures, be that as it may, a couple of trial and numerical examinations have been done on workmanship structures (Triantafillou and Fardis 1997; Ehsani et al. 1997, 2005). These investigations have focused on both

CFRP ligaments and cover and demonstrated that CFRP has an extraordinary potential to improve the quality and pliability of stone and block brick work structures.

For brick work structures fortified with CFRP covers, look into comes about have demonstrated that debonding of the CFRP overlay from the workmanship substrate is the controlling instrument of disappointment. This has been clear in workmanship dividers fortified to oppose either in-plane or out-of-plane burdens. For block and mud units, debonding may have an immediate association with the porosity of the stone work itself. Examinations ought to be done on various dividers worked with various and agent kinds of brick work units. The collaboration of reinforced dividers with the encompassing auxiliary components (i.e. shafts and sections) is of additionally critical since the adequacy of the fortifying might be hazardously overestimated because of untimely disappointments (e.g. pulverizing of brick work units at the limit districts). The flexural quality of URM dividers is fundamentally restricted by the rigidity of the mortar. Be that as it may, CFRP fortified divider can oppose huge minutes as it gives a huge tractable segment. It has additionally been demonstrated that the horizontal quality and solidness of the unreinforced workmanship dividers can be enhanced by giving appropriate CFRP retrofitting plans.

A standout amongst the most widely recognized utilizations for CFRP includes the repair and recovery of harmed or falling apart great structures. A few organizations over the world are starting to wrap harmed connect docks to anticipate fall and steel-fortified sections to enhance the basic trustworthiness and to forestall clasping of the support and strengthened segments.

Carbon Fiber Reinforced Polymers (CFRP) are in effect increasingly utilized for outer fortifying and seismic retrofitting of workmanship structures. Hence, portrayal of the short and long haul conduct of security between CFRP composites and brick work substrates in an administration situation is urgent for configuration purposes. A full collection of test and hypothetical examinations is required for solidness evaluation of CFRP reinforced structures. Be that as it may, the vast majority of the examination here has been committed to CFRP-solid examples, and the accessible information for FRP-fortified stone work parts are as yet missing and once in a while.

Mahfouz et al. (1998) tried square strengthened concrete columns restricted by FRP layer sunder uniaxial compressive load. The outcomes acquired demonstrate that huge increment in the malleability of square strengthened solid sections, alongside improvement in its definitive hub load could be accomplished.

Haggag et al. (1998) exhibited an exploratory assessment of the conduct of consumed strengthened concrete columns. A repairing strategy by a carbon fiber sheet (CFS) was used for repair of absconded segments because of salt-water presentation.

El-Kassabi (2017) displayed a few instances of archeological structures that utilized carbon fiber fortified polymers (CFRP) in the strengthening or fortifying of some noteworthy structures in Cairo, Egypt, similar to the vault of the mosque of Al-Aini, the Fatimid arch in Alazhar region, some stone lintels in the Qayaitbai organization in Algamalia territory and the Mohamed Ali Mosque in Shubra. As shown in Figs. 1, 2 and 3.

Carbon Fiber Reinforced Polymer (CFRP) is a moderately new class of composite material made from Carbon filaments and artificial resins (like Epoxy or polyesters) and has demonstrated effective and temperate for the advancement and repair of new and

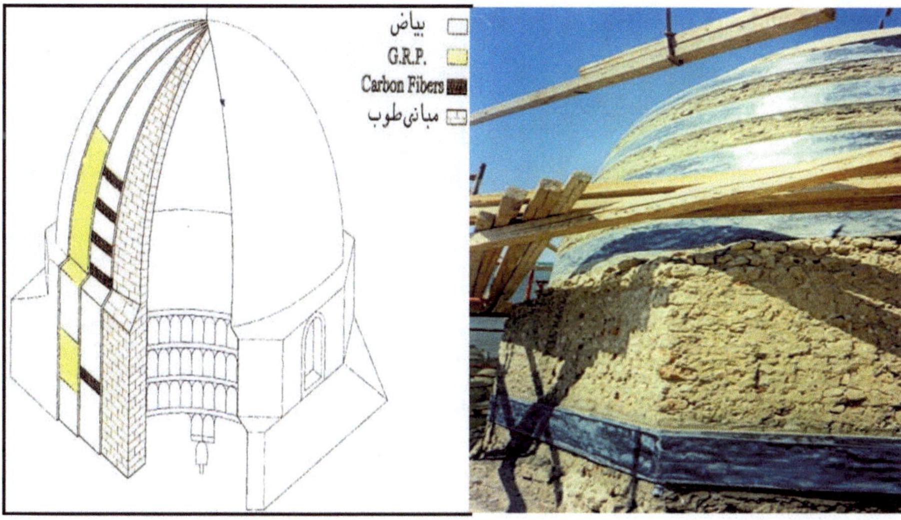

Fig. 1. The CFRP retrofitting technique for the dome of the mosque of Al-Aini in Cairo, Egypt (El-Kassabi 2017).

Fig. 2. The CFRP wrapping technique for the Sodon's dome in Cairo, Egypt (El-Kassabi 2017).

disintegrating brick work structures in structural designing. The mechanical properties of CFRP make them perfect for boundless applications in development and engineering conservation of memorable stone work structures around the world, as shown in Figs. 4, 5, 6 and 7.

Fig. 3. The CFRP external ties technique for the Column of the sultan Muayad sheikh mosque in Cairo, Egypt.

Fig. 4. An example of grid pattern application of CFRP, Villa Bruni, Padova.

Rochette and Labossiere (2000) played out a pivotal test, to analyze the conduct of roundabout, square, rectangular section models bound by composites. Then again, the impact of rehashed stacking and long haul mugginess presentation on flexural reaction of CFRP reinforced solid shafts, by Grace and Grace (2005).

Fig. 5. (a) An example of grid pattern application of CFRP, San Giorgio church, Italy. (b) An example of grid pattern application of FRP (Sika Limited 2003).

Fig. 6. Repair action in St. Fermo Church (Valluzzi et al. 2002).

It is imperative to call attention to that introduction of solid sections fortified with CFRP to sulfate assault for here and now, (Yousef et al. 2017). Expanded a definitive load, while for long haul comes about displayed that the proficiency of segment decreased somewhat and beneficial lessening in the attributes particularly for extreme load and pivotal strain flexibility proportion.

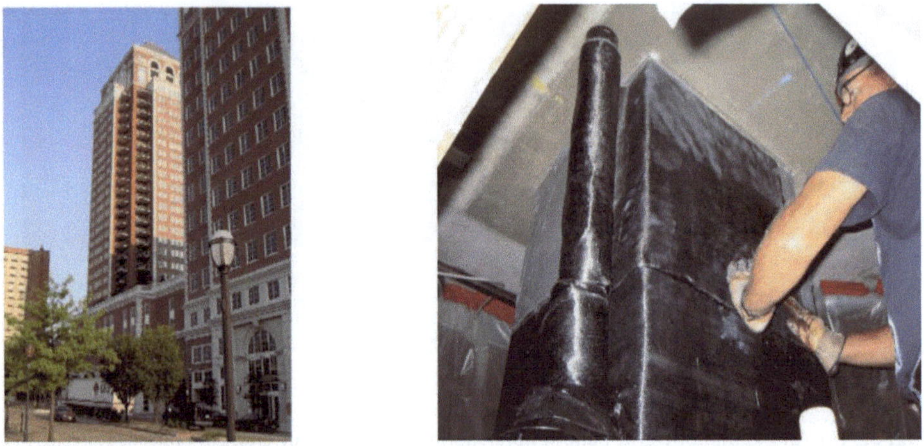

Fig. 7. Strengthening of columns in a 26-story residential tower in St. Louis, MO with CFRP (Ehsani 2005).

2 Sustainability of CFRP for Uses in Structural Restoration of Masonry Structures

The quality properties of CFRP aggregately make up one of the essential explanations behind which select them in the fortifying and seismic retrofitting of stone work structures. A material's quality is administered by its capacity to maintain a heap without intemperate misshapening or disappointment. At the point when a CFRP example is tried in pivotal pressure, the connected power per unit cross-sectional territory (push) is corresponding to the proportion of progress in an example's length to its unique length (strain). At the point when the connected load is evacuated, CFRP comes back to its unique shape or length. At the end of the day, CFRP reacts straight flexibly to pivotal pressure.

The reaction of CFRP to pivotal pressure is dependent on (1) the relative extent in volume of Carbon filaments, (2) the properties of the fiber and pitch, and (3) the interface bond quality. CFRP composite pressure disappointment happens when the strands show outrageous (frequently sudden and emotional) parallel or sides-way diversion called fiber clasping. CFRP reaction to transverse malleable pressure is especially subject to (1) the properties of the fiber and lattice, (2) the connection between the fiber and network, and (3) the quality of the fiber-framework interface. For the most part, in any case, elasticity toward this path is extremely poor.

Shear pressure is prompted in the plane of a territory when outside burdens tend to make two sections of a body slide more than each other. The shear quality of CFRP is hard to measure. For the most part, disappointment will happen inside the grid material parallel to the filaments.

Among CFRP high quality properties, the most significant highlights additionally incorporate amazing strength and erosion protection. Besides, their high quality to-weight proportion is of noteworthy advantage; a part made out of CFRP can bolster bigger live loads since its dead weight does not contribute fundamentally to the heaps that it must bear. Different highlights incorporate simplicity of establishment, flexibility, against seismic conduct, electromagnetic lack of bias, superb weariness conduct, and imperviousness to fire, as shown in Table 1.

Table 1. Characteristics of FRP composites and their constituents (CNR-DT 200/2004).

	Young's modulus (E)	Tensile strength (σt)	Strain at failure (εr)	Coefficient of thermal expansion (α)	Density (ρ)
	GPa	MPa	%	$[10^{-6}\ °C^{-1}]$	$[g/cm^3]$
E-glass	70–80	2000–3500	3.5–4.5	5–5.4	2.5–2.6
S- glass	85–90	3500–4800	4.5–5.5	1.6–2.9	2.46–2.49
Carbon (high modulus)	390–760	2400–3400	0.5–0.8	−1.45	1.85–1.9
Carbon (high strength)	240–280	4100–5100	1.6–1.73	−0.6–−0.9	1.75
Aramid	62–180	3600–3800	1.9–5.5	−2	1.44–1.47
Polymeric matrix	2.7–3.6	40–82	1.4–5.2	30–54	1.10–1.25
Steel	206	350–600 (failure)	20–30	10.4	7.8

Many of the CFRP components are characterized by high internal Damping properties. Thus leads to high absorption energy of vibrations induced by earthquakes through the materials and reduce its transmission through the adjacent buildings.

Nonetheless, as most basic materials, CFRP have a couple of downsides that would make some aversion in structural architects to utilize it for all applications like reclamation of landmarks: high cost, weak conduct, powerlessness to misshaping under long haul loads, UV corruption, photograph debasement (from introduction to light), temperature and dampness impacts, absence of configuration codes, and above all, absence of mindfulness.

3 CFRP Strengthening and Seismic Retrofitting of Circular Masonry Stone Columns

Numerous current notable historical and archeological structures not intended to withstand seismic impact have now turned out to be old because of advancement of more stringent outline codes and determinations. Besides, late tremors have incited a criticalness to repair and retrofit these seismic lacking noteworthy structures to lessen the harm and setbacks. Despite the fact that no such thing as completely quake confirmation structure can exist in genuine, appropriate retrofitting and recovery strategy can strikingly enhance the seismic execution of a structure.

Utilization and installation of CFRP has now turned into a typical option over steel to repair, retrofit and reinforce memorable and historical structures. CFRP materials may offer various favorable circumstances over steel plates and tubes which incorporate, 1. High particular firmness (E/ρ) 2. High particular quality (σult/ρ) 3. High consumption protection 4. As shown in Table 2. Simplicity of dealing with and establishment Moreover, its protection from high temperature and extraordinary mechanical and ecological conditions has settled on it a material of decision for seismic recovery.

Table 2. Some mechanical properties of Carbon Fiber Reinforced Polymer tendons.

Type	Ultimate tensile strength (N/mm^2)	Young's modulus in tension (N/mm^2)	Ultimate elongation (%)
Steel	483–690	200	6–12
CFRP	600–3960	120–580	0.5–1.7

At the point when a tremor and earthquakes hits, stone supporting columns and wharfs are thought to be the most helpless piece of a run of the mill noteworthy structures as they are the real load conveying component of the building. Least cross segment size and absence of fortification in under planned segments prompts a feeble section solid shaft development. It is critical to fortify the sections with the goal that the plastic pivots are framed in the pillars since it permits more viable vitality dispersal. In addition, segments ought to be satisfactorily intended to stay away from a delicate story fall of a working because of seismic activity.

Amid a quake, three methods of brick work stone segment disappointments that can occur because of cyclic pivotal and horizontal burdens are shear disappointment; flexural plastic pivot disappointment and lap graft disappointment. Absence of transverse fortification can bring about shear disappointment, which is both weak and calamitous in nature. Shear limit of insufficient segments can be fundamentally improved by furnishing remotely reinforced CFRP covers with filaments in the circle heading.

Investigates have demonstrated that an expansion in the thickness of CFRP coat relatively builds the shear quality of the redesigned segment or wharf. Trial thinks about (Kobatake et al. 1993; Saadatmanesh et al. 1996; Ehsani and Jin 1996) have demonstrated that appropriately outlined composite wrap for fortified concrete or stone section can expand shear quality to the degree that fragile shear disappointment mode is changed over to inelastic flexural misshapening mode and upgrades flexural pliability. Lap graft disappointments in fortified sections happen when the length of lap join in the segment is small to the point that the bond breaks amid seismic activity. As per an exploration led by Seible et al. (1997), the required CFRP wrapping thickness to secure the lap join area is straightforwardly corresponding to the compelling section distance across and contrarily relative to the modulus of versatility of the cover.

A few examinations have been led to think about the viability of CFRP in fortifying of roundabout, square and rectangular fortified solid sections. Haralji et al. (2006) revealed that keeping rectangular sections with CFRP brings about critical change in pivotal quality and pliability. For square segment segments without longitudinal support (plain concrete) the expansion in pivotal quality was observed to be 154, 213, and 230% for one, two, or three layers of CFRP wraps, separately. In any case, this expansion in quality because of FRP constrainment turns out to be less critical as the viewpoint proportion of the segment area increments. For square steel fortified solid sections, the expansion in pivotal quality coming about because of FRP retrofitting plan is 188, 255 and 310% with one, a few layers of CFRP wraps, individually (Haralji et al. 2006). Additionally, in fortified solid sections, FRP reinforcing counteracts untimely pressure disappointment of the solid cover and clasping of the longitudinal steel bars prompting enhanced execution of the segment under seismic stacking. This change is because of expanded strain limit of the bound cement, to upgraded restriction of bar clasping, and additionally to concealment of the impacts of shear on disfigurement limit (Bousias et al. 2004).

The shear disappointment of sections can make structures fall suddenly on account of the loss of hub stack conveying limit. In this manner, a retrofit strategy that considers the shear disappointment of segments is direly required.

A few retrofit choices were considered to add shear quality to the stone sections, including: outer edge fortifying, which would not be compelling without cross-supporting and fiber strengthened polymer (FRP) wrapping, which would expect testing to confirm its adequacy on cruciform-molded segments. There was an extensive open request to keep the cruciform state of the segments and the general tasteful interest of the structure in place. Along these lines, CFRP composite wrapping was retrofitted the segments.

In the recovery structure, the parallel reaction of the columns is unequivocally changed through retrofitting. Worldwide Intervention (GI) is a retrofit technique that outcomes in change of the worldwide auxiliary solidness. GI is important in frameworks with a high adaptability to influence, in torsional unequal frameworks, or where reinforcing of the current building is required.

It expressed that: Albeit neighborhood measures, (for example, CFRP jacketing) are an indistinguishable quality of the actualized restoration plans, they are not thought about piece of the fundamental system and don't go into the proportioning calculation of the worldwide recovery; they don't modify the firmness or flexural quality of the retrofitted individuals, however just influence their post yielding misshapening limit through restriction and by stifling untimely methods of disappointment (e.g., weak shear or lap-graft disappointment).

4 Carbon Nanotubes in Art Conservation

Since the time that their finding on 1991 (Helical 1991), carbon nanotubes (CNTs) require pushed analysts What's more engineers for future advances. This might be a result of those outstanding material properties of the little nanoscale tubes made about perfect carbon. CNTs are those most grounded furthermore stiffest materials alluded to in the planet and the simply material that may an opportunity to be fit the bill to setup and space lift that will not break by its personality or weight. They trademark an electrical conductivity similar to copper, a warm conductivity practically identical to precious stone, Furthermore A modulus more than one hundred times more astounding over steel. CNTs could an opportunity to be seen as sub-nuclear scale sheets about graphite (called graphene) moved ward upon settle on a tube What's more may make portrayed Likewise another part about carbon allotropes, building up the center of fullerenes additionally graphene. As for the most part perceived, singular divider Nanotubes (SWCNT) contain for single predictable moves of graphene (Helmut et al. 2013), same time Multi divider Nanotubes (MWCNT) include from asserting at least two moves (Dresselhaus et al. 1996; Yudianti et al. 2010).

5 Experimental Study

Two reference un-strengthened cylindrical stone specimens and five cylindrical stone specimens (with 110 mm length and 60 mm diameter) completely strengthened and wrapped with Carbon Fiber Reinforced Polymer laminates had been tested and investigated under the uniaxial compression test at the strength of materials laboratory, faculty of Engineering, Cairo University. The results indicated that ultimate load, increased by about (2.44–3.45%), respectively comparing with reference strengthened specimens. Also, effective confinement factor increased by about (3.67). Table 3 presents the experimental results for the reference un-strengthened and strengthened stone specimens.

Table 3. The experimental tests results for all tested stone specimens.

Specimens	Reference Specimens	SS_1	SS_2	SS_3	SS_4	SS_5
Loads (kN) Pu	360.76	653.4	504.3	622.3	663.02	545.1
Cracking load (kN) Pcr	202.1	345.5	355.1	455.2	455.1	357.0
Unconfined compressive strength MPa	10.12	11.0	14.0	15.9	15.0	16.1
Confined compressive strength MPa	9.8	12.2	14.6	15.6	16.6	15.3

S S = strengthened stone cylinder

6 CFRP Strengthening of the Eight Internal Masonry Stone Columns of St. George Church

The Church of St. George is situated in the Mar Girgis region. The congregation is exceptional and monstrous building complex brushing the basic and compositional components of various distinctive materials and auxiliary frameworks built in various timeframes: Roman period, 1909 development and 1941 recreation. It is a URM rotunda 27 m distance across at the base, 13 m arch measurement, 11 m vault and drum tallness and aggregate stature 22 m, worked toward the start of the twentieth Century upon the remains of a past Church. The rotunda depends on a three stories roman tower round about in design, referring in Figs. 8 and 9 (Sheehan 2010).

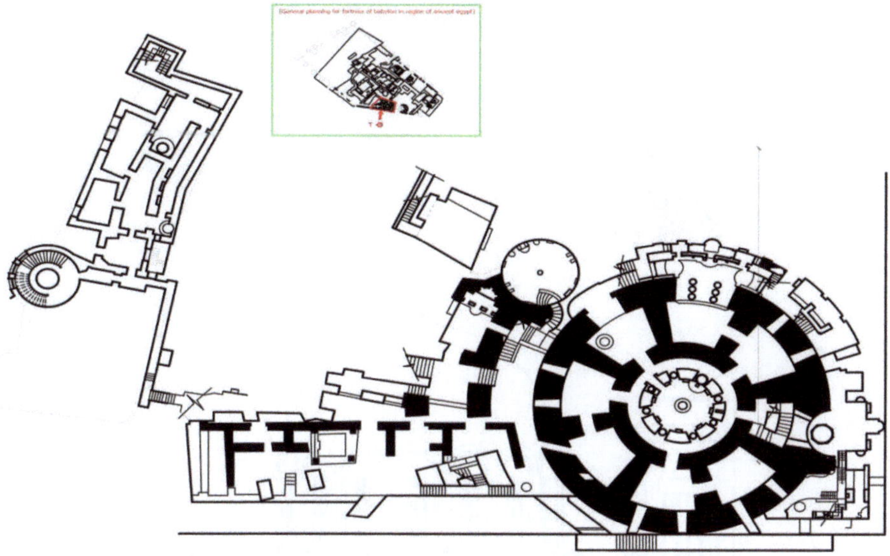

Fig. 8. The layout of the Greek St, George Church and roman tower underneath it.

After the end of the groundwater dewatering project (the agreement or project 102. 2000), differential settlement were seen between parts of the roman pinnacle and structures around the edge porch. Those settlements were short of what one centimeter and affirmed CDM's estimations (Johnson and Malhotra 2000) as well as the computations by (Antoniou and Lekkas 2015). The most extreme settlement changed from 0.7 to 0.9 cm (Penelis et al. 2014). The field observations demonstrated that the bases of the eight central columns have unsupported regions underneath them. The eight split stone segments in the focal well territory that are supporting the middle of the road time frame workmanship above give the presence of precariousness. At long last the eight inner round stone brick work columns, endured broad splitting at their base (spalling-off of mortar), as shown in Figs. 10 and 11.

Three intercession or interventions alternatives were talked about as indicated by the numerical examination to decide the basic zones: the mediations as following:

1. Strengthening of the outside border walls and the eight inner sections and columns with shotcrete.
2. Placing of pressure poles at the crotch vaults and between the sections and reinforcing of the eight inward segments with shotcrete or CFRP.
3. Introducing outside post tensioning with pressure poles at four levels remotely, and reinforcing and seismic retrofitting of the eight inward columns with CFRPs.

Amid the co-activity of all teaches, the third arrangement was chosen.

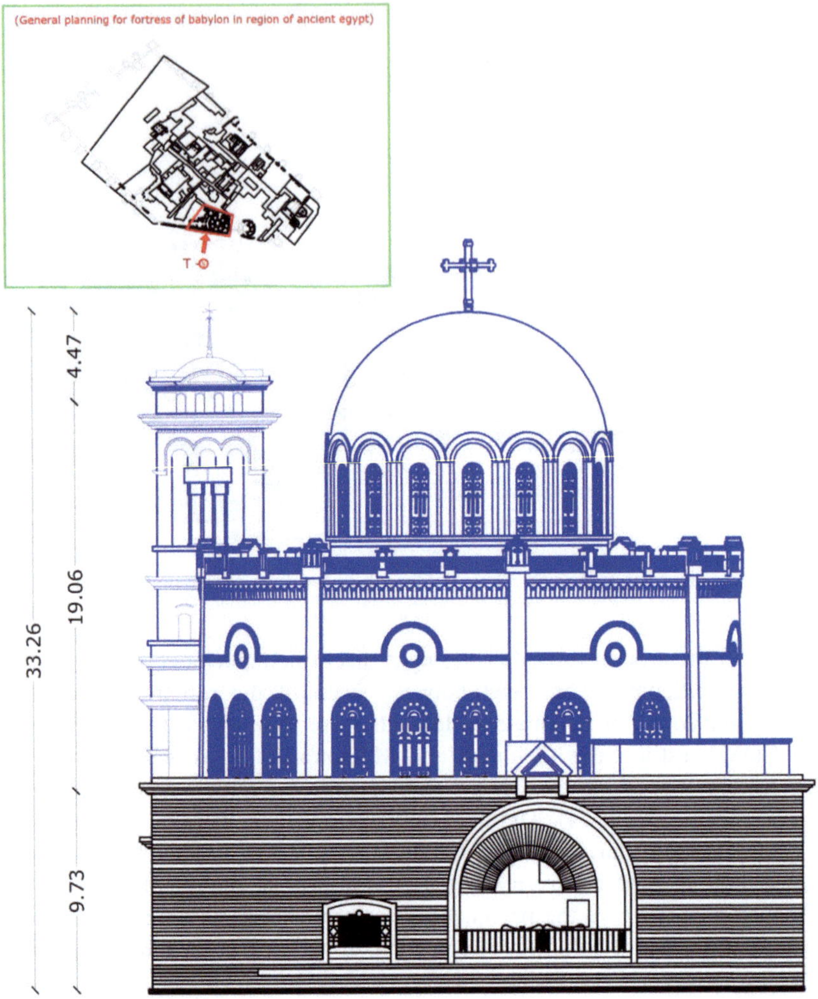

Fig. 9. The southern facade of the Greek St, George Church and roman tower underneath it.

7 CFRP Installation for Strengthening of the Eight Masonry Stone Columns

- Figures 12 and 13 present the preparation of Epoxy resin (hardener and monomer) and CFRP strips with inquire sizes and dimensions.
- Removing of the modern deteriorated oil painting and cement rendering layers on the columns. As shown in Figs. 14 and 15.
- Mechanical and chemical cleaning and treatment of the stone columns surfaces; crack injection, mortars repointing, stone consolidation with Wacker OH polymers. As shown in Fig. 16.

Fig. 10. External and interior views of Saint Geogre church in Mar Girgis area, old cairo, Cairo, Egypt.

Fig. 11. Extensive tensile and shear cracks and structural deficiency in the groin vaults of the ceiling of Saint Geogre church in Mar Girgis area, Cairo, Egypt.

- Installation of steel rings around the circular multiple stone columns as a supporting system for CFRP sheets. As shown in Fig. 17.
- Many laboratory mechanical testing like uniaxial compression test, splitting tensile strength, and pull-out test had been done to determine the efficiency of the CFRP technology in the retrofitting of the stone columns before the restoration project. The efficiency of the retrofitting techniques is depended on the cohesion between the CFRP laminates and the treated stone surfaces. As shown in Figs. 18 and 19.

Fig. 12. Preparation of the CFRP laminates or strips with the required dimensions.

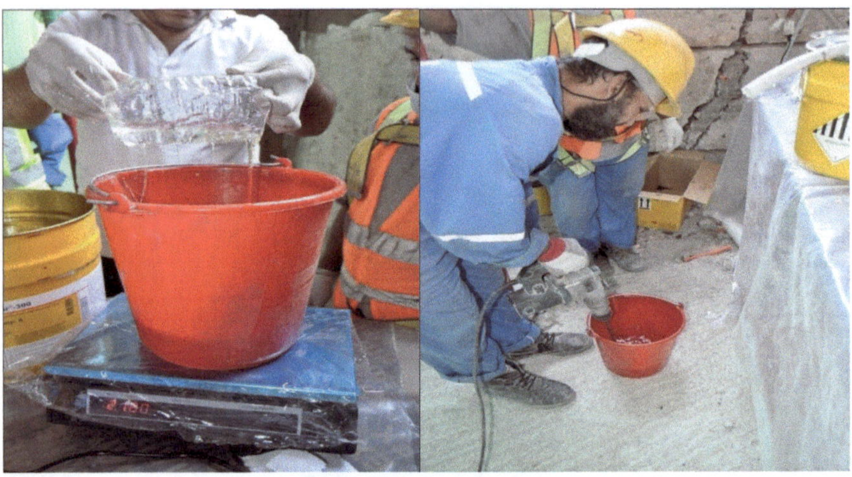

Fig. 13. Preparation of the epoxy resin (the adhesive matrix of the Carbon Fibers).

- Figures 20, 21 and 22 present the installation of ten laminates or levels of circular rings of CFRP with 50 cm width around the stone columns at different horizontal levels and perpendicular to the cracks directions on vertical levels.
- The eight circular multiple stone columns were demanded a specific retrofitting and strengthening technique, and it was installed in a passive CFRP wrapping system to achieve a horizontal stabilization for these columns.
- Figures 23 and 24 present the application of the new rendering and painting layers had been applied and covered the external surfaces of the CFRP laminates.
- Figures 25 and 26 present the eight stone masonry columns after the final restoration process.

Fig. 14. Mechanical removal of the deteriorated modern renders and oil painting layers.

Fig. 15. Removal of the cement rendering layers and discovering and polishing of the circular multiple stone columns.

Fig. 16. Removal of the cement rendering layers and discovering and polishing of the circular multiple stone columns

Fig. 17. Installation of steel rings around the circular multiple stone columns.

Fig. 18. In-situ investigation of the stone columns after discovery; roughness and moisture measurements and examinations.

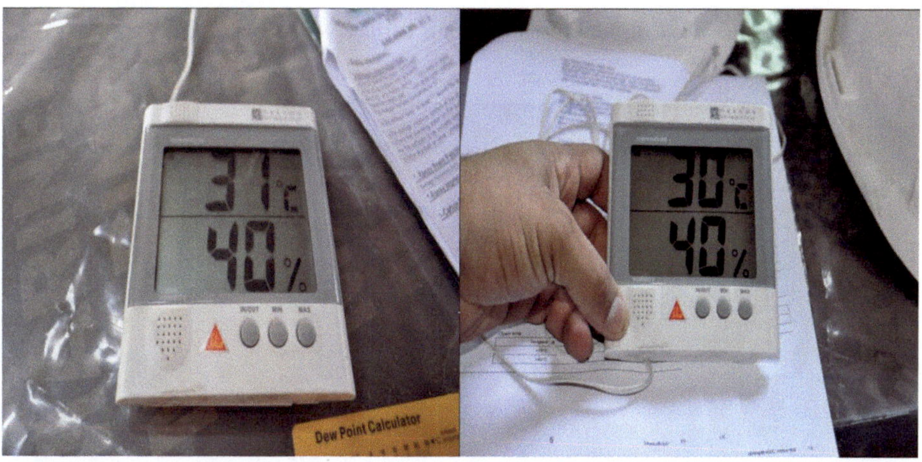

Fig. 19. In-situ measurements of temperature and humidity.

Fig. 20. Brushing of the columns surface with Epoxy resin.

Fig. 21. Installation of CFRP laminates on the surfaces of stone columns from top to bottom in ten levels.

Fig. 22. The eight stone columns after installation of the CFRP wrapping through the columns.

Fig. 23. The stone columns after the application of the new rendering cement layers covering the CFRP laminates.

Fig. 24. The stone columns after the application of the new rendering layers and painting layers covering the CFRP laminates.

Fig. 25. The eight stone masonry columns after the final restoration process.

Fig. 26. The stone masonry column after the final restoration process.

8 Conclusions

Look into effects show that retrofitting for flexural fortified stone masonry columns with composite fabrics of CFRP will be a powerful procedure for expanding those flexural Furthermore shear quality for these components. Clinched alongside these requisitions the measure and the quality of the fabric controls those mode about failure mechanism. Accepting that those epoxy doesn't fizzle prematurely, At lighter fabrics need aid used, those most extreme load may be that making strain disappointment of the fabric. At heavier fabrics need aid used, those parts will neglect by Possibly layering pulverizing alternately a shear disappointment In those plane passim through the longitudinal steel bars. Restriction for columns, also fortifying from claiming masonite dividers Furthermore cement pipes need aid different provisions the place CFRP items might be proficiently utilized. There will be minimal mistrust that the utilization from claiming FRPs will develop fundamentally in the nearing quite some time.

A few mediation retrofitting and fortifying strategies like expansion of new auxiliary components; outside post tensioning, steel plate holding and so on has been connected in the past with differing level of achievement. Among these techniques, fortifying and seismic retrofit with CFRP materials has increased outstanding acknowledgment from the structural building and archeological basic reclamation groups as of late. CFRP is portrayed with high quality to weight proportion, fantastic protection from crawl and weakness, amazingly sturdy and low stylish effect. Its advantages incorporate the include quality without including dead load, Withstands maintained and cyclic load conditions, greatly impervious to extensive variety of natural conditions, and simple to hide.

Acknowledgment. The Restoration and intervention retrofitting project of the historic church of ST. George and the roman tower underneath were achieved by the Egyptian company (the Arab Contractors (Osman Ahmed Osman & Co.)) under the Management of the Penelis, Consulting Engineering S.A, on behalf of the HellenicAid. The restoration project was funded by the Hellenic Ministry of Foreign Affairs-HELLENIC AID, Patriarchate of Alexandria and Whole Africa.

References

Antoniou, A., Lekkas, E.: Preservation of Saint George's church at Cairo, Egypt. In: Engineering Geology for Society and Territory, vol. 8 (2015)

Dresselhaus, M.S., Dresselhaus, G., Eklund, P.C.: Science of Fullerenes and Carbon Nanotubes. Academic Press, Waltham (1996)

Ehsani, M.R., Saadatmanesh, H., Al-Saody, A.: A shear behavior of URM retrofitted with FRP overlays. J. Compos. Constr. ASCE **1**, 17–25 (1997)

Ehsani. M.: Strengthening of Concrete and masonry Structures with FRP. In: 30th Conference on Our World in Concrete& Structures, Singapore (2005)

El-Kassabi, E.: Restoration of Structures, Cairo (2000)

El-Kassabi, E.: Restoration of historic buildings by carbon reinforced polymers. In: The First Arab Conference for Restoration and Reconstruction, Cairo (2017)

Helical, S.: Microtubules of graphitic carbon. Nature **354**(6348), 56 (1991)

Helmut, M., Saborwiski, K., et al.: Carbon nanotubes in art conservation. Int. J. Conserv. Sci. **16**, 1 (2013)

John, C.C., Malhotra, P.C.: (CCJM): Contract 102, Old Cairo Area Project. Works at Mar Girgis Church (in CDM's final report) (2000)

Mahfouz, I., Rizk, T., Sarkani, S.: Repairing and retrofitting of circular and square reinforced concrete columns using fiber reinforced plastic (FRP). In: Proceedings of the First Arabian Conference on Repairing and Rehabilitation of Structures, CICC-Nasr City, Cairo, Egypt (1998)

Sheehan, P.: Babylon of Egypt. Oxford University Press, Oxford (2010)

Penelis, G., Penelis, G., Papanikolaou, V.: Restoration of the historic church of ST. George in Cairo and the roman tower underneath. In: Proceeding of Second European Conference on Earthquake Engineering and Seismology, Istanbul (2014)

Sika Carbo Dur FRP Composites for Repair & Strengthening of Structures: Sika Poland Sp. z o. o. ul. Karczunkowska 89, 02-871, Warszawa, Polska (2003). www.sika.pl

Rochette, P., Labossiere, P.: Axial testing of rectangular column models confined with composites. J. Compos. Constr. ASCE **4**(3), 129–136 (2000)

Triantafillou, T.C., Fardis, M.N.: Strengthening of historic masonry structures with composite materials. Mater. Struct. **2**, 96–104 (1997)

Valluzzi, M.R., Tinazzi, D., Modena, C.: Shear behavior of masonry panels strengthened by FRP laminates. Constr. Build. Mater. **16**, 409–416 (2002)

Yousef, S., Abdel-Latif, H., Bahnasawy, H., Metawea, M.: Durability of reinforced concrete circular columns strengthened with carbon fiber reinforced polymers. Part 1, effect of sulfate attack. In: 9th International Conference on Nano-Technology in Construction, Sharm el-sheikh, Egypt (2017)

Yudianti, R., Indrarti, L., Onggo, H.: Thermal behavior of purified multi walled carbon nanotube. J. Appl. Sci. **10**(17), 1978–1982 (2010)

The Impact of Steel Fibers on the Properties of Self Compacting Concrete

Hassan Ghanem[1(✉)], Yehia Obeid[2], Ayman Trad[1], and Mohamed Dandachy[1]

[1] Department of Civil Engineering, Beirut Arab University (BAU), Tripoli, Lebanon
h.ghanem@bau.edu.lb
[2] Ferrovial, Al Ghubrah, Oman

Abstract. Self-compacting concrete (SCC) is a highly flowable, non-segregating concrete that has the ability to flow in every spot of the complex formwork and consolidate within that without any external compaction. Steel fiber reinforced self-compacting concrete (SFR-SCC) is a new mixed material that merge the advantages of the SCC with those of steel fibers in improving concrete mechanical properties. This paper is part of a study to analyse the effect of steel fibers on the rheological [J-ring test] and mechanical properties [compressive strength and four point bending test] of SCC. Five concrete mixtures were evaluated. The primary experimental variables are the type and aspect ratio of steel fibers. Test results have shown that the inclusion of fibers improves the compressive strength of SCC but it has a negative effect on the rheological properties of the SCC by reducing the slump flow and increasing the flow time, but better workability was obtained as aspect ratio of the steel fibers decreased. It was also found that the fiber geometry is a key factor affecting the mechanical performance in particular the toughness of the SFR-SCC material.

Keywords: SFR-SCC · J-ring · Aspect ratio · Rheological properties
Hardened properties

1 Introduction

Reinforced concrete was and still is one of the most extensively consumed construction materials due to the continuous growth of urban areas that are spreading all around the world. SCC is approximately a unique formation of high-performance concrete that is able to attain remarkable deformability and homogeneity in its fresh state, filling all space around the reinforcement, passing between dense steel reinforcement bars while consolidating under its own weight without any need of vibration (Daczko and Vachon 2006). During the last decade, SCC was highly enhanced because of the efforts of concrete technologists (Khayat 2000; Kwan and Ng 2010).

Concrete is known for being capable of attaining very high compressive strength; however its main weakness lies in its low tensile strength and low ductility due to its brittleness behavior. That's why steel fibers are being developed to enhance the mechanical properties of concrete. These fibers are discontinuous discrete entities and

H. Rodrigues and A. Elnashai (Eds.): GeoMEast 2018, SUCI, pp. 138–150, 2019.
https://doi.org/10.1007/978-3-030-01932-7_12

are distributed and oriented randomly (nominally uniformly) throughout the concrete matrix. SFR-SCC can be used by itself, or in conjunction with conventional reinforcing bars, depending on the application (Banthia 2001).

The use of steel fibers in SCC has been investigated by many researchers for use as a composite material. The majority of those studies found that the rhyological and mechanical properties of the SCC varies when steel fibers are added to the matrix (Ferrara et al 2012; Filiatrault et al. 1995; Hameed et al 2009; Deeb 2013). This variation is based upon several factors: (a) properties and types of the steel fibers, (b) aspect ratio L/D, (c) orientation of the fibers, etc. The above studies found that the addition of fibers to the SCC mixtures affects negatively its workability and the relationship between the steel fibers aspect ratio and its flowability is inversely proportional; i.e. when the aspect ratio increases, SCC workability decreases. Regarding steel fibers effect on the mechanical properties of SCC, previous work have shown that fibers had only minor effects on the ultimate compressive strength of concrete, slightly increasing or decreasing its magnitude, depending on the characteristics of the fibers themselves (ACI 544.1R-96 1996; ACI 544.3R-2 1998). However, their primary role in the SCC mixtues was to delay and limit the flexural cracking, as previous research indicated that steel fiber reinforced beams absorbe more energy and provide better ductility (Daniel and Loukili 2002).

The use of SFR-SCC is continuing to increase, but this development is being hindered by a general lack of theoretical knowledge and methods for its design, particularly under flexural loading (Jones et al. 2008). Thus, the mechanical properties of SFR-SCC must be studied empirically by means of standardized laboratory tests (Deeb 2013). The main objective of this experimental research was to inspect the interaction between SFR-SCC composites and SCC by marking the most sensitive variables in the SFR-SCC properties. This study compared the performance of steel fibers reinforcement systems used for strengthening self-compacting concrete structures.

2 Experimental Study

2.1 Materials

The cement used in this study is white Portland Cement identified as AALBORG WHITE® cement used in the preparation of concrete and mortar. The cement complies with all requirements according to EN 197-1 (2011) for Portland Cement CEM I 52, 5 R - SR5. Drinking-quality water was used in all concrete mixtures. To improve and maintain the workability of fresh concrete, a high range water reducing admixture (superplasticizer) commercially known as "Fosroc Auracast 200" that complies with EN 934 -2, 2012 was used.

The volume of fibers used is constant ($V_f = 0.5\%$), although four different types of fibers were used. The variation depends on the aspect ratio and shape of the fibers. The brand used is Dramix® 3D and 5D fibers. Dramix® are certified for structural use according to BS EN 14889-1, 2006. The properties of the steel fibers used are summarized in Table 1. Figure 1 describes the nomenclature present on the Dramix® Company packages for the 80/60 fibers.

Table 1. Fibers properties and shapes

Fiber type	45/50 BL	65/60 BG	65/60 BG	80/60 BG
Tensile strength (N/mm^2)	1.115	1.160	2.300	1.225
Young's modulus (N/mm^2)	±210.000	±210.000	±210.000	±210.000
Fiber family	3D	3D	5D	3D
Length L (mm)	50	60	60	60
Diameter D (mm)	1.05	0.90	0.90	0.75
Aspect ratio (l/d)	45	65	65	80
Minimum dosage (kg/m^3)	20	15	15	10
Sketch		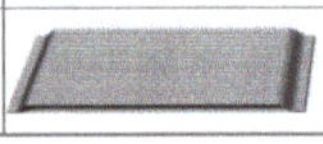		

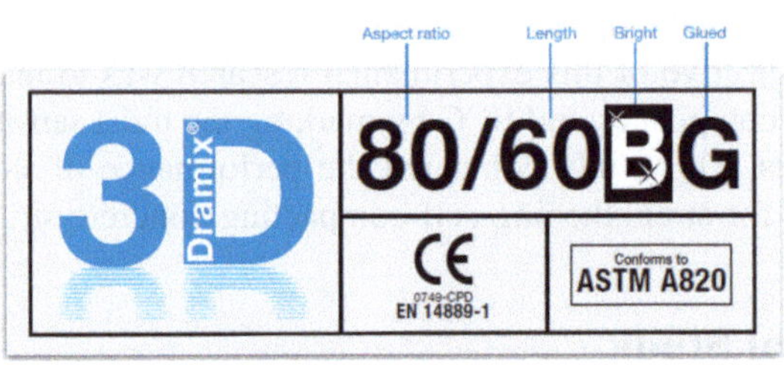

Fig. 1. Badge of steel fibers packages

2.2 Mix Designs

Five concrete mixes were evaluated in this study. Of the five concrete mix designs, four contain different types of steel fibers. The fifth mix is a plain SCC acting the role of a control sample. For each mix, three replicate specimens were cast for each test conducted. These mixes are labeled and presented in Table 2. The total number of specimens tested is thirty. The key mixture design information is presented in Table 3.

2.3 Sample Preparation

After measuring the material based on their proportions, they were placed in the concrete mixer in the following sequence. First, the gravel and sand were added and

Table 2. SCC mixes nomenclature

Mix 1	Plain SCC
Mix 2	SFR-SCC 45/50 3D
Mix 3	SFR-SCC 80/60 3D
Mix 4	SFR-SCC 65/60 3D
Mix 5	SFR-SCC 65/60 5D

Table 3. Mix design information

Materials	Amount (kg/m^3)	Mix dosage (kg)
Gravel	761.62	15.08
Sand	738.38	14.62
Added water	191.36	3.786
White cement	735.35	14.56
Superplasticizer	7.424	0.147
Fiber content	40	0.784

dry mixed for 3 min. Second, water and the superplasticizer were mixed together and then added to the drum and mixed for about 30 s. Cement was then added followed by another 3 min of mixing until the mix became homogenous. Finally, the steel fibers were added to the mix for a period of 1 min and mixed for another 2 min.

SCC in its fresh state should satisfy simultaneously the filling ability and passing ability. The J-ring test according to BS EN 12350, 2010 was carried out on all mixes to simulate simple field conditions. The J-ring test provides the most fundamental information regarding the flowability of SCC and SFR-SCC because it also includes the slump flow test in it. A slump flow diameter of 600 mm was adopted as a minimum requirement for SCC; otherwise the test had to be repeated. Visual inspection was used to observe if segregation occured.

Following mixing, the concrete was directly poured into the molds. For the compressive strength test, standard plastic cubic molds 100 × 100 × 100 mm were used according to BS EN 12390-3 2009. For the flexural test, standard beam specimen with internal dimensions 100 × 100 × 500 mm were used as shown in Fig. 2. After casting the specimens, they were sealed using polyethylene sheets until they were demolded after 24 h. Once demolded, they were placed in a lime-saturated bath for the required number of curing days which was 28 days according to BS EN 12390 2009.

3 Test Results

3.1 Fresh Properties of SCC and FR-SCC Mixes

As mentioned previously, the fresh properties of SCC and SFR-SCC mixes were evaluated using the J-ring test. No significant problems were observed in flowability, segregation resistance, filling ability, and passing ability of the SFR-SCCs mixes (Fig. 3).

Fig. 2. Casting of SCC cubes and beams

The test results are presented in Table 4. According to BS EN 12350-12 2010, the parameters mentioned in Table 4 are defined as follow: SF_J is the J-ring flow spread expressed to the nearest 10 mm and T500 is the J-ring flow time which is the time between elevating the cone from the base plate and the time SFR-SCC reaches a diameter of 500 mm.

Fig. 3. J-ring test - SFR-SCC (mix 5)

As shown in Table 4, the addition of steel fibers into the SCC mixes significantly affects its workability. This can be easily seen by making a few comparisons among SCC mixes. First, by comparing mixes 1 and 3, one can notice the difference in the flow spread and the flow time. For example, the flow spread for mix 1 [plain SCC mix] is 775 mm and it decreases with the addition of steel fibers to 720 mm for mix 3 [SFR-SCC]. Furthermore, the flow time is 4.1 for mix 1 and it increases to 9.7 s for mix 3. This indicates that the addition of steel fibers decreases SCC workability and its passing ability as the possibility of blocking increases with the addition of steel fibers. Results as well indicate that workability decreases as the aspect ratio of steel fibers increases; i.e. for mix 2, the flow spread and the flow time are 760 mm and 6.7 s respectively

Table 4. J-ring test results

Mix number	Fiber type	SF_J (mm)	T_{500} (s)
1	–	775	4.1
2	45/50 3D	760	6.7
3	80/60 3D	720	9.7
4	65/60 3D	755	8.1
5	65/60 5D	740	9.2

whereas they are 720 mm and 9.7 s for mix 3. This point indicates that the presence of long fibers is more challenging to handle during concrete pouring.

To illustrate the effect of the type of fibers used on SCC workability, a comparison is made between mix 4 [65/60 3D] and mix 5 [65/60 5D] who have the same aspect ratio but different number of hooks. Results indicate that the effect of multiple hooks [5D steel fibers] on SCC workability is similar to the effect of increasing the aspect ratio of fibers. For example, the flow time for mix 4 [3D] is 8.1 s. On the other hand, it is 9.2 s for mix 5 [5D]. The same effect can be seen on the flow spread as it decreases from 755 to 740 mm when using 5D steel fibers. This demonstrates the negative effect [from a workability perspective] of steel fibers with multiple hooks on SCC flowability. The above results are very consistent with what was mentioned in the literature.

3.2 Compressive Strength Tests

The results of 28 days compressive strength of 100 × 100 × 100 mm cubes are presented in Table 5. A few observations can be made. First, the strength of all SCC mixes is above 65 MPa indicating that those mixes can be considered as high strength concrete. Second, it was noticed that the compressive strength of SCC mixes improves with the addition of steel fibers. According to ACI, the compressive strength is slightly affected by the presence of fibers and a 15% increase in strength was observed with mixes having 1.5% by volume of fibers. However, in this study a much higher increase was achieved with much lower amount of fibers [0.5% of fibers]. This improvement ranges from 11% for SFR-SCC 45/50 3D mix to 26% for SFR-SCC 65/60 3D mix. This amelioration can be to the fact that the addition of steel fibers fibers increases strainability in compression failure and hence the compressive strength increases (Hannant 1978). Lastly, the effect of multiple hooks steel fibers on SCC compressive strength doesn't appear to be significant [13% increase]. This is expected as its contribution is anticipated to be on the flexural strength.

3.3 Flexural Strength Tests

Flexural strength is often needed as part of the design to check compliance with established specifications or to provide information necessary for the design of an engineering structure.

In this study, the flexural strength test was performed after 28 days of curing on SCC and SFR-SCC beam test specimens of size 100 × 100 × 500 mm with a span of

Table 5. Compressive strength test results

Concrete type	Average (3 samples) results of 28 days compressive strength (MPa)	% increase
SCC	69	0
SFR-SCC 45/50 3D	76.75	11
SFR-SCC 65/60 3D	87.43	26
SFR-SCC 65/60 5D	78.33	13
SFR-SCC 80/60 3D	83.76	21

450 mm. The beams were subjected to a four-point loading in flexure at a constant deformation rate control of 1 mm/min in accordance with BS 14488 2006 standards as shown in Fig. 4. Two symmetrically placed LVDTs placed on opposite sides of the specimens were used to get the mid-span deflections of the beams, and the net mid-span deflection was considered as the average between the two outputs of the LVDTs. In order to measure the pure beam deflection, a yoke was attached to the specimen to discharge additional supports. Steel rods were used as roller supports at 1/3 points with spacing of 450 mm. To avoid readings outside the linear range of the external LVDT, the tests on fibrous specimens were stopped at approximately 4.1 mm of deflection. Furthermore, the test was stopped after a load drop of 90%. These two adjustments were made for safety in order to prevent any over load that can break the system. The load and strain data were recorded through a data acquisition system. Ultimately, this system converts the voltage readings into actual stresses and strains.

3.3.1 Behavior of SCC Mixes

To illustrate the behavioral variations of SCC mixes, the stress deformation response is plotted and presented in Fig. 5. It should be noted that each plot is an average of three as each test was repeated three times to check the repeatability of the results.

Most of the plots display similar characteristics patterns. Three main stages occurs: (1) an elastic part of the ascending branch as the displacements increase linearly with the applied loads up to the first crack, (2) that part of ascending branch during which multiple cracks occur and (3) a descending branch well known as the post cracking stage dominated by the pull out of the fibers. The last step occurs when segments of the beams yield and start deforming plastically.

As shown from the plot, mix 1 [plain SCC] displays a brittle failure as the specimen reaches a stress of 10.24 Mpa and directly fails without any post-cracking resistance due to the absence of steel fibers. This is obvious, as the presence of steel fibers in a dispersed way in the SCC mix, limits the propagation of the cracks as the bonding effect of the steel fibers in the mix leads to a change in the failure mode from brittle to a more ductile mode. It also enhances the post-cracking load and toughness (energy absorption capacity).

Fig. 4. Flexural strength test setup for beam specimens

On the other hand, all SFR-SCC mixes deform plastically. With sufficient plastic deformation, a hairline crack develops near the midspan of the segments where the stresses are the highest. Previous research (Banthia 2001; Poh et al. 2009) indicated that when initial crack developed, the fibers began to tense up as the concrete experienced non-linear strain hardening. As shown, once the peak load is reached, the stress is supported by the fibers, which act as crack arresters, bridging the crack formation. The matrix is held together by the fibers as a long softening branch continues.

It can as well be seen from the figure that the (SFR-SCC 65/60 5D) specimen is the only one that exhibits a hardened behavior after the first crack, ultimately reaching a maximum compressive stress, followed by a softening branch, while all other SCC mixes have a softening behavior. This hardened behavior is the result of the multiple hooks in the 5D family steel fibers that made an extra bonding and bridging between the fibers and the concrete.

One of the objectives of this work is to study the effect of steel fibers [aspect ratio] on the mechanical properties of SCC specifically its flexural performance. It was expected from the literature that upon increasing the steel fibers aspect ratio, the flexural performance of SCC mixes should goes up as well. This was found to be partially correct. By comparing SFR-SCC 45/50 3D with SFR-SCC 65/60 3D, it is noticed that the stress rate decreases is much higher as the aspect ratio of steel fibers decreases. This is due to the fact that shorter fibers hold the stresses for a shorter period of time as you need less energy to pull out the fibers out of concrete. On the other hand,

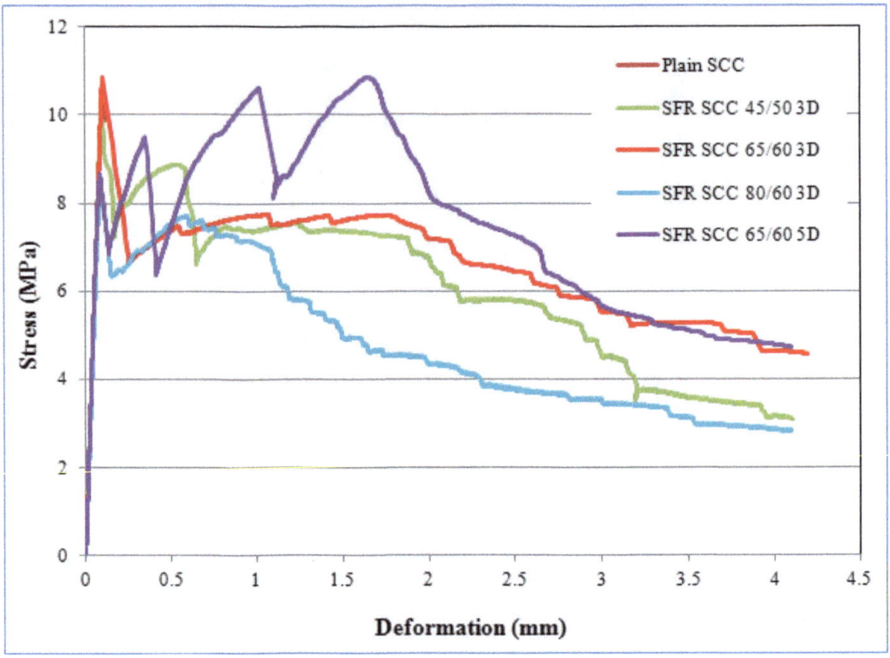

Fig. 5. Stress deformation curves for all mixes.

it is found that SFR-SCC 80/60 3D exhibits a lower flexural performance than SFR-SCC 65/60 3D. This result comes as a surprise because it is anticipated that longer fibers should hold the stress for longer period of time as they are expected to produce more residual strength and more toughness. This may be possibly due to the fact that high aspect ratio can lead to a balling effect during mixing, resulting in loss of flexural resistance. Therefore a threshold limit on the steel fibers aspect ratio should be placed.

3.3.2 Flexural Toughness of SCC Mixes

The addition of steel fibers in SCC mixes makes it distinguishable from plain SCC by its ability to absorb a large amount of energy and to withstand large deformations prior to failure. This preceding feature is referred to as toughness. In this work, flexural toughness is measured by taking the area under the load-deflection curve in flexure up to deflection of 4.1 mm. The area was calculated using AutoCAD software computer program. The results are presented in Fig. 6 using column bar chart.

It can be seen from Fig. 6 that plain SCC mix exhibits a very low flexural toughness of 4460 N-mm as plain SCC beam broke immediately into two pieces after the appearance of the first single crack in the middle of the beam. The introduction of steel fibers into SCC mixes significantly increases the beams toughness more than ten-fold as it enhances the post-failure strengths of the concrete after the formation of first crack. The bond between steel hook end fibers and concrete enables slower crack propagation and hence more energy is being absorbed by the specimen; this eventually allows the beams to deflect more before the commencement of first visible crack.

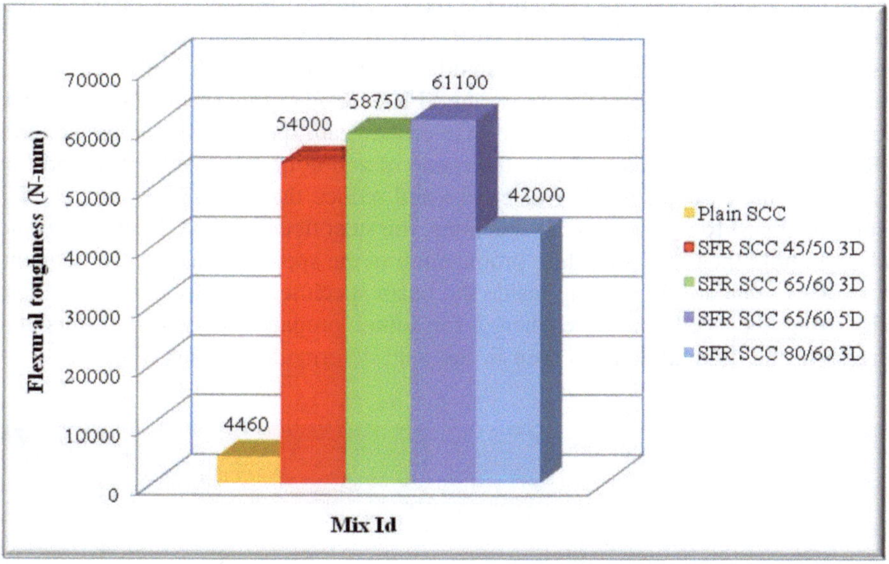

Fig. 6. Flexural toughness of all SCC mixes

As shown from the bar chart, the effect of steel fibers aspect ratio on the SCC flexural toughness is very dependent. Toughness increases as aspect ratio goes up. For example, the flexural toughness is 54000 N mm for SFR SSC 45/50 3D whereas it is 58750 N mm for SFR SSC 65/60 3D. This is due to the fact that longer fibers have the capability of absorbing more energy and are efficient in bridging macro-cracks resulting in an additional straining in the steel fibers. This mechanism will result in a larger area under the load deformation curve, eventually yielding a higher toughness value. On the other side, the flexural toughness decreases significantly as aspect ratio increases from 65 to 80; i.e. for SFR-SSC 65/60 3D, it is 58750 N mm while it is 42,000 for SFR SSC 80/60 3D. This is equivalent to a 38% decrease. This result appears shocking as it was expected that higher steel fibers aspect ratio yields higher flexural toughness as longer fibers are capable of bridging wider cracks and therefore delaying beams failure. A possible explanation for this phenomenon is that very long steel fibers are very difficult to handle during mixing leading to a balling effect and therefore fibers won't be uniformly distributed throughout the beams contributing to a loss of flexural resistance. Another possible reason is that longer steel fibers with smaller diameter [i.e. high aspect ratio] don't have the same energy absorption potential as those who have larger diameter.

The type of steel fibers have a big effect on the flexural toughness of SCC beams as SFR SSC 65/60 5D beam holds the highest toughness value 61100 N mm. This may be attributed to the excellent mechanical anchorage of multiple hooked 5D steel fibers leading to a hardening behaviour in the post cracking range.

The multiple hooking in the 5D family didn't only have a positive effect on the beam flexural toughness, but also on the cracking system. All SFR-SCC specimens using 3D family steel fibers displays a softening behavior in the load-deformation curves with a single cracking formation in the specimen surrounded by ramifications as shown in Fig. 7. On the other hand, SFR-SCC 65/60 5D beam exhibits a hardening behaviour with multiple cracks in the middle third of the beam. It can be clearly seen from Fig. 7 that the cracks at failure for the SFR-SCC 5D beam specimen are greater in numbers and more distributed over the span of the beam specimen in comparison to the SFR-SCC 3D beams. This is due to the mode of action of multiple hooked end fibers [5D] which increase the number of cracks and reduce the spacing between cracks at failure. In this SFR-SCC 65/60 5D specimen, the steel fibers around these micro cracks will try to hold them and stop their propagation in the specimen matrix. Consequently, micro-cracks that are developing inside the beam specimen have to take a meandering path, rising the need for more energy for future propagation, which results in an increase in the ultimate load as seen in the load-deformation curves (Fig. 5).

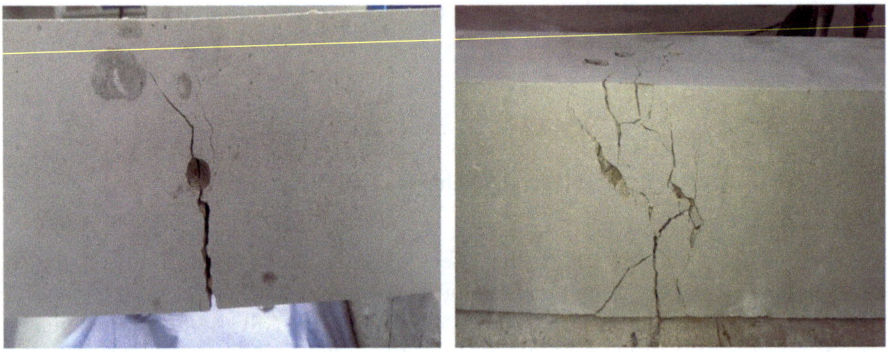

Fig. 7. Failure mode and crack pattern of SCC beams. (a) 3D family steel fibers (SCC 45/50 3D), (b) 5D family steel fibers (SCC 65/60 5D)

4 Conclusions

In this study, different types of steel fibers were used to produce fiber reinforced self-compacting concrete. The effect of steel fibers aspect ratio and geometry on fresh properties such as J-ring test, and on hardened properties such as compressive strength and four point bending tests were examined. It should be noted that the behavior of SCC structural members might be affected by other variables not intended for investigation in this study. However, the conclusions made herein would be an important step toward the development of design guidelines for SFR-SCC members.

(a) Regarding fresh properties, the J-ring test is used in this study to observe workability. Flow diameters above 600 mm are achieved in all cases without bleeding or segregation. It is observed that SCC workability [represented by the flow spread and flow time] increases as the aspect ratio of steel fibers decreases.
(b) The addition of steel fibers in the SCC mixtures improves the compressive strength. This increase ranges from 11 to 26% depending on the aspect ratio and type of steel fibers.

(c) From the flexural test, it is found that the addition of steel fibers to SCC mixes displays a deflection-softening response for all 3D types and a deflection-hardening response for the 5D type, both accompanied by single and multiple cracks respectively in the middle third of the specimens.
(d) Test results indicate that higher aspect ratio provides better flexural properties as longer steel fibers tend to bridge more cracks and absorb more energy. However, SFR- SCC 80/60 3D specimen yields the lowest flexural toughness. This may be attributed to the balling effect during mixing. Therefore, on can deduct that an upper limit [based in this study, it is 65] should be placed on the aspect ratio of steel fibers used in SCC mixes.
(e) By comparing flexural properties of SCC mixes, it was found that SFR-SCC 65/60 5D specimen has the highest flexural toughness. This indicates that 5D steel fibers provides better bond characteristics than 3D type steel fibers and has as well the highest energy absorption capacity.
(f) A strict mixing procedure is followed in this study, in terms of time of mixing, addition of components, and sequence of material addition. In order to achieve high quality SCC and SFR-SCC mixes, it is essential to strictly follow the recommended mixing procedure. This procedure with the given mix proportions has led to an SCC mix that was able to flow and fill the molds without any need of vibration.

References

ACI 544.3R-93: Guide for Specifying, Proportioning, Mixing, Placing and Finishing Steel Fiber Reinforced Concrete. American Concrete Institute, Farmington Hills (1998)

ACI 544.1R-96: State-of-the-Art Report on Fiber Reinforced Concrete. American Concrete Institute, Farmington Hills (1996)

Banthia, Nemkumar: Fiber Reinforced Cements and Concretes. Can. J. Civ. Eng. **28**(5), 879–880 (2001)

BS EN 14488-3: Testing Sprayed Concrete, Part 3: Flexural Strengths (first peak, ultimate and residual) of Fiber Reinforced Beam Specimens. British Standards Publication (2006)

BS EN 14889-1: Fibres for Concrete. Steel Fibres. Definitions, Specifications and Conformity (2006)

BS EN 12390-3: Testing Hardened Concrete, Part 3: Compressive Strength of Test Specimens. British Standards Publication (2009)

BS EN 12350-12: Testing Fresh Concrete, Part 12: Self-Compacting Concrete, J-Ring Test. British Standards Publication (2010)

BS EN 197-1: Cement. Composition, Specifications and Conformity Criteria for Common Cements (2011)

BS EN 934-2:2009+A1: Admixtures for Concrete, Mortar and Grout. Concrete Admixtures. Definitions, Requirements, Conformity, Marking and Labelling (2012)

Daczko, J.A., Vachon, M.: Self-Consolidating Concrete, pp. 637–645. Spon Press, Abingdon (2006)

Daniel, L., Loukili, A.: Behavior of high strength fiber reinforced concrete beams under cyclic loading. ACI Struct. J. **99**(23), 248–256 (2002)

Deeb, R.: Flow of Self-Compacting Concrete (Ph.D). Cardiff University, Cardiff (2013)

Ferrara, L., Bamonte, P., Caverzan, A., Musa, A., Sanal, I.: A comprehensive methodology to test the performance of Steel Fibre Reinforced Self-Compacting Concrete. Constr. Build. Mater. **37**, 406–424 (2012)

Filiatrault, A., Pineau, S., Houde, J.: Seismic behavior of steel-fiber reinforced concrete interior beam-column joints. ACI Struct. J. **92**(5), 543–552 (1995)

Hameed, R., Turatsinze, A., Duprat, F., Sellier, A.: Metallic fiber reinforced concrete: effect of fiber aspect ratio on the flexural properties. J. Eng. Appl. Sci. **4**(5), 67–72 (2009)

Hannant, D.J.: Fiber Cement and Fiber Concrete. Wiley-Interscience, Hoboken (1978)

Jones, P.A., Austin, S.A., Robins, P.J.: Predicting the flexural load-deflection response of steel fibre reinforced concrete from strain, crack-width, fibre pull-out and distribution data. Mater. Struct. **41**, 449–463 (2008)

Khayat, K.H.: Optimization and performance of air-entrained, self-consolidating concrete. Mater. J. **79**(5), 526–535 (2000)

Kwan, A.K.H., Ng, I.Y.T.: Improving performance and robustness of SCC by adding supplementary cementitious materials. Constr. Build. Mater. **24**(11), 2260–2266 (2010)

Poh, J., Tan, K.H., Peterson, G.L., Wen, D.: Structural Testing of Steel Fibre Reinforced Concrete (SFRC) Tunnel Lining Segments in Singapore. Land Transport Authority, Singapore (2009)

Assessing the Potential Value of a SHM Deployment on a Proposed Footbridge

David D. T. Nepomuceno[1(✉)], John Bennetts[2], Graham T. Webb[3], Matt Langhorne[2], Mike Johnson[4], John H. G. Macdonald[1], Theo Tryfonas[1], and Paul J. Vardanega[1]

[1] University of Bristol, Bristol, UK
{d.nepomuceno, john.macdonald, theo.tryfonas, p.j.vardanega}@bristol.ac.uk
[2] WSP, Bristol, UK
{john.bennetts, matthew.langhorne}@wsp.com
[3] WSP, London, UK
graham.webb@wsp.com
[4] South Gloucestershire Council, Bristol, UK
mike.johnson@southglos.gov.uk

Abstract. A fibre-reinforced polymer (FRP) cycle footbridge has been proposed for construction in Bristol, United Kingdom for South Gloucestershire Council. The superstructure will span 54 m, comprising a bowstring carbon fibre-reinforced polymer (CFRP) arch with a 5 m wide glass fibre-reinforced polymer (GFRP) deck supported by stainless steel hangers. Recently, a methodology has been proposed that provides a structured process to assess the value of a structural health monitoring (SHM) system for a bridge prior to deployment. This methodology outputs a simple metric that quantifies the likeliness of an SHM system to yield value to an asset owner. This FRP bridge is used as a case-study to 'road test' this process. Two possible systems were considered: a system of accelerometers and a system of strain gauges. From the resulting discussions, a deployment of accelerometers received a value-rating (VR) of 4.2. A strain gauge deployment received 3.7. The scores will contribute to a monitoring specification for the FRP bridge which is currently in the design phase. Expansions to the methodology have also been proposed to better capture the potential value of an SHM system which would be of interest to structural engineers and researchers, in particular to inform model validation and research activities.

H. Rodrigues and A. Elnashai (Eds.): GeoMEast 2018, SUCI, pp. 151–166, 2019.
https://doi.org/10.1007/978-3-030-01932-7_13

1 Introduction

1.1 Structural Health Monitoring

SHM systems can be classified into five categories: 'anomaly detection'; 'sensor deployment studies'; 'model validation'; 'threshold check' and 'damage detection' (Webb et al. 2015). When specifying an SHM system engineers should consider how the data obtained from the monitoring can be used to make decisions. Once the aims of the SHM system have been initially established, a structured process is needed to decide whether monitoring will yield actual value. Vardanega et al. (2016) have proposed a methodology that (prior to any deployment) brings together (arguably) the three key stakeholders: the 'SHM engineer', 'structural engineer' and 'asset manager' to participate in a facilitated discussion. Table 1 shows the nine questions that are proposed for the structured discussion, divided into three sections each to be answered by the 'SHM engineer', 'structural engineer' and 'asset manager' respectively. Each question is rated 1 to 5 by the relevant stakeholder (see Table 2) and these scores are averaged to provide a simple metric that relates to a project value statement. The framework provides structure to the discussion and is centered on the likeliness of the project 'to yield value to the asset owner/manager' (Vardanega et al. 2016). In this paper, this methodology is applied to the proposed FRP footbridge.

Table 1. Framework questions (adapted from Vardanega et al. 2016)

Stakeholder	Question	Ref.
SHM engineer	How likely is it that the required accuracy of the proposed measurements can be ascertained?	A1
	How likely is it that the specified system can be designed to be sufficiently robust such that it can be maintained over the life of the monitoring project?	A2
	How likely is it that the appropriate auditing will be conducted by a third party to certify that the gathered data are reliable? (e.g. monthly reports that give details of calibrations undertaken and 'sanity checking' of collected data.)	A3
Structural engineer	How likely is that relevant values of approximate threshold values will be assigned in consultation with the asset owner so as to ensure that the owner will act if these values are realized in the field?	B1
	How likely is it that the critical monitoring locations on the structure can be determined?	B2
	How likely is it that the data be used for validation (or falsification) of structural models which may assist with the design of more efficient future structures for the asset owner?	B3
Owner/asset manager	How likely is it that the necessary actions (or decisions) will be taken if threshold (trigger) values are exceeded/reached?	C1
	How likely is it that the data to be collected will be able to inform a maintenance regime?	C2
	How likely is it that there will be a secure budget necessary for the intended period of operation of the monitoring project?	C3

Table 2. Value ratings (adapted from Vardanega et al. 2016)

1	Very unlikely or not at all
2	Not likely
3	Neutral, i.e. may or may not occur
4	Likely
5	Very likely/certain

1.2 Project Background

A proposed FRP cycle footbridge is to be constructed in Bristol, United Kingdom for South Gloucestershire Council (SGC). The appointed designers are WSP. At the time of writing, the following description of the proposed design can be given:

- The superstructure will span 54 m, comprising a bowstring arch made of CFRP, with a 5 m wide GFRP deck supported by stainless steel hangers.
- The arch will have a triangular profile, inclined to the vertical.
- The arch material design will consist of unidirectional carbon fibres and biaxial glass fibres, bonded with an epoxy resin.
- The deck will be a modular construction, split into nine 6 m long modules.
- The internal diaphragms will provide strength and stiffness to the deck and also provide space for Tuned Mass Dampers (TMDs) and service ducts.

Preliminary drawings taken from the public planning application are shown in Figs. 1 and 2, with a visualization of the bridge shown in Fig. 3.

This project presents the opportunity to design and deploy an SHM system during the construction stage: allowing the collection of monitoring data both during construction and service, and offering insights into how monitoring systems can be embedded into a structure from the early design stages. Data of potential interest include: (a) long-term behaviour of FRP in service (i.e. observed durability);

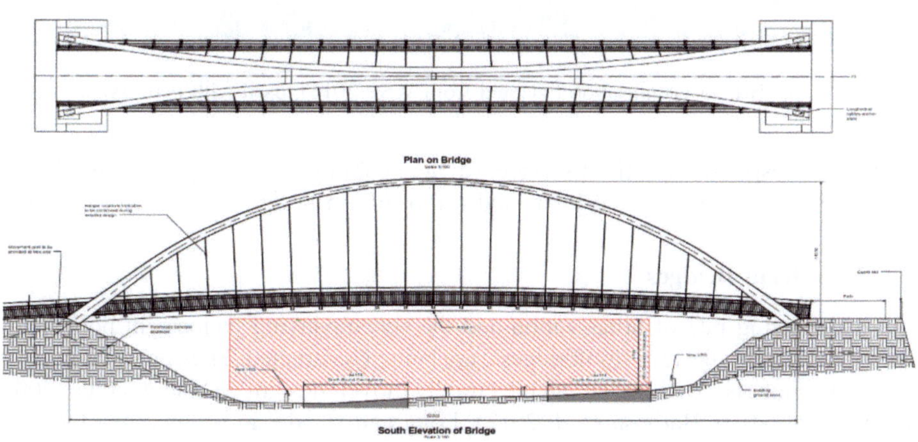

Fig. 1. Bridge plan and South elevation (adapted from WSP 2017), used with permission

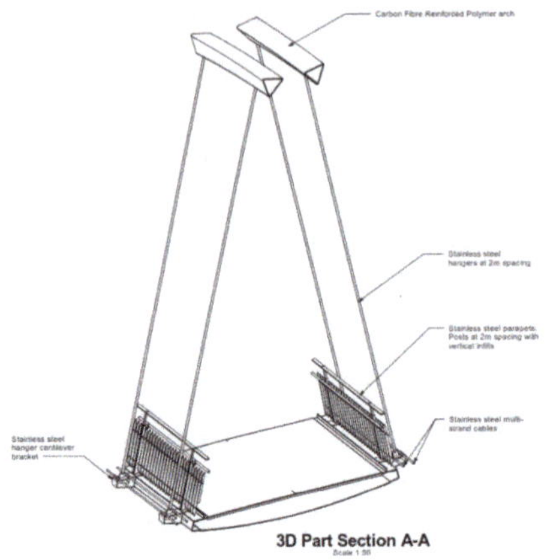

Fig. 2. 3D part section (adapted from WSP 2017), used with permission

Fig. 3. Bridge visualisation (adapted from WSP 2016), used with permission

(b) dynamic response of FRP in service; (c) applicability of existing deflection limits; (d) performance of constructed joints and (e) to inform bridge management practice.

1.3 Pedestrian Bridges

With the growing rate of urbanization and the accompanying increase in population, more pedestrian bridges (or 'footbridges') are being constructed. They are often used as statement symbols resulting in architecturally novel designs (Dallard et al. 2001; Caetano et al. 2010; Barbosa et al. 2013). Long and slender in profile, they typically use much less construction material than bridges that need to carry vehicles and are therefore much lighter. This low weight relative to the applied loading can make them susceptible to pedestrian-induced vibrations. Such vibrations can reach 'uncomfortable' levels. In recent times, this issue was brought into the public sphere during the closure of the London Millennium Bridge in 2000 (Dallard et al. 2001) just after its inauguration. Considerable retrofit costs were incurred to repair the situation, sparking many research efforts in the area of dynamic response of footbridges (Živanović et al. 2005). Numerous SHM systems have been deployed on footbridges and this is well documented in the literature (Parsekian et al. 2009; Caetano et al. 2010; Hu et al. 2013; Barbosa et al. 2013; Russell et al. 2017; Sá et al. 2017; Primi et al. 2017). Ingólfsson *et al* (2012) reported an extensive review of the state-of-the-art which focuses on cases of lateral bridge vibrations induced by pedestrians.

1.4 Fibre-Reinforced Polymers (FRP)

The deteriorating condition of infrastructure assets has been identified as a growing concern (cf. Thurlby 2013). Cost-effective and innovative solutions are needed to maintain these assets which are typically made of concrete and steel. Research into the use of FRP in construction has been on-going for many years (Bakis et al. 2002) leading to FRP being used as an alternative to these traditional materials. For instance, FRP has been reported to exhibit superior corrosion and fatigue resistance, high strength and stiffness-to-weight ratios, and can offer more rapid installation due to their pre-formed, modular construction and light weight (e.g. Bakis et al. 2002; Hollaway 2010). Altogether, these attributes contribute to lower maintenance costs over an asset's lifecycle (Kendall 2008; Mara et al. 2014).

Bridge decking is one of the more auspicious applications for FRP in infrastructure due to the benefits described above (Bakis et al. 2002; Hota and Hota 2002; Kumar et al. 2004; Sonnenschein et al. 2016). In the UK, the first major composite footbridge was the Aberfeldy Footbridge in Scotland, which was installed in 1992 (Skinner 2009). Canning and Luke (2010) carry out a comprehensive review of the many FRP bridges constructed in the UK. This includes the Mount Pleasant bridge in Lancashire, UK, which was the first FRP bridge installed on the UK motorway network by the Highways Agency in April 2006 (Canning 2008). Another 'pioneering' project is the St Austell FRP footbridge, which was designed by Parsons Brinckerhoff (Shave et al. 2010). Constructed in October 2007, it was the first structure built from FRP materials integrated into the UK rail network (Shave et al. 2009). Subsequently, many more successful installations have occurred across the UK transport network (Composites UK n.d.) as design guidance such as BD 90/05 has been developed (DMRB 2005; Shave and Bennetts 2012).

Several SHM deployments on FRP bridges have been carried out to investigate structural behaviour, such as dynamic response and creep deformation (Kumar et al. 2004; Sebastian et al. 2015; Votsis et al. 2017; Siwowski et al. 2018). Particular interest is held in the material's long-term behaviour (Sebastian 2016), as well as how it is affected by environmental factors (e.g. moisture ingress, UV exposure). More performance data on FRP bridges is needed for them to become more ubiquitous in future bridge construction efforts.

2 Instrumentation Overview

The sensor types which emerged as being most appropriate for this proposed deployment were *accelerometers* and *strain gauges*.

2.1 Accelerometers

Accelerometers are widely used for measuring dynamic responses of structures, including bridges (e.g. Russell et al. 2017). There are three main sensor types: piezoelectric, micro-electromechanical systems (MEMS) and piezoresistive (Middleton et al. 2016; Hanly 2016). Piezoelectric accelerometers are the most widely used for measurement applications (Hanly 2016) and typically contain a piezoelectric element supporting a mass, which induces a charge that is proportional to the force caused by any acceleration the sensor encounters.

Two key characteristics to consider are sensitivity and frequency response range (MMF 2001). The sensitivity of an accelerometer defines the magnitude of the electrical signal it generates in relation to the magnitude of acceleration. Higher sensitivity is more suitable when measuring small accelerations, as a cleaner signal (high signal to noise ratio) is obtained. The frequency response range gives the bandwidth of frequencies that a sensor can measure, usually specified as a tolerance band, relative to a reference frequency (range). Some typical performance characteristics for accelerometers are given in Table 3. For deployment on this bridge, accelerometers that can provide high accuracy readings in the 0.1–10 Hz range with high sensitivity (e.g. 10 V/g) would be ideal.

Table 3. Typical accelerometer performance (adapted from Midé 2017)

	Piezoelectric	DC MEMS	Piezoresistive
Frequency response (5% accuracy)	5–2000 Hz	0–1000 Hz	0–2000 Hz
Measurement range	±0.5 to ±2000 g	±16 to ±200 g	±100 to ±500 g
Sensitivity (±5%)	10.0 V/g	–	–
Resonant frequency	370–1220 Hz	–	–
Resolution	0.0008–0.06 g	0.004–0.05 g	0.003–0.015 g

$g = 9.81$ m/s^2 and V = volts

2.2 Strain Gauges

Webb (2014) describes a wide variety of sensors that can be used for both dynamic and static measurements, such as Electrical Resistance Strain Gauges (ERSs), Vibrating Wire Strain Gauges (VWSGs) and Demountable Mechanical Strain Gauges (DEMECs). Detailed descriptions of ERSs and VWSGs and some of their considerations can also be found in Middleton et al. (2016). ERSs are frequently used in SHM deployments (Sebastian et al. 2015; Hoult 2016; Siwowski et al. 2018). They consist of a thin metal wire that is bonded to the surface of the structure being monitored. As this surface experiences a strain, it changes the length of the wire which causes a proportional change in resistance. This resistance can then be measured with a suitable data-logger. The 'unit strain' output is dimensionless and can be used to infer changes in stress and quantify different stress states in the monitored parts of the bridge.

Several considerations should be made when deploying strain gauges. Temperature can affect the resistance of a strain gauge – which is a well-known issue (Webb 2014). To that end, most strain gauge manufacturers include temperature compensation capabilities in their sensors (National Instruments 2016). Another key thing to consider is the adhesive which is used to bond the strain gauge to the surface of interest. Adhesives experience creep and can therefore lead to inaccurate readings in the long-term. Also, ERSs only provide discrete measurements with relatively small gauge lengths, ranging from 0.5 mm to 100 mm. Therefore, one sensor is needed for each point of interest. For bridges, this number can get very large, which can raise the cost significantly and diminish aesthetics. With this in mind, one must be discerning when selecting points of interest.

3 Application of Value Assessment Methodology

3.1 Set-up and Agenda

As mentioned, this methodology involves facilitating a formal discussion between three key stakeholders: the 'SHM engineer', 'structural engineer' and 'asset owner'. For this case-study, these roles were filled by the persons shown in Table 4. Each participant has had at least four years' experience working in their respective field and was well-placed to have a meaningful discussion on this topic. The 2-letter references are used to signify spoken comments recorded from the meeting. The first author took the role of facilitator for the meeting. The third, fourth and fifth authors took on the key stakeholder roles.

Table 4. Roles occupied in discussion

Role	Job description	Experience	Ref.
SHM engineer	SHM expert, WSP	>4 years	SH
Structural engineer	Senior engineer, WSP	5 years	SE
Owner/asset manager	Structures team lead, SGC	22 years	AO

These three key stakeholders were to meet and discuss the questions/criteria shown in Table 1. Each criterion was value rated 1 to 5 by the relevant stakeholder based on its likeliness to be met (see Table 2). Participants were invited to write down comments to justify their score. These scores are averaged to give a simple metric that can be used to '*decide whether a particular SHM configuration may be worthwhile for a particular project*' (Vardanega et al. 2016). A more detailed description of the methodology development can be found in Vardanega et al. (2016).

Two SHM configurations were used for this case-study, one for accelerometers and the other for strain gauges, resulting in two pre-assessment matrices being filled out.

Whilst these two configurations were the primary focus, other sensors were also discussed. The meeting took place in the South Gloucestershire Council headquarters on 27th March 2018. The results are summarised in the following sections.

3.2 Results

3.2.1 Accelerometer Scores

Table 5 shows the assessment matrix filled out for the accelerometers scenario. Some entries in the comments column have been summarized for brevity. The resulting score of 4.2 indicates that deploying accelerometers on the bridge 'is very likely to yield value to the asset owner/manager'. This favourable score results from the relatively high ratings given in all the criteria, bar C3. The SHM engineer pointed out that "*it would be quite easy to make* [the system] *robust enough*" (SH) as accelerometers can be easily replaced. The structural engineer indicated maximum ratings for their criteria as "*expected levels of actual acceleration from the design phase and wind tunnel testing*" (SE) had been obtained which could be used to compare with the monitoring data for model validation.

Given the light-weight nature of the FRP footbridge, the asset owner expressed concern over adverse public relations if the bridge were to exceed the serviceability limits for acceleration: "*we would respond to those* [acceleration] *trigger levels, definitely*" (AO). The asset owner identified that the use of accelerometers would work in

Table 5. Filled in value assessment matrix for the accelerometer deployment

	1	2	3	4	5	Comments
A1					x	Accuracy easy to ascertain
A2				x		Could set-up a portable system; sensors easy to replace without losing information
A3			x			Depends on who is operating the monitoring. Sensors can be taken away for calibration
B1					x	Threshold would be up to the asset owner but could be based from code defined accelerations, varying comfort criteria, or the Wind Tunnel/dynamic analysis output
B2					x	Locations for peak expected accelerations are known from analysis therefore critical locations can be identified
B3					x	Testing in-situ, with the Tuned Mass Dampers (TMDs) locked and unlocked would allow the design output to be validated
C1					x	Could respond to concerns over vibration – aiding public relations. Could be used to change TMDs
C2				x		Would aid maintenance of TMDs. Would start with manufacturer maintenance regime then inform through monitoring
C3		x				Currently funded by Bristol University – v. likely. In 5 years' time, probably unlikely. Would need to be in Operation & Maintenance manual. Finding funding for short time period activities easier. Short-term cost a stronger driver than long- term
AVG	4.2					

tandem with the maintenance of the TMDs included in the bridge design. On the subject of potential user complaints about uncomfortable bridge vibrations they mentioned that "*we could do a 'before and after' and show that we've actually improved the situation by changing the* [frequency of] *the dampers*" (AO). Whilst funding for the project was secure for the next five years, the asset manager stated that from SGC's point of view long-term funding would not be likely as the next incumbent of their position might "*not put the same importance*" (AO) to the monitoring system. Therefore, a score of 2 was given for question C3.

3.2.2 Strain Gauge Deployment Scores

The assessment matrix for a strain gauge deployment is found in Table 6. The average score of 3.7 indicates that deploying strain gauges on the bridge 'is likely to yield value to the asset owner/manager', making it potentially less valuable than the accelerometer deployment. From the SHM engineer's point of view, the robustness of strain gauges in the long-term was a concern: "*Strain gauges are delicate… experience shows that when you stick them* [on a structure] *for 20 years, they're not going to last*" (SH). It is also pointed out that if you replace a strain gauge "*it's very hard to re-baseline… and you start getting more and more uncertainties in your absolute stress and strain*" (SH). To mitigate this, embedding the sensors into the structure during manufacture was suggested. This takes away most of the potential environmental deterioration and is much less intrusive aesthetically. However, the practicalities of embedding sensors are "*not as straightforward as attaching accelerometers*" (SH). Despite strain measurement being relatively easy from a sensor point of view, there is difficulty in interpreting the data because "*the magnitude of strain measurements is so much smaller than other things you try to correlate it with*" (SH). In other words, as strain changes are often small, it is challenging to distinguish meaningful data points from background noise. Furthermore, strain changes can be induced by various factors such as temperature, pedestrian loading and wind speeds, and it hard to separate the effects of each of these. This resulted in a lower score for ascertaining the required accuracy (A1).

The structural engineer was confident that "*given the level of modelling that's been or will be undertaken, at certain locations you'd be able to put a threshold* [strain] *value*" (SE). Critical monitoring locations could be determined from the design models, with connections most likely to be targeted. However, developing specific load cases and condition factors to obtain threshold values and critical locations would be a more difficult process, resulting in slightly lower scores compared to the accelerometer deployment.

The asset manager expressed they would "*take a decision if a* [strain] *trigger value is exceeded*" (AO), such as deciding whether to "*close the bridge or not*" (AO). However, they expressed that it would be less likely to inform a maintenance regime because FRP design is currently so conservative that any excessive deflections were unlikely. This is also based on observations made on a previously built FRP bridge within South Gloucestershire Council's stock in Frampton Cotterell (Sebastian et al. 2015): "*we were surprised that we got about half the deflection we anticipated so it performed really well, meaning it was a lot stiffer*" (AO). Similar to the accelerometer deployment, the same concerns over the lack of a secure budget from the asset manager's organization were stated, resulting in a lower score.

Table 6. Filled in value assessment matrix for the strain gauge deployment

	1	2	3	4	5	Comments
A1				x		Measuring strain is relatively easy from a sensor point of view. Required accuracy would depend on the structural engineer, so is closely interlinked to B1
A2				x		Sensor replacement may increase data uncertainty as it's difficult to re-baseline data. Can mitigate by embedding the sensors into the structure, reducing risk of a damage
A3			x			Depends on who is operating the monitoring. Strain gauges are much harder to calibrate
B1					x	Consideration should be given to long-term and short-term monitoring. The threshold values could be established from the design models; however, we wouldn't expect these to be reached in practice in the short term given the various loading and material factors that are considered in FRP design. They could still be considered useful threshold for **long-term** monitoring and management of the structure For **short-term** monitoring, a specific load case could be developed, run in the design model and replicated in practice on the structure. Threshold values could be established for this and strain measurements reviewed at regular intervals
B2				x		The anticipated strain can be identified at any location in the structure from the design models. Expected locations of peak strains/stresses can be determined and areas of high utilisation could be targeted The behaviour around bolt groups would be of interest
B3				x		In order to influence future design, the data and lessons learned would need to be fed into the development of future design guidance and standards. It is envisaged that data from other structures and wider considerations would be needed for this to happen The matrix answer has been based on the assumption that a particular load case is developed and run in the mode and on the structure for validation. We would consider the answer to be not likely/neutral if this was not undertaken
C1				x		Damage detection at specific places would be valuable – e.g. vandalism/fire – give alert to prompt inspection, should target non-visual issues Don't want to do lots of data analysis – needs a user interface
C2			x			Less likely to inform maintenance regime than accelerometer
C3		x				Currently funded by Bristol University – v. likely. In 5 years' time, probably unlikely. Would need to be in O&M manual. Finding funding for short time period activities easier. Short-term cost a stronger driver than long- term
AVG	3.7					

4 Further Considerations

Given the assessment results, the following implications and considerations emerge. A deployment of accelerometers would be expected to provide value to the asset owner. A system of these sensors would: (1) be simple to make robust, (2) be able to collect data that can inform the validation of design models, and (3) have the potential to inform the bridge's maintenance regime. The value obtained from a deployment of strain gauges would likely be lower, because they are harder to make robust and are perceived to have less value to the asset owner. However, the strain data can be used for model validation by structural engineers and inform research activities. To ensure long-term robustness, it is highly recommended they be embedded into the structure. This would involve liaising with the eventual manufacturers to co-ordinate installation.

The asset owner stated the desire for a layman's interpretation of the data: "*we wouldn't want to have to do any processing of the data, we'd want a system to go 'we've got an issue at hanger 12', we need to go out and do something.*" (AO). An appropriate user-interface may have to be developed to ensure insights are given to the asset owner. This leads to other considerations pertinent to an SHM system such as cabling, power supply, data transmission and data storage. These aspects require appropriate planning ahead of deployment and perhaps should be captured in the SHM criteria of this value assessment methodology.

Other potential instrumentation was also discussed in the meeting. It was noted that sensors that could be used for other projects were perceived to be more valuable to this asset owner. A weather station was brought up as an example: "*they would be very useful for multiple disciplines and then* [that means] *being able to fund those easier because I could share the funding with other teams*" (AO). It was also in the asset owner's intentions to install "*automatic pedestrian counters on the approaches to the bridge*" (AO). The structural engineer also described pedestrian counters as a "*useful*" (SE) sensor to investigate user-induced vibrations. This could be integrated into the monitoring system to provide context to accelerometer data.

5 Reflections on the Methodology

Overall, the participants found this methodology a worthwhile exercise in assessing the potential value of the monitoring for this project: "*I think challenging all the proposals would be useful for the project*" (SE), to which the other participants agreed. Indeed, having this structured process to critique sensor proposals allowed for constructive discussions leading to a meaningful quantification of value. However, it became widely accepted as the discussion drew on that the framework was not entirely suitable to this project, as it was asset owner focused. As pointed out by the asset manager, the wider aim of the project was not for "*South Gloucestershire Council to have an all-singing, all-dancing bridge that could be monitored every minute of the day*" (AO), but it "*was for industry and for academics to learn from… and to help other asset owners get more efficient, cost-effective structures*" (AO). The SHM engineer expounded on this point: "*it's not just South Gloucestershire doing all this monitoring for their own purposes, but it's wider – using that information for other purposes which may or may not lead to value to the asset owner*" (SH).

From this, it can be said that any monitoring on this bridge would be predominantly for the purposes of 'model validation' and 'sensor deployment study' (cf. Webb et al. 2015); rather than obtaining data that would directly drive decision-making by the asset owner. This shifts most of the potential value away from the asset owner, and towards the structural engineers and researchers involved in the project. Therefore, the framework (Vardanega et al. 2016) has been modified for this scenario (cf. Table 7 and Fig. 4). A fourth stakeholder – the research engineer – who may or may not be relevant for future projects, has been added to the framework. Additional questions have been included for the SHM engineer, structural engineer and research engineer (see Table 7). It is hoped these modifications better capture the likeliness of value for a monitoring system perceived to have a primary purpose of a 'model validation' deployment, where these four stakeholders are involved.

6 Summary and Conclusions

The value assessment methodology proposed by Vardanega et al. (2016) was tested for a proposed SHM deployment. The methodology proved effective at facilitating discussion between the key stakeholders at an early stage in the project. A fourth (optional) key stakeholder has been included in the methodology (research engineer) and extra questions have been added (see Table 7).

The following recommendations for the footbridge are made:

(a) The system of accelerometers would probably be valuable for the asset owner as it has potential to inform a maintenance regime;
(b) A robust long-term monitoring system can be deployed;
(c) The collected data can be used to investigate the bridge dynamic response;
(d) The system of strain gauges would be of less value to the asset owner but are potentially useful for structural engineers and research engineers informing future designs and understanding FRP behaviour/performance.

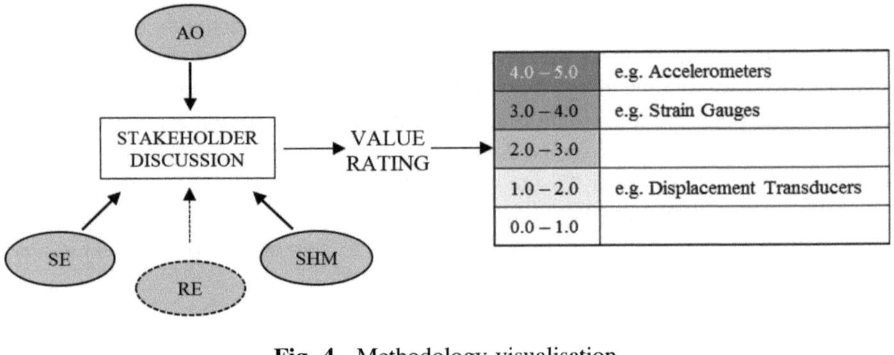

Fig. 4. Methodology visualisation

Table 7. Revised methodology questions (adapted from Vardanega et al. 2016)

Stakeholder	Question
SHM engineer	How likely is it that the required accuracy of the proposed measurements can be ascertained?
	How likely is it that the specified system can be designed to be sufficiently robust such that it can be maintained over the life of the monitoring project?
	How likely is it that the appropriate auditing will be conducted by a third party to certify that the gathered data are reliable? (e.g. monthly reports that give details of calibrations undertaken and 'sanity checking' of collected data.)
	*How likely is it that the system can be designed as (part of) a network (considering power supply, cabling, data storage etc.) to enable effective data transmission to key stakeholders?
Structural engineer	How likely is that relevant values of approximate threshold values will be assigned in consultation with the asset owner so as to ensure that the owner will act if these values are realized in the field?
	How likely is it that the critical monitoring locations on the structure can be determined?
	How likely is it that the data be used for validation (or falsification) of structural models which may assist with the design of more efficient future structures for the asset owner?
	*How likely is it that the findings from the monitoring will be disseminated to relevant authorities to develop standard guidance?
	*How likely is it that the findings from the monitoring will inform standard practice 'in-house'?
Owner/asset manager	How likely is it that the necessary actions (or decisions) will be taken if threshold (trigger) values are exceeded/reached?
	How likely is it that the data to be collected will be able to inform a maintenance regime?
	How likely is it that there will be a secure budget necessary for the intended period of operation of the monitoring project?
*Research engineer (if applicable)	*How likely is it that data collected from the monitoring system will be useful for current research activities?
	*How likely is it that the monitoring of this bridge will be shared to other research organisations to contribute to related topics?
	*How likely is it that there will be sufficient resources (e.g. labour, funding) available to operate and maintain the monitoring system?
	*How likely is it that support will be received from the asset manager to monitor the bridge in the long-term?

*Denotes newly added questions to the original framework cf. Table 1

7 Data Availability Statements

The work reported in this study has not generated new experimental data.

Acknowledgments. The first author would like to acknowledge support from the Engineering and Physical Sciences Research Council (EPSRC) through the National Productivity Investment Fund (Grant Number EP/R51245X/1).

References

Bakis, C., Bank, L., Brown, V., Cosenza, E., Davalos, J., Lesko, J., Machida, A., Rizkalla, S., Triantafillou, T.: Fiber-reinforced polymer composites for construction-state-of-the-art review. J. Compos. Constr. **6**(2), 73–87 (2002). https://doi.org/10.1061/(ASCE)1090-0268(2002)6:2(73)

Barbosa, R., Magalhães, F., Caetano, E., Cunha, Á.: The Viana footbridge: construction and dynamic monitoring. Proc. Inst. Civ. Eng. Bridge Eng. **166**(4), 273–290 (2013). https://doi.org/10.1680/bren.10.00048

Caetano, E., Cunha, Á., Moutinho, C., Magalhães, F.: Studies for controlling human-induced vibration of the Pedro e Inês footbridge, Portugal. Part 1: assessment of dynamic behaviour. Eng. Struct. **32**(4), 1082–1091 (2010). https://doi.org/10.1016/j.engstruct.2009.12.034

Canning, L.: Mount pleasant FRP bridge deck over M6 motorway. In: Fourth International Conference on FRP Composites in Civil Engineering (CICE 2008), Switzerland (2008)

Canning, L., Luke, S.: Development of FRP bridges in the UK—an overview. Adv. Struct. Eng. **13**(5), 823–835 (2010). https://doi.org/10.1260/1369-4332.13.5.823

Composites UK: BRIDGES – Case Studies (n.d.). https://compositesuk.co.uk/composite-materials/applications/construction/bridges. Accessed 19 April 2018

Dallard, P., Flint, A., Le Bourva, S., Low, A., Smith, R.M.R., Willford, M.: The London millennium footbridge. Struct. Eng. **79**(22), 17–35 (2001)

Design Manual for Roads and Bridges DMRB: BD 90/05: Design of FRP Bridges and Highway Structures, vol. 1 (Section 3) (2005). http://www.standardsforhighways.co.uk/ha/standards/dmrb/vol1/section3/bd9005.pdf. Accessed 26 June 2018

Hanly S.: Accelerometers: Taking the Guesswork Out of Accelerometer Selection. Engineering Solutions, Midé (2016). https://blog.mide.com/accelerometer-selection. Accessed 21 April 2018

Hollaway, L.C.: A review of the present and future utilisation of FRP composites in the civil infrastructure with reference to their important in-service properties. Constr. Build. Mater. **24**(12), 2419–2445 (2010). https://doi.org/10.1016/j.conbuildmat.2010.04.062

Hota, G.V.S., Hota, S.R.V.: Advances in fibre-reinforced polymer composite bridge decks. Prog. Struct. Mat. Eng. **4**(2), 161–168 (2002). https://doi.org/10.1002/pse.113

Hoult, N.A.: Case study: Black River Bridge. In: Middleton, C., Fidler, P., Vardanega, P.J. (eds.) Bridge Monitoring: A Practical Guide, pp. 69–71. ICE Publishing, London (2016)

Hu, W.H., Caetano, E., Cunha, Á.: Structural health monitoring of a stress-ribbon footbridge. Eng. Struct. **57**, 578–593 (2013). https://doi.org/10.1016/j.engstruct.2012.06.051

Ingólfsson, E.T., Georgakis, C.T., Jönsson, J.: Pedestrian-induced lateral vibrations of footbridges: a literature review. Eng. Struct. **45**, 21–52 (2012). https://doi.org/10.1016/j.engstruct.2012.05.038

Kendall, D.: The business case for composites in construction. Reinf. Plast. **52**(7), 20–27 (2008). https://doi.org/10.1016/S0034-3617(08)70239-X

Kumar, P., Chandrashekhara, K., Nanni, A.: Structural performance of a FRP bridge deck. Constr. Build. Mater. **18**(1), 35–47 (2004). https://doi.org/10.1016/S0950-0618(03)00036-9

Mara, V., Haghani, R., Harryson, P.: Bridge decks of fibre reinforced polymer (FRP): a sustainable solution. Constr. Build. Mater. **50**, 190–199 (2014). https://doi.org/10.1016/j.conbuildmat.2013.09.036

Metra Mess- und Frequenztechnik (MMF): Piezoelectric Accelerometers – Theory and Application (2001). http://www.gracey.co.uk/downloads/accelerometers.pdf. Accessed 21 April 2018

Middleton, C.R., Fidler, P.R.A., Vardanega, P.J.: Bridge Monitoring: A Practical Guide. ICE Publishing, London (2016)

Midé: Slam Stick: Shock and Vibration Data Loggers. Engineering Solutions, Midé (2017). https://info.mide.com/data-loggers/download-slam-stick-datasheets-page. Accessed 21 April 2018

National Instruments: Measuring Strain with Strain Gauges (2016). http://www.ni.com/white-paper/3642/en/. Accessed 21 April 2018

Parsekian, G., Shrive, N., Brown, T., Kroman, J., Seibert, P., Perry, V., Boucher, A., Ghoneim, G.: Full-scale testing of a fibre-reinforced concrete footbridge. Proc. Inst. Civ. Eng. Bridge Eng. **162**(4), 157–166 (2009). https://doi.org/10.1680/bren.2009.162.4.157

Primi, S., Acero, R.L., Paulotto, C., Herrera, I.C.: Delivery of a 40 m long fibre-reinforced polymer composite footbridge in Madrid, Spain. Proc. Inst. Civ. Eng. Struct. Build. (2017). https://doi.org/10.1680/jstbu.16.00233

Russell, J., Wei, X., Živanović, S., Kruger, C.: Dynamic response of an FRP footbridge due to pedestrians and train buffeting. Procedia Engineering **199**, 3059–3064 (2017). https://doi.org/10.1016/j.proeng.2017.09.411

Sá, M.F., Guerreiro, L., Gomes, A.M., Correia, J.R., Silvestre, N.: Dynamic behaviour of a GFRP-steel hybrid pedestrian bridge in serviceability conditions. Part 1: experimental study. Thin Walled Struct. **117**(May), 332–342 (2017). https://doi.org/10.1016/j.tws.2017.05.013

Sebastian, W., Ross, J., Johnson, M., Twyman, C., Henderson, J.: Continuous monitoring and lorry testing of the Frampton Cotterell FRP road bridge. In: Proceedings of the 2015 ACIC Conference, Cambridge, pp. 21–26 (2015)

Sebastian, W.: Case study: instrumentation and monitoring of the Frampton Cotterell FRP traffic bridge. In: Middleton, C., Fidler, P., Vardanega, P.J. (eds.) Bridge Monitoring: A Practical Guide, pp. 95–98. ICE Publishing, London (2016)

Shave, J., Bennetts, J.: Some principles for designing safe and robust FRP structures. In: Halliwell, S., Whysall, C. (eds.) FRP Bridges: Conference Proceedings Full Papers from the 1st FRP Bridges Conference, 13th–14th September 2012, London, UK. NetComposites Limited, Chesterfield, UK, pp. 4–14 (2012)

Shave, J., Denton, S., Frostick, I.: St Austell Footbridge: The first fibre reinforced polymer structure on the UK rail network. In: Proceedings of the 2009 ACIC Conference, Edinburgh, pp. 116–126 (2009)

Shave, J., Denton, S., Frostick, I.: Design of the St Austell fibre-reinforced polymer footbridge, UK. Struct. Eng. Int. **20**(4), 427–429 (2010). https://doi.org/10.2749/101686610793557780

Siwowski, T., Kaleta, D., Rajchel, M.: Structural behaviour of an all-composite road bridge. Compos. Struct. **192**(March), 555–567 (2018). https://doi.org/10.1016/j.compstruct.2018.03.042

Skinner, J.M.: A critical analysis of the Aberfeldy Footbridge, Scotland. In: Proceedings of Bridge Engineering 2 Conference 2009 (April) (2009)

Sonnenschein, R., Gajdosova, K., Holly, I.: FRP composites and their using in the construction of bridges. Procedia Eng. **161**, 477–482 (2016). https://doi.org/10.1016/j.proeng.2016.08.665

Thurlby, R.: Managing the asset time bomb: a system dynamics approach. Proc. Inst. Civ. Eng. Forensic Eng. **166**(3), 134–142 (2013). https://doi.org/10.1680/feng.12.00026

Vardanega, P.J., Webb, G.T., Fidler, P.R.A., Middleton, C.R.: Assessing the potential value of bridge monitoring systems. Proc. Inst. Civ. Eng. Bridge Eng. **169**(2), 126–138 (2016). https://doi.org/10.1680/jbren.15.00016

Votsis, R.A., Stratford, T.J., Chryssanthopoulos, M.K., Tantele, E.A.: Dynamic assessment of a FRP suspension footbridge through field testing and finite element modelling. Steel Compos. Struct. **23**(2), 205–215 (2017). https://doi.org/10.12989/scs.2017.23.2.205

Webb, G.: Structural Health Monitoring of Bridges. Ph.D. thesis, University of Cambridge, Cambridge, UK (2014)

Webb, G.T., Vardanega, P.J., Middleton, C.R.: Categories of SHM deployments: technologies and capabilities. J. Bridge Eng. **20**(11), 04014118 (2015). https://doi.org/10.1061/(ASCE)BE.1943-5592.0000735

WSP: Bridge visualisations. In: PK16/6500/F|Erection of an Composite Pedestrian and Cycle Bridge Linking Emersons Green East (Gateway) Development and the Existing District Centre, Across the A4174 Avon Ring Road (2016). http://developments.southglos.gov.uk/online-applications/applicationDetails.do?activeTab=summary&keyVal=OH7M8AOKISZ00. Accessed 27 June 2018

WSP: Bridge cross-sections and elevations. In: PK16/6500/F|Erection of an Composite Pedestrian and Cycle Bridge Linking Emersons Green East (Gateway) Development and the Existing District Centre, Across the A4174 Avon Ring Road (2017). http://developments.southglos.gov.uk/online-applications/applicationDetails.do?activeTab=summary&keyVal=OH7M8AOKISZ00. Accessed 27 June 2018

Živanović, S., Pavic, A., Reynolds, P.: Vibration serviceability of footbridges under human-induced excitation: a literature review. J. Sound Vib. **279**, 1–74 (2005). https://doi.org/10.1016/j.jsv.2004.01.019

Seismic Demands Assessment of Tall Buildings: Theoretical Approach and Applications

Benazouz Chikh[1(✉)], Mustapha Remki[1], Abdelkader Benyoucef[1], Youcef Mehani[1], Mohamed Hadid[2], and Abdelmounaim Mechaala[3]

[1] Earthquake Engineering Division, National Earthquake Engineering Research Center, Hussein Dey, Algiers, Algeria
cheikhpbsd@gmail.com
[2] National School of Built and Ground Works Engineering (ENSTP), 01 Rue Sidi Garidi, Vieux Kouba, 16003 Algiers, Algeria
[3] Civil Engineering Department, University of Abbés Laghrour, Khenchla, Algeria

Abstract. In recent years, a new interest has in fact arisen in the displacement based seismic design, especially in tall buildings. Several contributions have already been submitted for the development of different procedures based on displacements with the aim to simplify the evaluation of seismic demands of structures. In this paper, an improved procedure based on displacements applicable to the analysis and design of tall buildings is presented and illustrated by examples. It is considered as a new procedure for estimating the seismic deformation of Multi-Degree-of-Freedom systems. An example of a calculation that highlights the steps involved in applying this inelastic response evaluation approach is included to demonstrate the simplicity of the method. The accuracy of the proposed procedure is verified against the Nonlinear Time History Analysis (NL-THA) results and Uncoupled Modal Response History Analysis (UMRHA) of a 9-story steel building subjected to El-Centro 1940 (N/S) as a first application. In the second application the accuracy of the improved procedure is verified against the nonlinear time history analysis results of two tall buildings. The comparisons showed that the new method gives accurate deformations.

1 Introduction

Several high-rise buildings have recently been designed and are being built in areas of high seismicity. As many of these buildings use high performance materials or exceed the height limits specified in current building codes, they are designed according to an alternative performance procedure, permitted in the alternative clause of the codes. For such projects, Nonlinear Time History Analysis NL-THA is now the most used to determine Engineering Demand Parameters (EDPs) for a set of multi-component ground motions. The selection and scaling of ground motions required for the NL-THA has been the subject of many researchs in recent years.

Several simple evaluation methods have been proposed as an alternative to complex NL-THA for estimating EDPs (Gülkan and Sözen 1974; Freeman et al. 1975;

H. Rodrigues and A. Elnashai (Eds.): GeoMEast 2018, SUCI, pp. 167–179, 2019.
https://doi.org/10.1007/978-3-030-01932-7_14

Newmark and Hall 1982; Fajfar and Fischinger 1988; Kowalsky 1994; Sasaki et al. 1998; Fajfar 1999; Gupta and Kunnath 2000; Albanesi et al. 2000; Priestley and Kowalsky 2000; Miranda 2001; Chopra and Goel 2001; Lin and Chang 2003; Maja and Fajfar 2012; Chikh et al. 2014; Zerbin and Aprile. 2015; Kazaz 2016). The basic idea of these methods is to relate the structural capacity to the physical basis of elastic or inelastic demand spectra, as the Capacity Spectrum Method (ATC–40 1996) and its different implementations.

According to the Nonlinear Static Procedures (NSP) described in FEMA-440 (FEMA 2005), ASCE/SEI 41-06 (ASCE 2007) and Eurocode 8 (British Standards 2004), EDPs can be calculated by a nonlinear static analysis of the structure known as Pushover analysis. These procedures are now standard in the practice of structural engineering since they provide a better assessment of the actual capacity and expected performance of the structure compared to the code-based NSP procedure. Traditionally they require much less computational effort compared to rigorous nonlinear dynamic analysis. NSP procedures are attractive for the additional reason that their implementation only requires the spectral design; it is necessary to have records of ground motions.

This paper is divided into two main sections. A presentation of the Non Linear Multimodal Spectral Analysis (Nl-MSA) developed by Chikh et al. (2016) in terms of an inelastic response spectrum to obtain a good estimate of seismic demands which considers the ductility factor as the inelastic response of the system which the contribution of higher vibration modes is introduced. In the second part, this procedure is applied to two (02) tall buildings.

2 Nonlinear Multimodal Spectral Analysis (NL-MSA)

An approximate spectral method called Nonlinear Multimodal Spectral Analysis (NL-MSA) is investigated in this paper. This method was developed by Chikh et al. (2016) and is suitable for estimating the inelastic seismic response of Multi-Degree-Of-Freedom (MDOF) systems. However, it is based on the inelastic modal decomposition, and will be simplified when using an inelastic response spectrum in terms of ductility demand. In this procedure the maximum ductility demand for the n considers modes are not generally attained at the same time t. Therefore the value of ductility demand of the structure should be calculated as a combination. A summary of the implementation of this procedure is presented below. In this method, the peak responses of inelastic MDOF systems can be calculated from the inelastic response spectrum. The results of this procedure are quite accurate for structural design applications.

2.1 Design Spectra

The seismic demand in this method is determined by using the inelastic design spectra developed by Benazouz et al. (2012), which have been called ductility demand response spectrum (DDRS). These spectra depends on the normalized yield strength coefficient (η) (defined as the yield strength coefficient divided by the Peak Ground Acceleration of the earthquake) which is fixed. The ductility demand is calculated

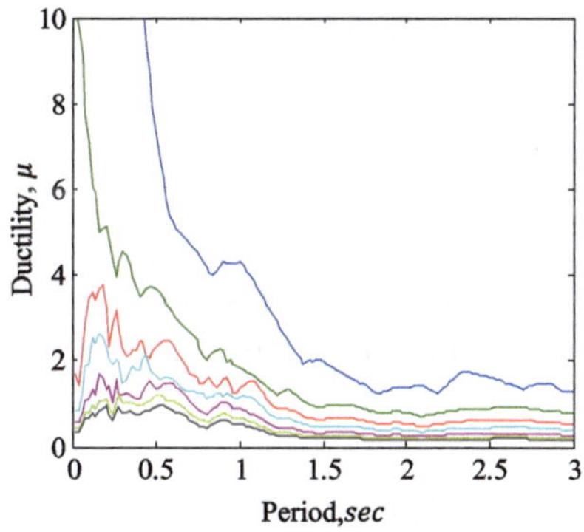

Fig. 1. DDRS for inelastic system computed for El Centro 1940 (N/S) component (η = 0.25, 0.5, 0.75, 1, 1.5, 2, 2.5 from top line to bottom line).

directly by drawing a vertical line passing through the natural period of the nth mode of the structure (see Fig. 1). On the other hand, DDRS spectra are obtained when maximum ductility demands are assessed for a constant value of normalized yield strength coefficient (η) defined as (Benazouz et al. 2012; Chikh et al. 2016; Lin and Mahin 1984):

$$\eta_n = \frac{q_n g}{PGA} \tag{1}$$

Where:

q_n is the yield strength coefficient of the nth mode.
g is the gravity.
PGA stands for the Peak Ground Acceleration.

The ground acceleration has been normalized such that its value varies from −1 to 1. The NL-MSA implies that for a MDOF inelastic system, if α_n (post-to-preyield stiffness ratio) and η_n of the nth mode are fixed, the intensity of the ground motion has no effect on the normalized peak ductility factor (Chikh et al. 2016). This permits the construction of the ductility response spectrum for an ensemble of ground motions with common frequency content. The DDRS is constructed also for El Centro 1940 ground motion (N/S) component and is shown in Fig. 1 in function of the parameter (η).

Simplified equations for ductility demands μ would obviously facilitate the estimation of the ductility and deformation of an inelastic SDOF system has been developed by (Benazouz et al. 2012). Such an equation for μ, could be used to determine the ductility demand for a new or rehabilitated structure with known normalized yield strength coefficient (η) and post-to-preyield stiffness ratio (α).

The spectrum is divided into three period regions according to the procedure described by Benazouz et al. (2012), where the limit period T = 0.6 s marks the transition from the acceleration-sensitive region to the velocity sensitive region which ends at T ~ 3 s. Such an equation for μ has been derived in terms of η, α, and T, and they are tabulated in Table 1 for different ranges of η and α.

Table 1. Parameters in Eqs. (2 and 3) for each value of η and α.

Parameters	$\eta < 1$				$\eta \geq 1$ (all values of α)
	$\alpha(\%)$				
	0	3	5	10	
a	1.24	1.12	1.08	1.04	1.23
b	0.98	0.94	0.92	0.88	0.85
c	1.69	1.65	1.68	1.68	$1.21 + \alpha$

$$\mu = \frac{a(b)^{1/T}(T)^{\eta - c}}{\eta} \quad \text{for All values of } T \tag{2}$$

$$\mu = \frac{\omega^2(1-\alpha)}{\eta} 0.027T^{0.84} \quad T \geq 0.6\,\text{s} \tag{3}$$

2.2 Step by Step NL-MSA Procedure

For convenience we summarize the nonlinear multimodal spectral analysis (NL-MSA) as a series of steps used to estimate the peak inelastic response of MDOF systems:

1. Calculate the natural vibrating period T_n, the natural frequencies ω_n and the mode shapes ϕ_n, for linearly elastic vibration of the structure (Fig. 2a, b).
2. For the nth mode, develop the diagram of force-deformation relationship between the base shear and the top displacement $V_{bn} - x_{rn}$ (capacity curve) (Fig. 2c).
3. Idealize the obtained capacity curve as a bilinear system.
4. Transform the idealized Pushover curve to the capacity diagram $S_{an} - D_n$ (Acceleration – Displacement format) (Fig. 2d).
5. Define the ground motion $\ddot{x}_g(t)$ (Fig. 2e).
6. Compute the post-to-preyield stiffness ratio α_n and the normalized yield strength η_n with the known q_n and PGA, and select the damping ratio ξ_n of the design structure (Fig. 2f).
7. Calculate the ductility demand for the nth mode by using the DDRS spectra.
8. Calculate peak roof displacement and base shear of the original structure (Fig. 2g).
9. Calculate the floor displacement, story drifts, etc, from the capacity curve database obtained in the second step of the procedure.
10. Repeat Steps 3 to 8 many times (for different modes) as required for sufficient accuracy.
11. Determine the total response (demand) by combining the peak "modal" responses using the SRSS rule.

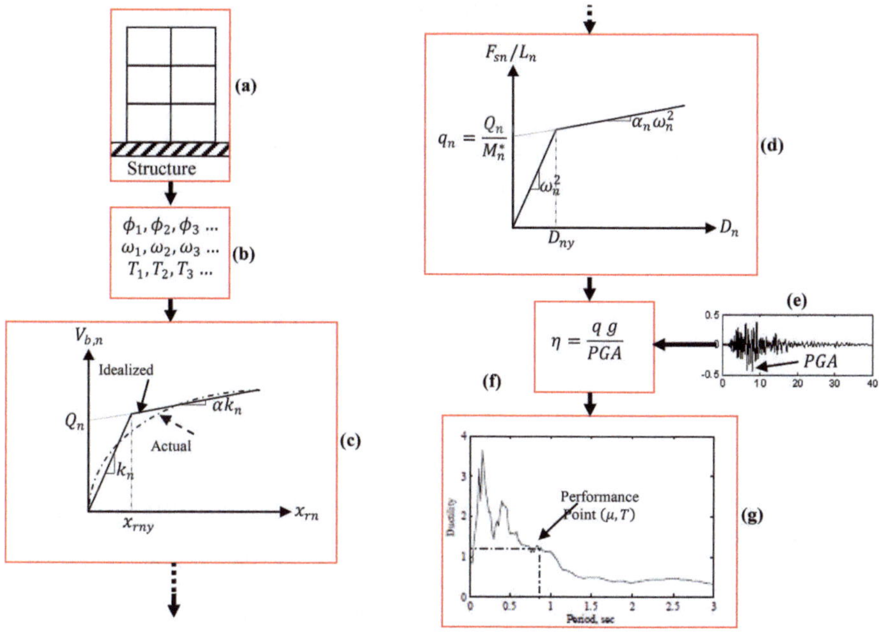

Fig. 2. Flow chart of the NL-MSA procedure.

2.3 Application

In recent years, Chopra and Goel (2002) assessed the strength variation of several procedures including the modal Pushover analysis (MPA), that they developed. The MPA analysis is based on structural dynamics theory. Its accuracy and reliability in estimating the peak response of inelastic MDOF systems has been evaluated extensively by the authors. Chopra and Goel (2004) analyzed and evaluated the response of several procedures for nonlinear static analysis, including Pushover analysis where only fundamental mode was taken into account.

The accurate of the proposed procedure is evaluated for a 9-story SAC steel building (Chopra and Goel 2001). The 'exact' response of a rigorous nonlinear time history analysis (NL-THA) is compared with the response obtained by the NL-MSA.

The 9-story structure meets the seismic code requirements and represents typical medium-rise buildings designed for the Los Angeles, California region. The Pushover curves of this structure presented in (Chopra and Goel 2001) are sufficient for the objectives of this study. The selected structure is tested and detailed in this section when subjected to one time and half (1.5) El Centro 1940 ground motion in order to compare our results with those of Chopra and Goel (2001). The properties of the first three modes of vibration are summarized in Table 2.

The capacity curves of the three first modes are shown in Fig. 3. Next, the Pushover curves are transformed to equivalent SDOF systems (see Fig. 3). The conversion of the

Table 2. Properties of modal inelastic SDOF systems.

Properties	Mode 1	Mode 2	Mode 3
L_n (kg)	2,736,789	− 920,860	696,400
Γ_n	1.36	− 0.5309	0.2406
M_n^* (kg)	3,740,189	488,839.1	167,531.5
$D_{n,y}$ (cm)	26.51	18.65	19.12
T_n (s)	2.2671	0.8525	0.4927
α_n	0.19	0.13	0.14
k_n (kN/cm)	210.3867	500.2020	1132.6086
ξ_n (%)	1.948	1.103	1.136
Q_n (kN)	6168.977	4374.343	4414.347
q_n (g)	0.168	0.912	2.685

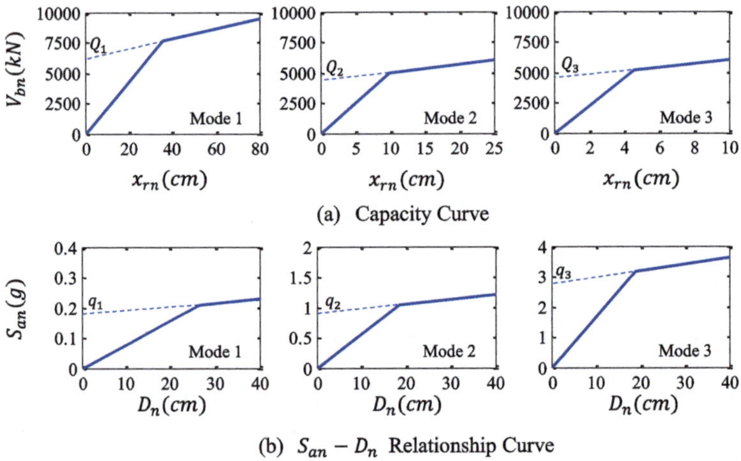

Fig. 3. Modal Pushover curves and capacity diagrams for the first three modes.

idealized Pushover curve to the force-displacement, (see Fig. 3b) for the nth–mode of inelastic SDOF system is obtained by using $\left(F_n^* - D_{n,y}\right)$:

$$S_{an} = \frac{V_{bn}}{M_n^*} = F_n^*, \quad D_n = \frac{x_{rn}}{\Gamma_n \phi_{rn}} \tag{4}$$

In which S_{an} is the spectral acceleration, V_{bn} the base shear, ϕ_{rn} the amplitude of ϕ_n and x_{rn} the roof displacement.

The NL-MSA of the structure starts with obtaining the multimodal Pushover curves of the MDOF system subjected to lateral forces distributed over the building height. In the proposed procedure, the movements will be decomposed in the form of a series of normal modes in terms of the ductility demand. The obtained response in terms of

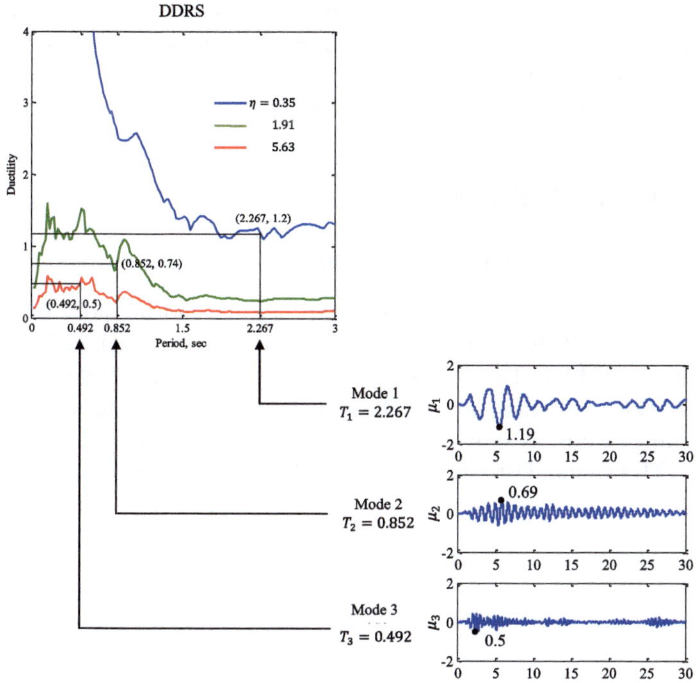

Fig. 4. Peak response of ductility demand from the DDRS spectra and the NL-THA for 1.5 × El Centro ground motion: first three modal responses.

ductility demand for the three first modes of the selected building subject to 1.5 times El Centro ground motion (N/S) component (PGA = 0.32 g, PGV = 36.14 cm/s, and PGD = 21.34 cm) are shown in Fig. 4.

Table 3. Peak values of floor displacements from the NL-MSA and NL-THA for 1.5 × El Centro ground motion.

Procedure	Combined (NL-MSA)			NL-THA	Errors (%)		
	1 Mode	2 Modes	3 Modes		1 Mode	2 Modes	3 Modes
DDRS spectra	44.12	44.38	44.45	44.6	− 1.08	− 0.49	− 0.33
DDRS design spectra	34.54	34.90	34.91		− 29.12	− 27.79	− 21.39

Following the NL-MSA procedure aforementioned, the total response is determined using the NL-THA ("exact"). Figure 4 shows the ductility demand and ductility time histories. It is clear from the comparison shown in this figure that the NL-MSA gives results in good agreement with the NL-THA.

Table 3 presents the results of the combined response of the first application (9-story SAC steel building) obtained by the DDRS spectra considering one, two, and three "modes," respectively, and the errors in these estimates relative to the exact response from NL-THA (Chopra and Goel 2001).

3 Application to Tall Buildings

3.1 Structural Systems and Ground Motions

The structural systems considered in this study are 48 and 62 story buildings taken from the project (Tall Building Initiative at the Pacific Earthquake Engineering Research Center (PEER). The lateral resisting system of the buildings is a ductile concrete core wall consisting of two channel-shaped walls connected to rectangular walls by coupling beams, as shown in Figs. 5 and 6. The total height of the tall building of 48 stories (TB48) and TB62 buildings is 143.56 and 192.02 m, respectively.

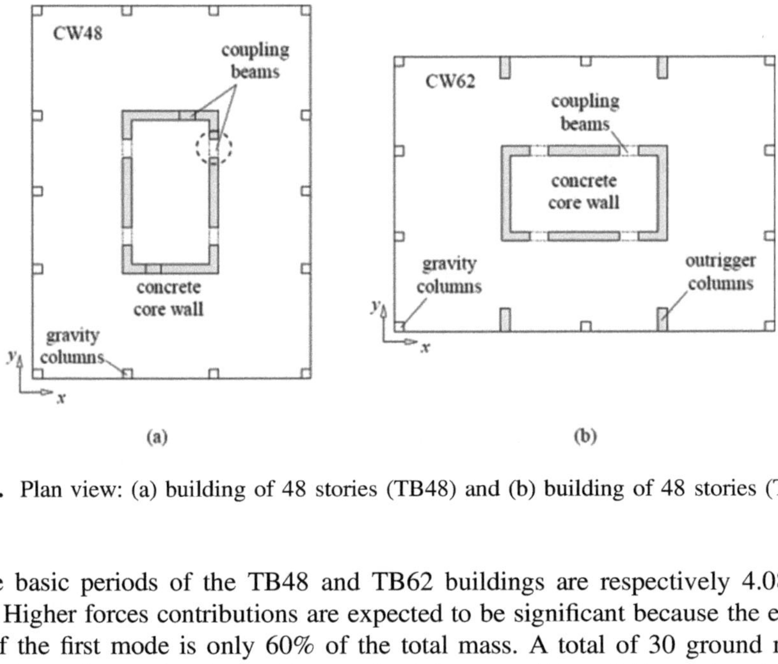

Fig. 5. Plan view: (a) building of 48 stories (TB48) and (b) building of 48 stories (TB62).

The basic periods of the TB48 and TB62 buildings are respectively 4.08 s and 4.50 s. Higher forces contributions are expected to be significant because the effective mass of the first mode is only 60% of the total mass. A total of 30 ground motions records from nine earthquakes with magnitude between 7.3 and 7.9 used and selected by Reyes and Chopra (2011) are used in this investigation.

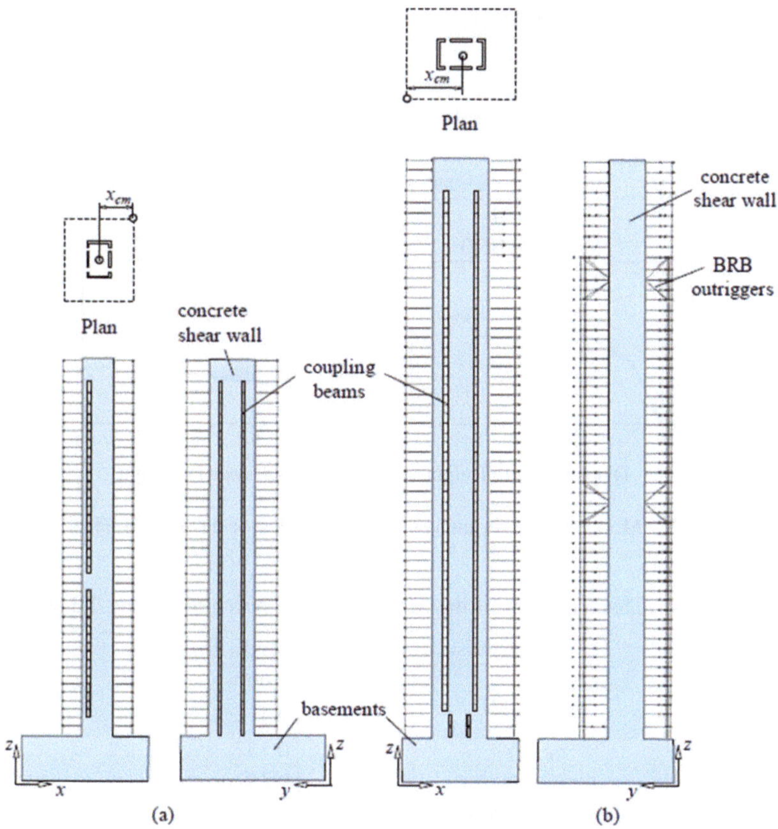

Fig. 6. Elevations: (a) building of 48 stories (TB48) and (b) building of 48 stories (TB62) (Reyes and Anil 2011).

3.2 Analyse Spectrale Multimodale Non Linéaire (NL-MSA)

As a first step in the evaluation of the NL-MSA procedure, it is necessary to present the Pushover curves and the reference displacement response of each building. Figure 7 shows capacity curves for each building associated with their first and second "modes" in the x and y directions, respectively, and Tables 4 and 5 summarize the calculations of the 1DDL inelastic equivalent system for the two selected buildings (Fig. 8 and Tables 6, 7).

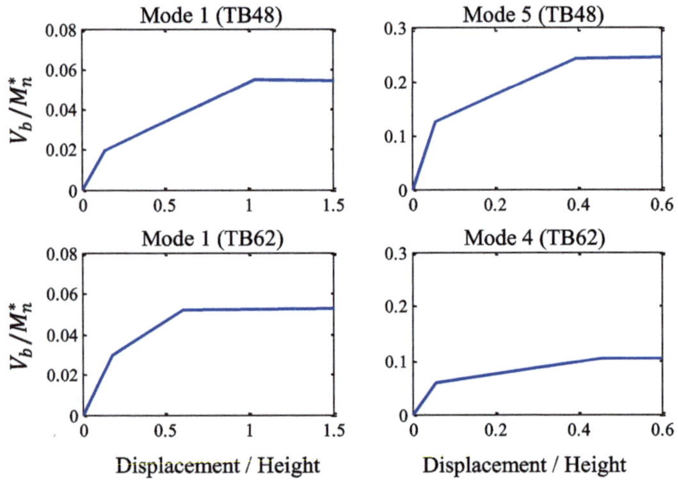

Fig. 7. Modal Pushover curves for two modes of TB48 and TB62.

Table 4. Characteristics of the SDOF inelastic system (building TB48).

Characteristics	Mode 1	Mode 5
T (s)	4.08	0.99
α_n (%)	6.1	9.4
q_n (g)	0.375	0.137
η_n	1.23	0.45
x (NL-MSA)	0.485 m	0.276 m
x (NL-MSA Eq)	0.18 m	0.055 m

Table 5. Characteristics of the SDOF inelastic system (building TB62).

Characteristics	Mode 1	Mode 4
T (s)	4.50	1.22
α_n	2.5	3.3
q_n (g)	0.046	0.075
η_n	0.15	0.245
x (NL-MSA)	0.326 m	0.141 m
x (NL-MSA Eq)	0.14 m	0.049 m

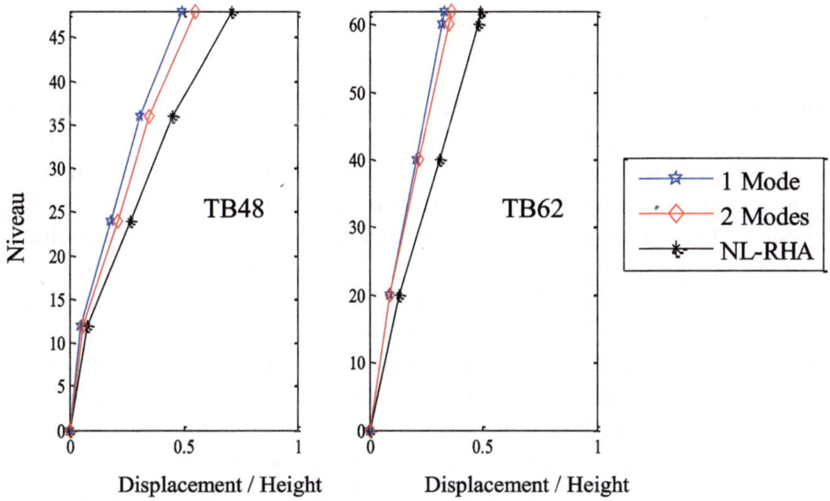

Fig. 8. Comparison of the displacement coefficient of TB48 and TB62.

Table 6. Maximum values of roof displacements obtained by NL-MSA and NL-THA for 1.5 × El Centro 1940 (N / S) for TB48.

Procedure	NL-MSA (combined)		NL-THA	Errors (%)	
	1 Mode	2 Modes		1 Mode	2 Modes
Spectre DDRS	0.485	0.55	0.71	− 31.69	− 22.53
Spectre DDRS Eq (5.20)	0.18	0.19		− 74.64	− 73.23

Table 7. Maximum values of roof displacements obtained by NL-MSA and NL-THA for 1.5 × El Centro 1940 (N/S) for TB62.

Procedure	NL-MSA (combined)		NL-THA	Errors (%)	
	1 Mode	2 Modes		1 Mode	2 Modes
Spectre DDRS	0.326	0.355	0.49	− 33.46	− 27.51
Spectre DDRS Eq (5.20)	0.14	0.148		− 71.42	− 69.79

4 Conclusions

An approximate procedure for seismic demands assessment of MDOF system has been developed and its accuracy verified by examples. An inelastic modal decomposition in terms of ductility has been developed to construct the NL-MSA. The base shear–roof displacement ($V_{bn} - x_{rn}$) curve is developed from a Pushover analysis. This Pushover curve is idealized as a bilinear force–deformation relation for the nth mode of inelastic SDOF system. This idealization is used to determine the yield strength coefficient q_n and the post-to-preyield stiffness ratio α_n to estimate the ductility demand.

The efficiency of the NL-MSA procedure is evident; the designer needs only to have the Pushover curve of the structure and the design earthquake(s) to determine peak response of any structure, namely, displacement and base shear. This method is applicable to a variety of uses such as a rapid evaluation technique for a large inventory of buildings, a design verification procedure for new construction, and evaluation procedure for an existing structure to identify damage states.

The results of the NL-MSA method are acceptable while compared to the nonlinear dynamic analysis results. For structures that the first mode contributes significantly to the response, the NL-MSA method will generally give good estimates of demand for global deformations. However, inelastic dynamic response may differ significantly from a selected or from the response distributions with constant lateral loads. For example, we can expect significant differences in structural response due to the influence of vibration modes at higher frequencies.

References

Albanesi, T., Nuti, C., Vanzi, I.: A simplified procedure to assess the seismic response of nonlinear structures. Earthq. Spectra **16**(4), 715–734 (2000)

ASCE/SEI Seismic Rehabilitation Standards Committee: Seismic Rehabilitation of Existing Buildings (ASCE/SEI 41-06). American Society of Civil Engineers, Reston (2007)

ATC: Seismic Evaluation and Retrofit of Concrete Buildings. Applied Technology Council, California (1996)

Benazouz, C., Moussa, L., Ali, Z.: Ductility and inelastic deformation demands of structures. J. Struct. Eng. Mech. **42**(5), 631–644 (2012)

Chikh, B., Leblouba, M., Mehani, Y., Kibboua, A., Hadid, M., Zerzour, A.: Ductility spectrum method for design and verification of structures: single-degree-of-freedom bilinear systems. In: Proceedings of the 15th European Conference on Earthquake Engineering, Istanbul, Turkey, August (2014)

Chikh, B., Mehani, Y., Leblouba, M.: Simplified procedure for seismic demands assessment of structures. Struct. Eng. Mech. **59**(3), 455–473 (2016)

Chopra, A.K., Goel, R.K.: A Modal Pushover Analysis Procedure to Estimate Seismic Demands for Buildings: Theory and Preliminary Evaluation. PEER, Berkely (2001)

Chopra, A.K., Goel, R.K.: A modal Pushover analysis procedure for estimating seismic demands for buildings. Earthq. Eng. Struct. D **31**(3), 561–582 (2002)

EN, C. (1998). 1: EuroCode 8: Design of Structures for Earthquake Resistance: Part 1: General Rules, Seismic Actions and Rules for Buildings. European Committee for Standardization, Bruxelles (2004)

Fajfar, P., Fischinger, M.: N2 a method for nonlinear seismic analysis of regular structures. In: Proceedings of 9th World Conf on Earth Eng, Tokyo-Kyoto, Japan, vol. 5, pp 111–116 (1988)

Fajfar, P.: Capacity spectrum method based on inelastic demand spectra. Earthq. Eng. Struct. D **28**(9), 979–993 (1999)

FEMA, A. 440, Improvement of Nonlinear Static Seismic Analysis Procedures. FEMA-440, Redwood City (2005)

Freeman, S.A., Nicoletti, J.P., Tyrell, J.V.: Evaluation of existing buildings for seismic risk-a case study of Puget Sound Naval Shipyard. In: Proceedings of 1st U S National Conference on Earthquake Engineering, Berkeley, USA, pp. 113–122 (1975)

Gülkan, P., Sözen, M.: Inelastic response of reinforced concrete structures to earthquake motions. ACI J. Proc. **71**(12), 604–610 (1974)

Gupta, B., Kunnath, S.K.: Adaptive spectra-based Pushover procedure for seismic evaluation of structures. Earthq. Spectra **16**(2), 367–392 (2000)

Kazaz, L.: Seismic deformation demands on rectangular structural walls in frame-wall systems. Earthq. Struct. **10**(2), 329–350 (2016)

Kowalsky, M.J.: Displacement-based design—a methodology for seismic design applied to RC bridge columns. Master's thesis, University of California at San Diego, La Jolla, California (1994)

Lin, Y.Y., Chang, K.C.: "An improved capacity spectrum method for ATC-40. Earthq. Eng. Struct. D **32**(13), 2013–2025 (2003)

Lin, J., Mahin, S.A.: Response of subsystems on inelastic structures. ASCE, United States (1984). https://inis.iaea.org/search/searchsinglerecord.aspx?recordsFor=SingleRecord&RN=16061353

Maja, K., Fajfar, P.: The extended N2 method considering higher mode effects in both plan and elevation. Bull. Earthq. Eng. **10**(2), 695–715 (2012)

Miranda, E.: Estimation of inelastic deformation demands of SDOF systems. J. Struct. Eng. **127** (9), 1005–1012 (2001)

Newmark, N.M., Hall, W.J.: Earthquake Spectra and Design. Earthquake Engineering Research Institute, Berkeley (1982)

Priestley, M.J., Kowalsky, M.J.: Direct displacement-based seismic design of concrete buildings. Bull. N. Z. Soc. Earthq. Eng. **33**(4), 421–444 (2000)

Sasaki, K.K., Freeman, S.A., Paret, T.F.: Multimode Pushover procedure (MMP)—a method to identify the effects of higher modes in a Pushover analysis. In: Proceeding of the 6th U.S. National Conference on Earthquake Engineering, Seattle, Washington (1998)

Zerbin, M., Aprile, A.: Sustainable retrofit design of RC frames evaluated for different seismic demand. Earthq. Struct. **9**(6), 1337–1353 (2015)

Verification of Mechanism on Improvement of Drying Shrinkage or Air Permeability on Concrete Using Blast Furnace Slag Sand Based on Pore Structure

Ryo Yoshida[1(✉)], Kazuhide Saito[2], and Chiaki Yoshizawa[3]

[1] Nagoya Institute of Technology, Aichi, Japan
yoshida.ryo@nitech.ac.jp
[2] Construction Chemicals Div., TAKEMOTO OIL & FAT CO., LTD., Aichi, Japan
[3] Nihon University Research Institute, Chiba, Japan

Abstract. In this study, the differences in the pore structures densified by the blast furnace slag (BFS) sand and powder were investigated by using the mercury intrusion test and removing the fine particles contained in the sand. By using BFS sand, the pores with a diameter of 400 nm or higher were densified. The characteristics of the transition zone detected in natural sand were not observed with the elemental analysis around the BFS sand. A strong correlation was found between the volume of the pores densified by mercury intrusion method and drying shrinkage and the air permeability coefficient. It was verified that the densification and adhesion strengthening of cement paste and sand by BFS sand improve the drying shrinkage and air permeation resistance. Using BFS sand, concrete infrastructures can be more durable.

1 Introduction

Despite the remarkable development of the steel industry, the issue of conversion of the by-products of BFSs into resources still remains. Today, the industry is promoting the effective use of BFS sands by emphasizing on alleviation of environmental burdens, including the restrictions on the use of natural aggregates and regulation of CO_2 emissions. In addition, as the focus of the business landscape shifts toward improving the maintenance efficiency and durability of infrastructure, the conversion of BFS sands into highly durable materials is now being promoted.

The physical properties of the hardened concrete bodies obtained from BFS sands are difficult to understand. The raw materials used for both high furnace slag fine powder and sand are the same, which have common latent hydraulic properties and high long-term compressive strength. However, they have different physical properties. For example, the drying shrinkage for BFS fine powders increases when the degree of fineness or the hardening temperature is increased, whereas it reduces for the BFS sands under the same conditions. Furthermore, the neutralization speed is slightly increased by using the BFS fine powders and decreased by using the BFS sands. A previous study demonstrated that the size of the BFS influences the difference in the

H. Rodrigues and A. Elnashai (Eds.): GeoMEast 2018, SUCI, pp. 180–193, 2019.
https://doi.org/10.1007/978-3-030-01932-7_15

physical properties of the hardened concrete. With respect to this, however, the pore structure formed by the BFS fine powders and the BFS sands have not been fully determined, preventing an understanding of the related mechanism.

The purpose of this study is to clarify the difference in pore structure densified by the BFS sand and the fine powder. In addition, the results were used to understand the mechanism based on the pore structure that improves the drying shrinkage and air permeation resistance of the BFS sand.

2 Outline of the Experiment

2.1 Materials

Table 1 shows the physical properties of the materials used. The specimens were prepared using ordinary Portland cement (C), BFS fine powder (SgP), JFE mineral (SgS1) and Nippon Steel Sumikin (SgS2) as BFS sands, Oigawa sand (S1), Okazaki crushed stone (G), high performance AE water reducing agents, and AE reducing agents. Gypsum was not added to the BFS fine powder. All the BFS sands were sprayed with an anti-caking agent.

Table 1. Properties of materials

Symbol	Physical properties
C	Density: 3.16 g/cm^3
SgP	Specific surface area: 4325 cm^2/g, non gypsum type, density: 2.91 g/cm^3
S1	(1) Density in saturated surface-dry condition: 2.57 g/cm^3, (2) Percentage of water absorption: 2.44%, (3) Fineness modulus: 3.01, (4) Solid content: 67.1%, (5) Content of materials finer than 75 μm Sieve: 1.5%
SgS1	(1) 2.74 g/cm^3, (2) 0.62%, (3) 2.50, (4) 59.7%, (5) 2.6%
SgS2	(1) 2.73 g/cm^3, (2) 0.51%, (3) 2.35, (4) 59.0%, (5) 3.1%
G	(1) 2.66 g/cm^3, (2) 0.78%, (3) 61.0%

The formulations are shown in Table 2. The water–cement (W/C) ratio was 50% and the amount of unit water (W) was constant. The BFS sand was replaced with 0 and 100 vol%, and the additive rate of fine powder was 0, 2, and 5 vol%. In addition, the samples for washed BFS sands after removing the fine particles are labeled as SgS1w and SgS2w; the fine particles of 0.15 mm or less were washed off on the day before the test and the suspension water was received with a 0.15 mm square sieve. After drying the sand, the surface water was adjusted and used for preparing the test specimens.

The slump and the air volume fell within the ranges of 9.5–13.0 cm and 5.3%–6.2%, respectively.

Table 2. Mix-proportion

Specimen	W/C	[kg/m^3]						[C × %]	[C × 0.001%]
		W	C	S1	SgS	SpP	G	SP	AE
S1	0.50	175	350	671	0	0	1072	0	10.0
S1-SgP2				658		15		0.20	
SgS1				0	715	0			4.0
SgS1w									1.5
SgS1w-SgP2					701	15			1.0
SgS1w-SgP5					680	38			
SgS2					713	0			
SgS2w									

3 Physical Properties Tests of the Hardened Concrete

3.1 Compressive Strength Test

The size of the specimen was φ 100 × 200 mm and standard curing was performed. The test was conducted according to the Japanese Industrial Standard JIS A 1108 with a 28-day aged specimen.

3.1.1 Drying Shrinkage and Mass Decreasing Rate Test

The specimen was obtained from a 100 × 100 × 400 mm steel mold. After curing in water at 20 °C for 7 days, the base length was measured. The test was conducted at 20 °C with a relative humidity (RH) of 60% in accordance with JIS A 1129-3.

3.1.2 Air Permeability Test

Disk specimens of φ 100 × 50 mm were cured in water at 20 °C for 28 days, successively immersed in acetone for 35 days, and finally dried at 20 °C and RH 50% until a constant weight was achieved.

To prevent air inflow and outflow from the surface of the specimens, they were coated with urethane resin and overlaid epoxy resin. For the test, the amount of air passed through the cross section of the specimen was measured at the outlet. The loading pressure was in the range of 0.2–0.3 MPa and the air permeability was measured using the downward displacement of water method.

3.2 Pore Structure Analysis

3.2.1 Porosity Test (Archimedes Method)

Disk specimens of φ 100 × 50 mm were cured in water at 20 °C for 28 days, and then their volume and wet mass were measured using the Archimedes method. Thereafter, they were immersed in acetone for 24 h and dried at 20 °C and RH 50% until constant weight was achieved. Then, the mass of the specimens was measured again and the porosity of the concrete was calculated from the difference between wet mass and dry mass.

3.2.2 Mercury Intrusion Test

Mortar samples of 8 × 8 × 8 mm were obtained from those used for the air permeability test using a chisel and a hammer. These samples were immersed in acetone for 24 h and vacuum dried for 48 h in a desiccator filled with silica gel. A Pascal 140/240 porosimeter was used for the mercury intrusion porosimetry test.

3.3 Electron Microscopy and Elemental Analysis

The surfaces of the mortar samples used in the mercury intrusion test were roughly polished using a water-resistant P600 abrasive paper, and then polished with a 9 μm diamond slurry. The samples were pretreated by vacuum drying with silica gel for 4 days. Carbon was used for vapor deposition. A JXA-8530F electron microscope (JEOL Co., Ltd.) was used for observation. Energy dispersive X-ray spectrometry (EDS) was performed on a straight line straddling the boundary between sand and cement paste at 15.0 kV and a 200 × magnification.

3.4 Differences in Pore Structures Densified by BFS Sand and Fine Powder

In this chapter, we describe the pore structure of the concrete separated by washing from the fine particles with a size of 0.15 mm or less contained in the BFS sand. In addition, we discuss the change in the pore structure caused by the BFS fine powder as a result of displacement by its internally splitting into natural and BFS sand.

Figures 1 and 2 show the cumulative pore volume curves obtained by the mercury intrusion test. The comparison of S1 and SgS1 (Fig. 1) confirmed the differences in the pore structure depending on the type of sand, as the pore structure influenced by the BFS sand was denser than that obtained with the natural sand in the entire range (diameter of 10 nm or more) measured by the mercury intrusion test. In addition, the comparison between S1 and S1-SgP2 confirmed the densification of the pore structure when 2% of the BFS fine powder was substituted with natural sand. The degree of densification by the blast furnace fine powder was smaller than that of the sand, but it could be due to the pore structure of the cement paste substituted in large quantities with the binder. Moreover, since the curves of S1 and S1-SgP2 overlapped for pore diameters of 400 nm and higher, with practically no change, the pores densified by the BFS fine powder had a diameter lower than 400 nm. Hence, the BFS sand can be effectively used with a high w/c which is likely to form a transition zone, in order to densify large-diameter pore structures (transition zones).

Figure 2 confirms the densification of the pore structure when the fine powder is substituted by BFS sands. One of the reasons why the cumulative pore volume of SgS1w without fine particles was smaller than that of SgS1 containing fine powder is the washing off of the sand while removing the fine particles. Maybe, the anti-caking agent was removed by washing and the latent hydraulic property of the sand surface was activated, resulting in a pore structure denser than that in case of SgS1. All the curves of the w-series, subjected to washing, overlapped for pores with diameter of 200 nm or more and were smaller than the SgS1 curve. This further suggests that the latent hydraulic property of the surface of BFS sand was activated. However, in light of

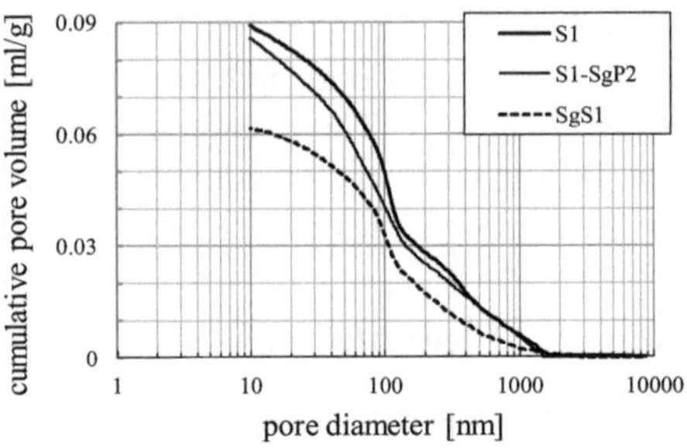

Fig. 1. Cumulative pore volume curves (S1, S1-SgP2, SgS1)

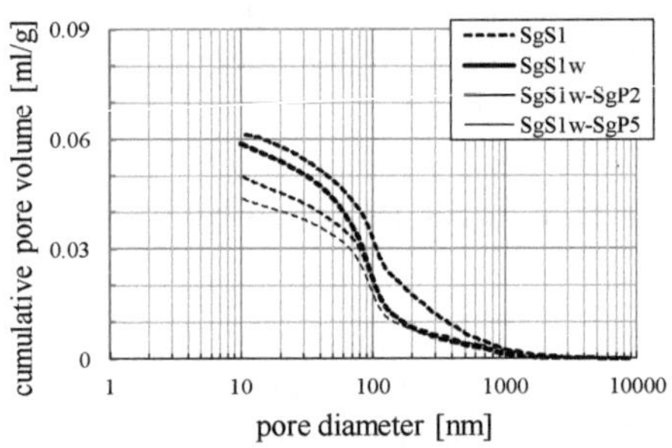

Fig. 2. Cumulative pore volume curves (SgS1, SgS1w, SgS1w-SgP2, SgS1w-SgP5)

improving the physical properties of the hardened body discussed later, even with the anti-caking agent on the surface of the aggregate, the BFS sands would have brought sufficient densification of the pore.

In the w-series, the effect of the substitution rate of fine powder was confirmed. As a result, the pore volume of 200 nm or less diameter decreased as the substitution rate increased from SgS1w to SgS1-SgP2 to SgS1-SgP5.

In the concrete using BFS sands, it was not possible to clearly separate the range of densification caused by sand and fine powder. However, based on the results obtained, it is reasonable to infer that the densification caused by both sand and fine powder acts on pores with diameters lower than 400 nm, while that resulting only by sand manifests on larger pores.

Figure 3 shows a comparison of the pore volumes obtained by the mercury intrusion test. Based on the abovementioned results with regards no densification by BFS fine powder for pore diameters of 400 nm or larger, the pore volume data were divided into two groups, pores with diameters of 10–400 nm and those larger than 400 nm. The pore volume for the second group was less than 0.02 ml/g in case of natural sand, and it was reduced to less than 0.01 ml/g when BFS sand was used. The effect of the substitution rate of the fine powder was difficult to recognize. In contrast, for the group of smaller pores, the pore volume decreased by approximately 0.020 ml/g when using BFS sand. The decrease in the pore volume when the fine powder was substituted by 5% was about 0.015 ml/g because of the non-substitution effect.

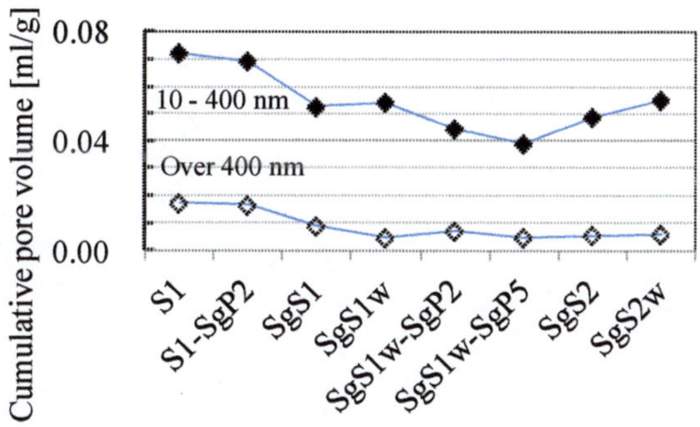

Fig. 3. Cumulative pore volume of mortar part (Mercury intrusion porosimetry test)

Figure 4 shows the porosity of the concrete specimens obtained by the Archimedes method. In this case, the porosity ranged between 0.10 and 0.13 ml/ml. With regards to the effect of the different types of sand, the porosity of the specimens using natural sand was about 0.13 ml/ml and that of the specimens using BFS sand was 0.10–0.12 ml/ml, with a resulting difference of 0.01–0.03 ml/ml that is almost the same as the effect of fine powder (0.02 ml/ml). The differences in the porosity of the mortar part according to the type of sand, as seen in Fig. 3, could have been averaged by the pores around the coarse aggregate that were included in the concrete specimens. By comparing the effect on the porosity of the mortar section and the concrete, we can infer that the densification by BFS sand does not extend to the transition zone around the coarse aggregate.

Fig. 4. Total pore volume of concrete (Archimedes' method)

3.5 Verification of the Transition Zone in Pore Structures Densified by BFS Sand

Electron microscope observation was conducted to analyze the pore structure with diameters of 400 nm or more potentially densified only by BFS sand and mercury intrusion test. Photo 1 shows the secondary electron images of the areas around the sand. Even with a magnification higher than 200 times, it was impossible to visually confirm a typical pore structure around the sand. Therefore, EDS was used to detect the

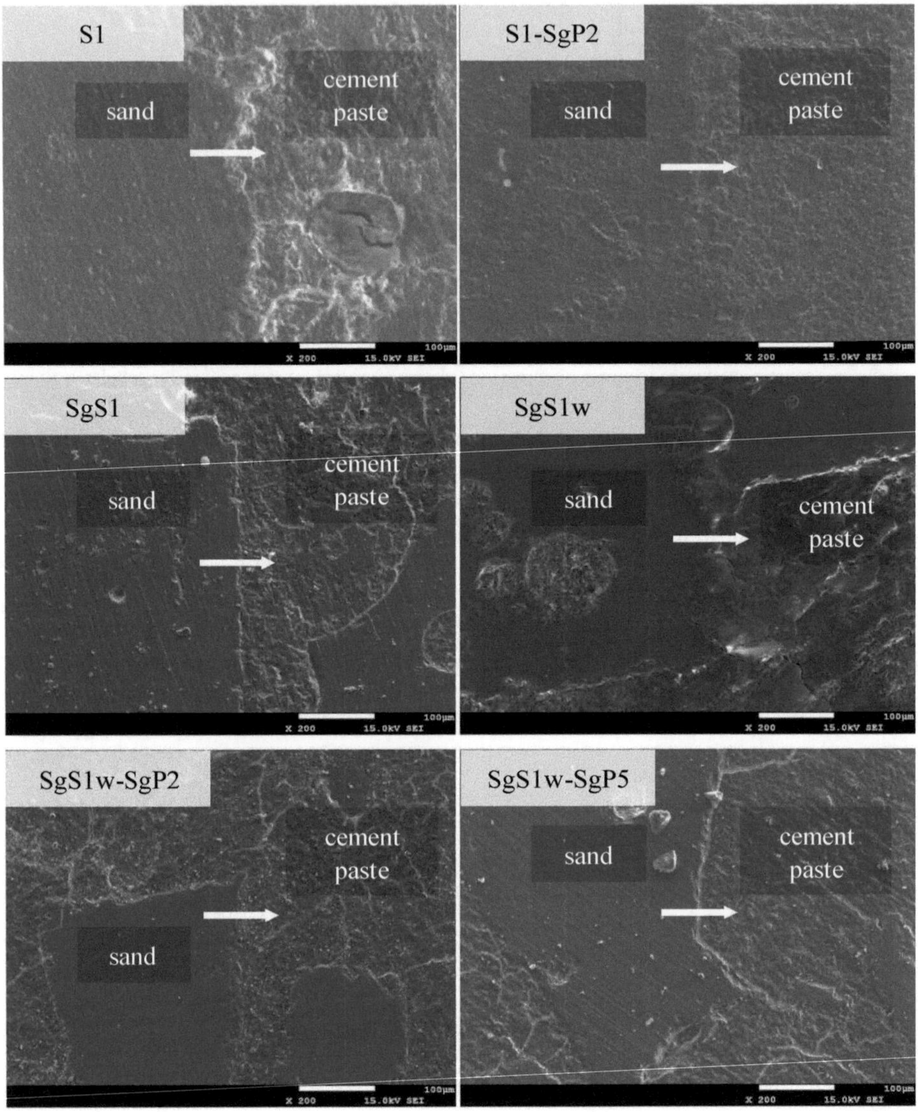

Photo 1. Secondary electron image

characteristic presence of hydrate around it. EDS was performed on the straight lines, as shown by white arrows in the picture, directed from the sand portion toward the cement paste section and with a length of about 100 μm. The investigated elements were O, Ca, and Si.

Figure 5 shows the EDS results. In S1 and S1-SgP2 using natural sand, at the boundary between the sand and the cement paste Ca increased and Si decreased. The width of this boundary region was about 50 and 25 μm, respectively, for the two

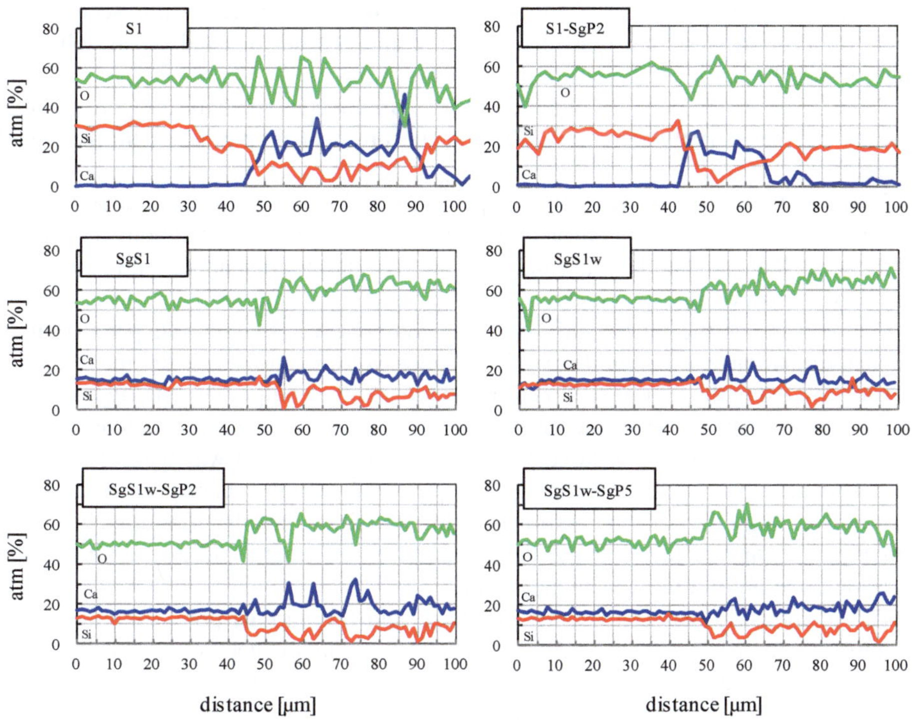

Fig. 5. Result of EDS

samples. Parameters such as the quantitative relationship between Ca and Si, and the position and width of the boundary region were consistent with the features of the transition zone reported in previous studies. In the transition zone, a coarse space rich in Ca(OH) and poor in Si was detected.

In contrast, around the aggregate of the specimen using BFS sand, proportions of O, Ca, and Si similar to the cement paste part were observed, suggesting that a hydrate close to CSH was formed as the main component of the cement paste. The fact that the aforementioned region rich in Ca and poor in Si could not be detected around the BFS sand indicates that the transition zone, which was a coarse space around the natural sand, was densified by the latent hydraulic nature of the BFS sand. Furthermore, the effects of the substitution rate of the fine powder and washing (activation of surface) could not be explained as a difference of atomic percentages.

3.6 Relation Between the Densified Pore Structure and the Various Physical Properties of the Hardened Concrete

In the previous section, the EDS results suggested that as no transition zone feature such as that observed in natural sand was found around the BFS sand, in the second case, the transition zone was densified by the latent hydraulic property. The correspondence between the pore structure measured by the mercury intrusion test and the transition zone detected by EDS should be verified in the future. In this section, the effects of the pore structure densified by BFS fine powder or sand on the physical properties of the hardened body are discussed.

Table 3 shows the porosity obtained by the Archimedes method, the pore volume obtained by the mercury intrusion test, and the physical properties of the hardened concrete.

Based on these values, the relationships between the pore volumes obtained by the mercury intrusion test and the physical property values of the hardened body are shown in Figs. 6, 7, 8 and 9. The natural sands and the BFS sands are marked with empty and solid scatters markers, respectively.

Figure 6 shows the relationship between compressive strength and pore volume. The determination coefficient (R^2) of the pore volume measured by the mercury intrusion test using the mortar section as a sample and the compressive strength was not high. As previously reported, the determination coefficient between concrete porosity and compressive strength measured by the Archimedes method was 0.74. The correlation between the porosity, including the pores around the coarse aggregate, and compressive strength was successfully confirmed. These results supported the data from Fig. 4 about the fact that the densification by BFS sand did not extend to the transition zone around the coarse aggregate.

The relationship between drying shrinkage and the various pore volumes in the concrete specimens is shown in Fig. 7. The determination coefficient for any pore volumes was about 0.90, whereas that of the porosity of concrete obtained by the Archimedes method was 0.70. The pore volume measured in the mortar part showed a higher R^2. Among others, a very high determination coefficient of 0.95 was obtained with the pore volume of 400 nm or more in diameter. Regarding the drying shrinkage, the aggregate contributed to suppress the shrinkage of the cement paste, but this result suggests that the adhesion characteristics (transition zone structure) of sand and cement paste had a significant impact on it.

From the large diameter side, the determination coefficient was the highest for diameters of 400 nm or larger and the lowest for a diameter of 100 nm. Thereafter, the smaller the value of the minimum pore diameter, the higher the coefficient of determination became. We suppose that this happened because the moisture dissipation property, which was the main mechanism of the drying shrinkage phenomenon, was dominated by the pore structure with a smaller diameter. This theory is supported by the relation with the mass decreasing rate described later.

Figure 8 shows the relationship between mass decreasing rate and pore volume measured for the same specimen used in the drying shrinkage test. A correlation was identified for the pore volume, and R^2 increased as the pore diameter decreased. In particular, the determination coefficient between the pore volume for 10 nm or more and the mass decreasing rate, at which BFS fine powder also influenced the densification, was as high as 0.97. In contrast, that between the pore volume for 400 nm or more in diameter was 0.83, which was smaller than for 10 nm or more in diameter. This result suggests that not only the large pores around the sand but also the small pore structures, such as capillary voids in the cement paste part, should be considered for the moisture dissipation phenomenon.

Table 3. Properties of hardened concrete specimens

	Archimedes' method pore volume [ml/ml]	Mercury intrusion porosimetry test pore volume [ml/g]				Compressive strength [N/mm^2]	Drying shrinkage [10^{-6}]	Mass loss rate [%]	Air permeability coefficient * 10^{-17} [m]
		Over 10 nm	Over 40 nm	Over 100 nm	Over 400 nm				
S1	0.128	0.089	0.074	0.049	0.017	40.1	490	1.83	0.00
S1-SgP2	0.127	0.086	0.066	0.041	0.016	36.2	509	1.87	0.00
SgS1	0.112	0.061	0.051	0.033	0.009	43.3	359	1.45	0.00
SgS1w	0.118	0.059	0.047	0.022	0.005	37.9	334	1.47	0.00
SgS1w-SgP2	0.116	0.052	0.042	0.023	0.007	40.0	324	1.29	0.00
SgS1w-SgP5	0.105	0.044	0.036	0.018	0.005	47.2	316	1.11	0.00
SgS2	0.109	0.054	0.042	0.020	0.006	45.9	340	1.32	0.00
SgS2w	0.117	0.061	0.047	0.025	0.006	41.8	328	1.37	0.00

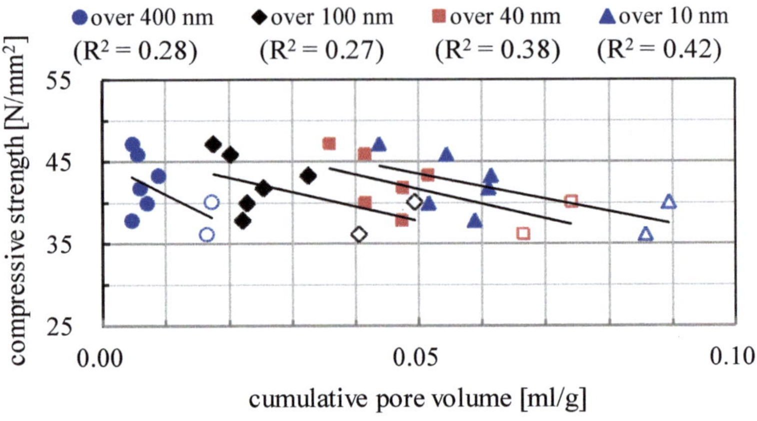

Fig. 6. Relationship between compressive strength and cumulative pore volume (MIP)

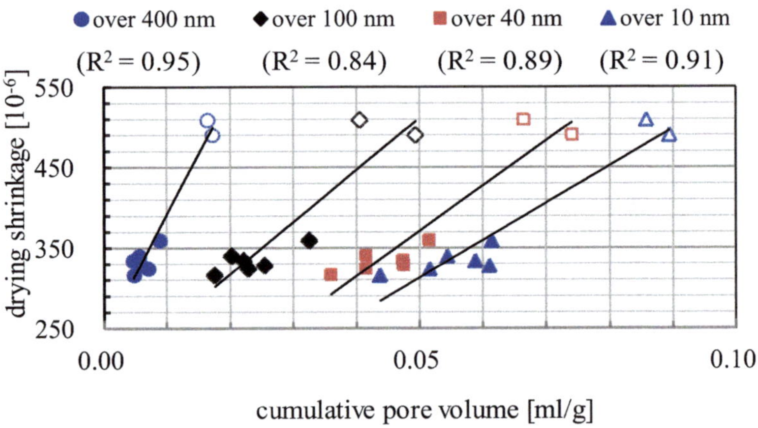

Fig. 7. Relationship between drying shrinkage and cumulative pore volume (MIP)

Based on the result of the mass decreasing rate (Fig. 8), the drying shrinkage (Fig. 7) was re-examined. It was confirmed that by using BFS sands, the amount of drying shrinkage was reduced by about 170 μ from that of natural sands concentrated around 330 μ. The high coefficient of determination obtained for a diameter of 400 nm or more and a pore volume of 10 nm or more supported such high drying shrinkage resistance. It was also due to the adhesion property to the sand attributable to the transition zone around it and the moisture dissipation property of the pores in the cement paste portion.

By using BFS sands, the adhesion is improved by strengthening the interface between the sand and the cement paste. By improving the adhesion, the sands can efficiently constrain the shrinkage of the cement paste in the drying shrinkage. The hardness of BFS sands can be demonstrated in drying shrinkage, which can be reduced because the transition zone becomes strong and adhesion is improved.

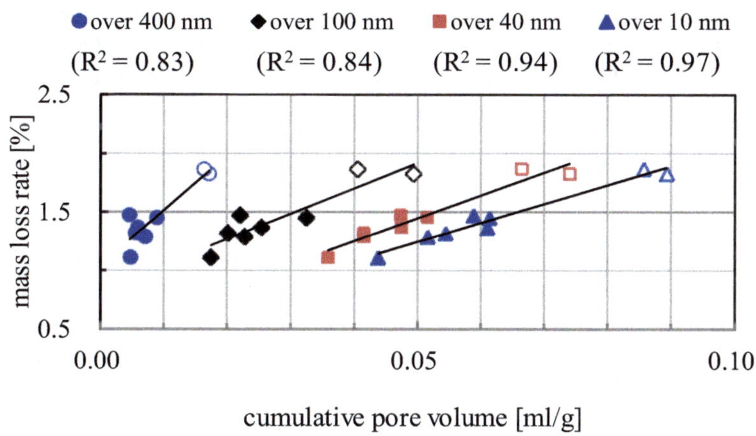

Fig. 8. Relationship between mass loss rate and cumulative pore volume (MIP)

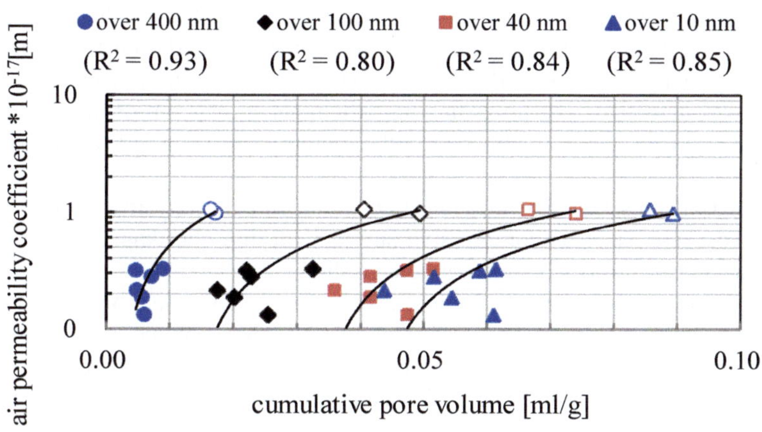

Fig. 9. Relationship between air permeability coefficient and cumulative pore volume (MIP)

At the same time, by densifying the small diameter pore structure in the cement paste part, it is possible to suppress the dissipation of moisture from the cement paste. Therefore, concrete using BFS sand exhibits high resistance to drying shrinkage.

Figure 9 shows the relationship between the permeability coefficient and the pore volume. In both pore volumes ranges with a diameter of 10 nm or more and a diameter of 400 nm, the coefficient of determination with the porosity of concrete was reported as $R^2 = 0.70$, obtaining a higher determination coefficient. In particular, the relationship between the porosity of 400 nm or more diameter and the permeability coefficient had a high coefficient of determination (0.93), and the approximation line also headed toward the origin. This suggests that the air permeability of concrete was significantly influenced by large size pores such as the transition zone around sand in the mortar part. By using BFS sands, the permeability coefficient of concrete can be reduced by 101 m compared to natural sands. Such high air permeation resistance obtained is likely to be attributable to the densification of the transition zone around the sands (pores having a diameter of 400 nm or more).

4 Summary

The purpose of this study was to understand and explain the effects of the pore structure on the physical properties of a hardened concrete body by clarifying the differences in the pore structures densified by BFS sands and fine powders. Concrete specimens with sands, from which the fine particles contained in the original BFS sand have been removed, were analyzed.

Concrete with BFS fine powder showed no change in pores with a diameter of 400 nm or larger and the densification was confirmed only in capillary voids having a diameter lower than 400 nm. In contrast, in the case of BFS sand, the densification was confirmed in both in large pores, having a diameter of 400 nm or more, and capillary voids.

The elemental analysis by EDS around the sand showed the presence of a transition zone around the aggregate in natural sand even when replaced with BFS fine powder. However, by using the BFS sands, the result suggested that the transition zone around the sands was densified. The pores having a diameter of 400 nm or more, which were densified using BFS sands as obtained in the mercury intrusion test, correspond to the transition zone.

The relationships between the densified pores and the physical properties of the hardened concrete, such as compressive strength, drying shrinkage, mass decreasing rate, and air permeability coefficient, were studied.

No correlation between the compressive strength and the porosity of concrete (Archimedes method) could be confirmed by the pore volume in the mortar part (mercury intrusion test). Although the compressive strength of concrete increased by using BFS sand, the densification by BFS sand did not reach the transition zone around the coarse aggregate.

A stronger correlation was confirmed between the drying shrinkage and the pore volume in the mortar part. In particular, a high coefficient of determination was obtained for pore volumes of 400 nm or more diameter and 10 nm or more in diameter, suggesting that the drying shrinkage was affected by the adhesion characteristics of the sand and the cement paste (due to the transition zone around the sand) and by the moisture dissipation characteristics of the mortar part (due to small diameter pores such as capillary voids). The moisture dissipation property was also supported by the correlation between the mass decreasing rate and the pore volume of the mortar part.

The relationship between the permeability coefficient and the pore volume of the mortar part also showed a stronger correlation than for the concrete porosity. Especially, a high coefficient of determination was obtained for pores of 400 nm or more diameter.

Considering the abovementioned results, it was verified that the BFS sand material strengthens the adhesion with cement paste, improves the drying shrinkage resistance, and effectively increases the air permeation resistance by densifying pores having a diameter of 400 nm or more detected by the mercury intrusion test (transition zone around the sand). Using BFS sand, reinforced concrete structure can be more durable. BFS sand can save the cost of infrastructure maintenance.

Effect of Wet-Mat Curing Time on Chloride Permeability of Concrete Bridge Decks

Hassan Ghanem(✉), Ayman Trad, Mohamed Dandachy, and Adel ElKordi

Department of Civil Engineering, Beirut Arab University (BAU), Tripoli, Lebanon
h.ghanem@bau.edu.lb

Abstract. This study presents findings related to chloride permeability of several widely used Texas Department of Transportation bridge deck concrete mixtures. The rapid chloride permeability test (RCPT) was selected and seven mix designs were tested; five were evaluated using cores taken from field-cast model bridge decks and the other two from laboratory-cast specimens. Mix designs were selected based on their extent of use around the state, with differences in terms of type of cement, coarse aggregate type, and supplementary cementitious materials. Wet-mat curing durations of 0, 2, 4, 8, 10 and 14 days were evaluated for each mix. Experimental results showed that no significant benefit for resistance to chloride permeability beyond 2, 4 or 8 days depending on the mixture, beyond which the benefits appeared to diminish. A comparison of results obtained from the RCPT and its alternative test; the ponding test, showed a correlation between their results for some mixtures.

Keywords: Concrete permeability · Curing · Bridge deck · Mix design

1 Introduction

Concrete durability can be approximately determined by the amount of time of maintenance-free service it provides. It can be designed to last for its service life if the right prediction is made of the environmental conditions that will be subjected to [1]. Concrete durability is closely associated with its permeability. Concrete permeability is the main factor that affect the rate of aggressive agents that will diffuse through the concrete matrix and the steel rebars [2].

During extreme cold weather, de-icing chemicals, such as sodium and calcium chloride are applied to the top surface of bridge decks in order to remove the snow and ice that has been accumulated. These salts with time eventually diffuse throughout the concrete and attack the reinforcement leading the steel to corrode [3].

Previous studies indicate that the two parameters that have the largest influence on concrete penetrability are the water cement ratio and the curing duration [4, 5]. The water cement ratio is the variable that singularly influences the pore permeability of concrete and is often the cause of much of the concrete deterioration that takes place. Curing is the process of preserving moisture content and temperature in concrete so that

H. Rodrigues and A. Elnashai (Eds.): GeoMEast 2018, SUCI, pp. 194–208, 2019.
https://doi.org/10.1007/978-3-030-01932-7_16

the desired properties are achieved. Lack of sufficient moisture will lead to cracks and reduce concrete long term durability [6, 7].

In the standard specification, item [420], the Texas Department of Transportation allows the contractors to use "wet mats" to help in the curing of concrete [8, 9]. The 1995 specification stipulates that a minimum of 14 days wet mat curing is required to allow construction vehicles not exceeding three-quarter tons and a minimum of 21 days of curing before opening to traffic.

Although concrete mix designs have evolved towards achieving high early strengths, there are still insufficient laboratory and field data on the durability properties of early loaded concrete. TxDOT main concern is that any delay in the opening of bridge deck to traffic will worsen traffic congestion and increase projects costs. The objective of this work is to analyze field and laboratory data in order to update current guidelines as to when bridge decks can be open to traffic without compromising concrete durability.

2 Construction of Model Field Bridge Decks

Seven concrete mixtures currently used in different districts of the State of Texas were included in this study. These were selected based on regional climatic conditions and other local conditions including coarse aggregate type used in concrete and other mix constituent materials. The key design information for all mixtures is presented in Table 1.

Of the seven mixture designs, five were actually tested in the field in five different TxDOT districts: New Lubbock, El Paso, Fort Worth, San Antonio and Houston. The remaining mixture designs had specimens cast at Texas Tech University laboratories using raw materials brought in from two additional regions (Pharr and Atlanta districts). At each field site, a relatively small-scale, though full-size (typically 12′ (3.65 m) × 32′ (9.75 m)), bridge deck was constructed. Construction of the field test sites was designed to mimic actual TxDOT bridge design and construction procedures as closely as possible and each model field deck was cast to accommodate six panels, with each one subjected to wet-mat curing durations of 0, 2, 4, 8, 10 and 14 days. The first phase of the field site work involved construction of the footings and of the two abutment walls (Fig. 1a). The second construction phase involved setting the beams and pre-cast panels on top of the abutment walls (Fig. 1b and c). The third phase of site construction consisted of forming the bridge deck, tying top mat reinforcement, and setting the grade (Fig. 1d). Reinforcement for the bridge was provided to mimic a real bridge deck. The 4.5 in. (114 mm) thick in situ slab casted on the precast panels gave the deck a total thickness of 8.5 in. (216 mm).

After the bridge deck was poured, curing compound was sprayed on the surface of each section to prevent moisture loss (Fig. 1e). This process was usually done within the first half hour after the pour. Once the concrete achieved initial set, cotton mats were laid on the 2-, 4-, 8-, 10-, and 14-day cure sections of the concrete surface and

Table 1. Mixture designs information for the bridge decks used in the State of Texas.

Variables	Districts						
	El-Paso	Fort Worth	San Antonio	Houston	New Lubbock	Atlanta	Pharr
Cement type	I/II	I/II	I/II	I/II	I/II	I	I
Cement factor	3.2	4.8	5.2	4.5	4.2	4.9	6.5
Mineral	Slag	Fly ash F	Fly ash C	Fly ash C	Fly ash F	Fly ash F	None
Admixture	50%	22%	20%	27%	31%	21%	
CA	LS	LS	LS	LS	LS	Sil	Sil
W/C	0.42	0.43	0.43	0.46	0.45	0.46	0.45
Air	5%	6%	6%	5%	6%	6%	6%
Admixture	Retarder	Retarder	Retarder	Retarder	Retarder	None	None
Casting location	Field	Field	Field	Field	Field	Lab	Lab
Nb. of RCPT samples	18	18	18	18	18	18	18
Ponding duration	180	90	180	180	90	180	180
Time of casting	July	September	August	June	May	November	November
Ambient temperature (°C)	35	34	36	31	25.3	Room Temp	Room Temp
Slump (in)	4	3	4	3	3.5	4	3.5
Unit weight (lb/ft^3)	148	140	151	145	147	142	146
Compressive strength (psi) at 7 days	5919	4911	6070	6803/5205	6305	4770	4010
Split Tensile strength (psi) at 7 days	473	344	547	510	475	401	375
Modulus of elasticity (ksi) at 7 days	4461	3713	4527	4511	4615	3799	3620

CA = coarse aggregate, *LS* = limestone, *Sil* = siliceous gravel
10000 psi = 68.9 MPa; 1in = 2.54 cm

wetted to provide the necessary moisture for hydration continuation. An automated system consisting of a water tank, a pump with a timer, and a series of soaker hoses was employed to maintain moisture in the cotton mats. Since periodic supervision was not possible at the field site, evaporation of moisture from the cotton mats was minimized by placing plastic sheeting on top of the cotton mats. Cotton mats and plastic sheeting were removed from the corresponding slab panel at the end of each curing duration being evaluated; 0-, 2-, 4-, 8-, 10-, and 14-days wet-mat curing. Curing periods were based on actual calendar days. The 0-day cure section of the bridge deck did not receive any wet-mat curing, but curing compound was applied to its surface.

For Pharr and Atlanta districts, 4″ × 8″ [10 × 20 cm] concrete cylinders were cast using raw materials brought from both sites. After 24 h of casting, specimens were stripped from their molds and moist cured for 0-, 2-, 4-, 8-, 10-, and 14-days.

Fig. 1. Construction of a two-bay model bridge deck slab. (a) construction of footings and abutment walls, (b) setting the beams, (c) setting the pre-cast panels on top of the abutment walls, (d) setting the grade, (e) spraying the surface of section to prevent moisture loss, (f) drilling the cores, (g) cores retraction, (h) slabs after coring, (i) casting the RCPT and ponding samples.

3 Test Methods

Two tests that are commonly used to determine concrete permeability to chloride ions are the rapid chloride permeability test (RCPT) and the ponding test.

The ponding test which has the designation of AASHTO T 259/260 [10, 11], is well known as either the "Salt Ponding Test" or "Soaking Test". It is a long-term test to measure the chloride ion penetration into concrete. This test is often used to validate the results of the RCPT. This test requires a minimum of 3 slabs, each having a thickness of at least 3 in. (75 mm) and a surface area of 12 in. (300 mm^2) square. The top surface of the slab is ponded for 90 days using a 3% sodium chloride solution. Following this period, the chloride concentration at different depths throughout the slab is determined. This procedure is very liked by concrete corrosion specialists as it determined accurately the chloride penetration through the concrete. However, the length of the test is a major drawback. For that reason, the RCPT has become popular among owners and specifiers.

4 Rapid Chloride Permeability Test (RCPT)

4.1 Sample Preparation

Cores were taken from the five bridge decks at 56 days with a portable coring machine using a four inch inside diameter coring bit (Fig. 1f). The coring depth for each core was approximately 4.5–5 in. [114–127 mm], usually just below the precast panel. Once the coring bit reached this depth, the bit was drawn back and the core was retracted (Fig. 1g). The cores were kept moist until cut into 2-in. thick discs (i.e. each disk is 4 in. [101.6 mm] diameter and 2 in. [50.8 mm] thickness). The side of the cylindrical samples was coated with epoxy, and following drying period, it was placed in a vacuum chamber for 3 h. The sample was vacuum saturated for 1 h and then soaked for 18 h to make the cores fully saturated.

4.2 Test Procedure

A RLC Instrument Company 164A 4-cell model was used to determine the electrical resistivity of the disk samples. One side of the cells was filled with 3.0% NaCl solution and the other side is filled with 0.3 N (Normal) NaOH solution. Current and charge readings were taken during 6-h duration of the test at 30-min intervals including the initial reading. As the current and time history are known, the total charge in coulombs that passed through the samples can be calculated using Simpson's integration rule. ASTM C1202 provides a qualitative rating based on the charge that passes through the specimens (Table 2 [12, 13]).

Although the above guidelines are acknowledged by practitioners, commonly accepted in practice, they can be sometimes misleading if other variables such as cement content, aggregate type and size, and concrete density are not considered [14, 15]. The set-up of the RCPT is presented in Fig. 2. RCPT data reflects the electrical resistivity of the concrete being tested rather than the resistance of the concrete to chloride penetration. Therefore, this procedure is an indirect way of measuring concrete permeability.

5 Analysis of Test Results

The RCPT results [a set of relations between the total charge in coulombs that passed through the specimens and the curing durations] for the seven mixtures are presented in Fig. 3. The coulomb charges appear to be related to the curing duration for most mixtures. As curing duration increases, the coulomb charges decrease, indicating that concrete becomes less permeable. This may suggest that concrete microstructure has been changed, thus altering the electrical conductivity of concrete.

Table 2. RCPT ratings [12, 13].

Charge passed (Coulombs)	Chloride permeability	Type of concrete	Total integral chloride to 40.6 mm (1.6 in.) depth after 90 day ponding test
>4000	High	High water cement ratio, conventional (>0.6) PCC*	>1.3
2000–4000	Moderate	Moderate water cement ratio conventional (0.5 to 0.5) PCC*	0.8–1.3
1000–2000	Low	Low water cement ratio, conventional (<0.4) PCC*	0.55–0.8
100–1000	Very Low	Latex modified concrete, internally sealed concrete	0.35–0.55
<100	Negligible	Polymer impregnated concrete, polymer concrete	<0.35

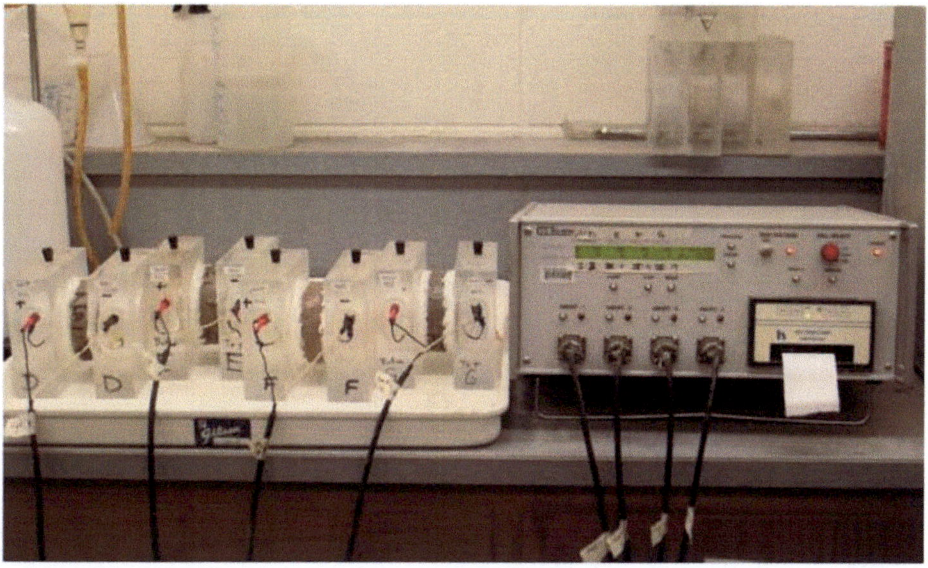

Fig. 2. Test setup: RCPT

As an example, for El-Paso mixture, it is observed that concrete permeability reduces to a large extent. From Fig. 3a, the charges decrease from 3170 coulombs (moderate RCPT rating) for 0 day curing to 770 coulombs (very low RCPT rating) for 2 days curing. This sudden drop in coulomb charges may be due to the following reason. The use of ground granulated blast furnace slag (GGBS) involved pozzolanic reaction in concrete. The higher the replacement percentage of ordinary Portland cement by GGBS, the lower the pH pore solution. In the case of El Paso mixture

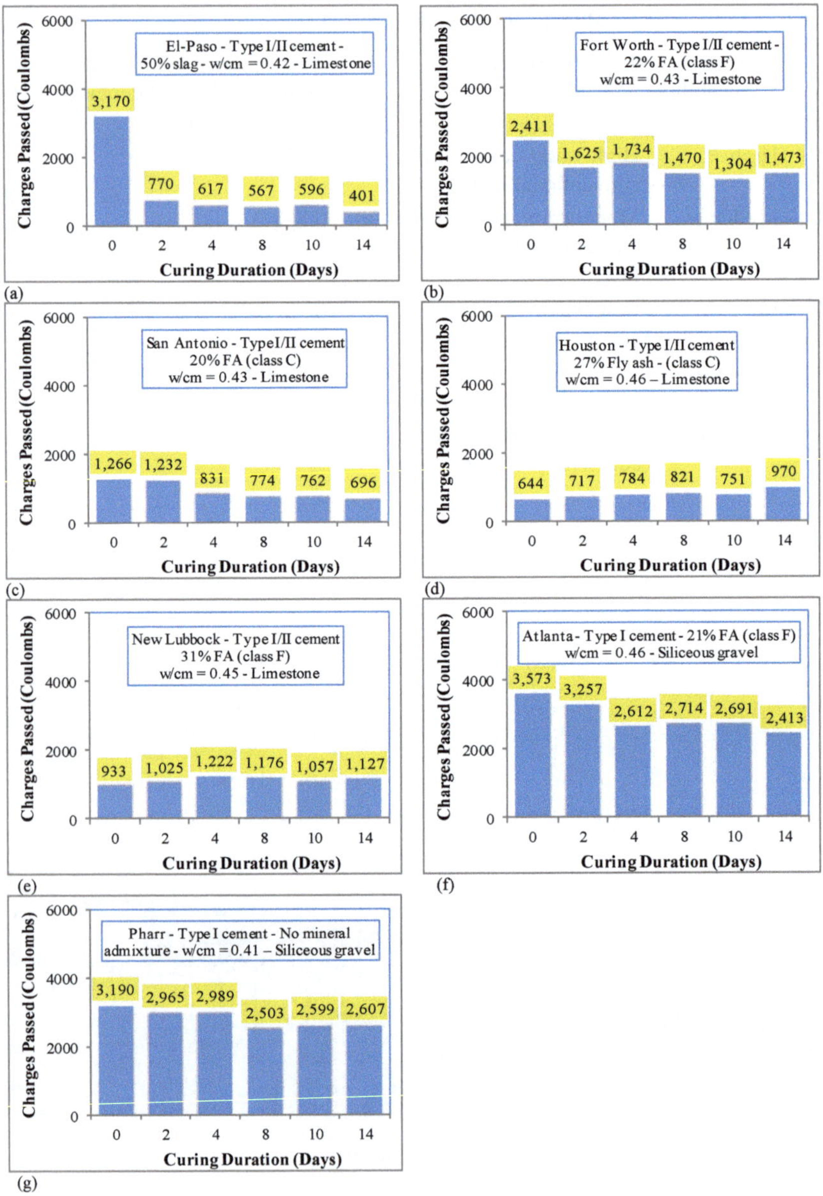

Fig. 3. RCPT test results. (a) El-Paso, (b) Fort Worth, (c) San Antonio, (d) Houston, (e) New Lubbock, (f) Atlanta, (g) Pharr

design, 50% of cement was replaced by GGBS. During RCPT, the pH pore solution influences the total charge passed through GGBS concrete. In addition to that, the influence of the pH pore solution on the charge passed is governed by the curing regimen as the OH^{-} ions in the pore fluid deplete with the pozzolanic reaction [16]. That's may explain the reason for the great reduction in coulomb charges.

It is also interesting to notice that coulombs charges continue to go down as curing increases but only up to a specific point past which there is no significant change in coulomb charges. This threshold curing duration looks significantly different for each concrete mixture; 2 days for El Paso mix, 4 days for San Antonio, and 8 days for Pharr, Atlanta and Fort Worth mixes.

Fort Worth and San Antonio RCPT results are presented in Fig. 3b and c respectively. As seen from Fig. 3b, 0-cure day for Fort Worth mixture showed moderate permeability (2411 coulombs), whereas 14-cure day concrete displayed low permeability (1473 coulombs). On the other hand, San Antonio mixture showed surprisingly low permeability (1266 coulombs) even at 0-day curing, whereas at 14-cure day, the concrete falls in the very low permeability range (696 coulombs). The above results illustrate the beneficial effect of curing on concrete permeability. As the curing duration increases, more calcium silicate hydrate likely is formed. Thus, voids in the concrete structure tend to become discontinuous and smaller and the permeability of the concrete is reduced [17]. From the experimental test results of Fort Worth and San Antonio mixture, it can as well be stated that fly ash is very effective in reducing the permeability of concrete even at a lower rate of replacement level (22% Class F fly ash for Fort Worth and 20% Class C fly ash for San Antonio). This could be explained by the pozzolanic reaction caused by fly ash. Fly ash is capable of reacting with the calcium hydroxide produced during cement hydration and forms calcium silicate hydrate. This silicate hydrate fills large capillary voids and thus void continuity is disrupted. Consequently, the blocking of capillary voids reduces the permeability of concrete [1].

Unlike previous mixtures, the Houston mixture does not follow the expected pattern (Fig. 3d). For example, the 0 day curing charges (644 coulombs) for Houston mixture are higher than those at 14 days curing (970 coulombs) which can be translated that non cured concrete has a better permeability than a well cured concrete and this is unrealistic. This may be attributed to the following. The ready-mixed concrete used for this test slab came from two different concrete trucks. No noticeable difference was observed between the mixes at the time of casting and both truckloads were ordered for the same standard mix design. However, subsequent tests indicated that concrete from the second truck that went to cast the 0-day cured panel and a part of the 2-day cured panel was stronger than the concrete from the first truck. It appears that this anomaly in the concrete mixes is reflected in the RCPT current flow results.

Similar to Houston mixture, New Lubbock results do not abide by the same trend (Fig. 3e). The 0-day cure core specimens indicated the lowest current flow (thus lower permeability) when compared to other curing durations. Furthermore, the cores from the 4-day cure panel showed the highest current flow. The discrepancy between the results of New Lubbock mixture is probably due to construction problem (bad compaction) in addition to environmental issues (very dry and windy conditions).

For Atlanta district, no test slab was constructed. Instead, test specimens for the RCPT were taken from laboratory cast 4-in. (101 mm) diameter, 8-in. (203 mm) long cylindrical specimens that were cast using local materials from the district. Test results are presented in Fig. 3f. An interesting point is observed. All coulombs charges values for all curing durations fall in the moderate range of permeability (2000–4000 coulombs) although Atlanta contains 21% Class F fly ash. However, 0-day cure concrete displays permeability near the upper limit of moderate permeability range

[3573 coulombs] whereas the 14-days curing shows permeability near the lower limit of moderate permeability range [2413 coulombs]. Consequently, the increase in curing duration helps reduce the coulomb charge values and therefore reduces concrete permeability but its beneficial effect is not as apparent as that for El-Paso, Fort Worth and San Antonio samples. This phenomenon could be due to the following reason. In Atlanta, siliceous river gravel is used in the mixture whereas El-Paso, Fort Worth, San Antonio, New Lubbock and Houston mixtures contain limestone. Siliceous river gravel is well known for its smooth surface texture, which is attributed to the establishment of a weak bond between the cement paste and the aggregate [18–20]. This weak bond leads to the formation of a weak interface zone, which increases the penetrability of concrete by supplying channels for the chloride solution to pass. This fact may be the primary reason why site such as Atlanta with siliceous gravel coarse aggregate is showing higher RCPT results at all curing durations compared to other sites.

Similar to Atlanta, Pharr specimens are casted and tested in the laboratory. Interesting, the curing duration appears to play only a minor role in reducing concrete permeability [from 3190 at 0 day curing to 2607 coulombs at 14 days curing] and all RCPT values are in the moderate permeability range (Fig. 3g). This can be due to the fact that Pharr mixture does not contain any mineral admixture and siliceous river gravel was used. It can be as well noted that no improvement whatsoever is noticed after 8 curing days which one can deduct that 8 days of curing is enough for the Phar site. By comparing Pharr and Atlanta RCPT results, one can conclude that the effect of mineral admixture in reducing concrete permeability is limited when siliceous river gravel is used in the mixture. On the other side, in other mixtures where the type of aggregate is limestone, it can be seen that the presence of fly ash significantly decreases concrete permeability as curing duration increases, as fly ash changes both the microstructure and the pore solution composition of the concrete, resulting in a finer pore structure in both the cement paste matrix and the paste aggregate interface in concrete.

6 Correlation Between the RCPT and the Ponding Test

As mentioned previously, the RCPT and the ponding test are generally used to determine concrete permeability to chloride ions. While establishing the coulomb-based classification system (Table 2), Whiting determined chloride concentrations for the same concrete mixtures in his study. He introduced the parameter of "Total Integral Chloride, *I*", which is a dimensionless value representing the integral of chloride concentration versus unit depth increments of 5.1 mm (0.2 in) [rather than actual depth] from the surface of the ponded specimen to a depth of 1.6 in (41 mm). "*I*" was used to represent the results of the 90/180 days ponding test in only one value. In equation form, the total integral chloride is represented by:

$$I = \frac{\textit{Total area under the line representing chloride content vs. Depth}}{5.1\,\text{mm}\,(0.2\,\text{in.})\ \textit{depth increments}}$$

Using a linear regression analysis, Whiting developed a correlation between the 90/180 days ponding test and the rapid 6 h permeability test with an R^2 coefficient of 0.708. The correlation was based upon tests conducted on a single core of each of the concrete mixtures in Table 2. None of the concrete mixtures tested contained mineral admixtures.

In a previous work, the ponding test was conducted on the same seven mixtures in this study for all curing durations, and chloride concentration was calculated at four different depths; 1/16″–0.25″ (1.5–6.3 mm), 0.25″–0.75″ (6.3 to 19 mm), 0.75″–1.25″ (19–31.7 mm) and 1.25″–1.75″ (31.7–44.5 mm) [21]. Using the permeability definition presented above, the mixtures were classified based upon their water soluble integral chloride and this is illustrated in Table 3. As shown in the table, 2 days of curing for El-Paso mixture is very beneficial for reducing the total integral chloride. For San Antonio mixture, 4 days of curing was necessary to achieve significant reduction in integral chloride while 8 days is needed for Fort Worth, Houston, Atlanta and Pharr sites.

Table 3. Classification of sites based on water soluble integral chloride.

Curing duration	El Paso	Fort worth	San Antonio	New lubbock	Houston	Atlanta	Pharr
0	1.18	1.04	0.96	0.66	0.86	1.27	1.19
2	0.91	0.95	0.79	0.62	0.76	1.04	1.13
4	0.90	0.79	0.80	0.50	0.65	0.86	1.05
8	0.93	0.61	0.80	0.48	0.63	0.85	0.95
10	0.92	0.58	0.79	0.48	0.61	0.83	0.94
14	0.90	0.54	0.78	0.42	0.58	0.81	0.88

To make comparison with Whiting's original research, all mixtures with 14 days of curing are classified based on their water soluble integral chloride and their coulomb values, as shown in Table 4. Three sites (Pharr, Atlanta and Fort Worth) exhibit identical water soluble integral chloride and coulomb classifications while the other four sites showed some discrepancies. For example, the El Paso and San Antonio mixtures exhibit moderate water soluble integral chloride while exhibiting very low coulomb values. The New Lubbock mixture exhibits very low water soluble integral chloride values, while exhibiting low coulomb values. The Houston mixture exhibits low integral chloride values, while exhibiting very low coulomb values. As a result, it is concluded that the two tests are not always consistent in predicting the permeability levels defined in the original test development of AASHTO T277 [22]. This is probably due to two reasons: (a) current coulomb measurements are related to the movement of all ions (not just chloride ions) and can be influenced by increasing specimen temperatures resulting from the test [23]; (b) The use of supplementary cementing materials such as ground blast furnace slag and fly ash have alter the chemistry of the pore solution of concrete. However, it is important to note that in this study, most of the AASHTO T277 results ended in a lower classification than the ponded test results and this is for sites ponded for 90 and 180 days. Most discrepancies arose from the fact that the total integral chloride resulted in a higher classification.

Table 4. Classification of sites based on water soluble integral chloride and Coulomb values.

	El Paso	Fort worth	San Antonio	New lubbock	Houston	Atlanta	Pharr
Water soluble integral chloride	0.9	0.54	0.78	0.42	0.58	0.81	0.88
	Moderate	**Low**	Moderate	Very Low	Low	**Moderate**	**Moderate**
Coulomb values	401	1473	696	1127	970	2413	2607
	Very Low	**Low**	Very Low	Low	Very Low	**Moderate**	**Moderate**

The correlation between AASHTO T277 and AASHTO T259/260 for this experimental program is investigated further by performing a linear regression analysis on three sets of data. Three cases are considered: (1) the water soluble integral chloride values, *I* values, were coupled with their corresponding coulomb values for the sites subjected to 180 days of ponding for 14 days of curing, (2) the 180 day chloride results in % weight of concrete for the top ¼ in (6.3 mm) were coupled with their corresponding coulomb values for 14 days of curing; we should note here that El Paso results were not included in the correlation for both methods because of the inconsistent classification shown in Table 4; and (3) the water soluble integral chloride values, *I* values, were coupled with their corresponding coulomb values for all sites and for all curing durations. Results of these three regression results are shown in Figs. 4 and 5. As shown, the uppermost ¼ in (6.3 mm) resulted in an R^2 correlation coefficient of 0.625 (Fig. 4a) and exhibit a better correlation to the T277 results than the water soluble integral chloride which displayed an R^2 coefficient of 0.502 (Fig. 4b). Whiting in his study used the total integral chloride and reported an R^2 correlation coefficient of 0.70 [13]. However, the correlation is slightly lower in this case and possibly due to the fact that mixtures tested include mineral admixtures while all the mixes tested in Whiting's research do not contain mineral admixtures.

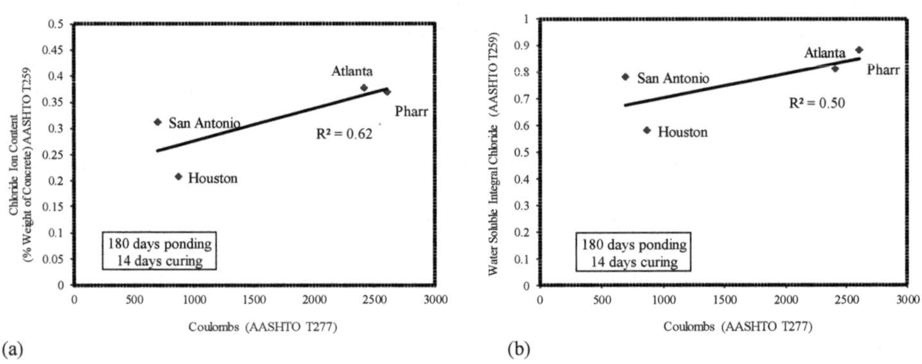

Fig. 4. Correlation data. (a) Water soluble chloride content by % weight, (b) Correlation data water soluble chloride ion content by integral

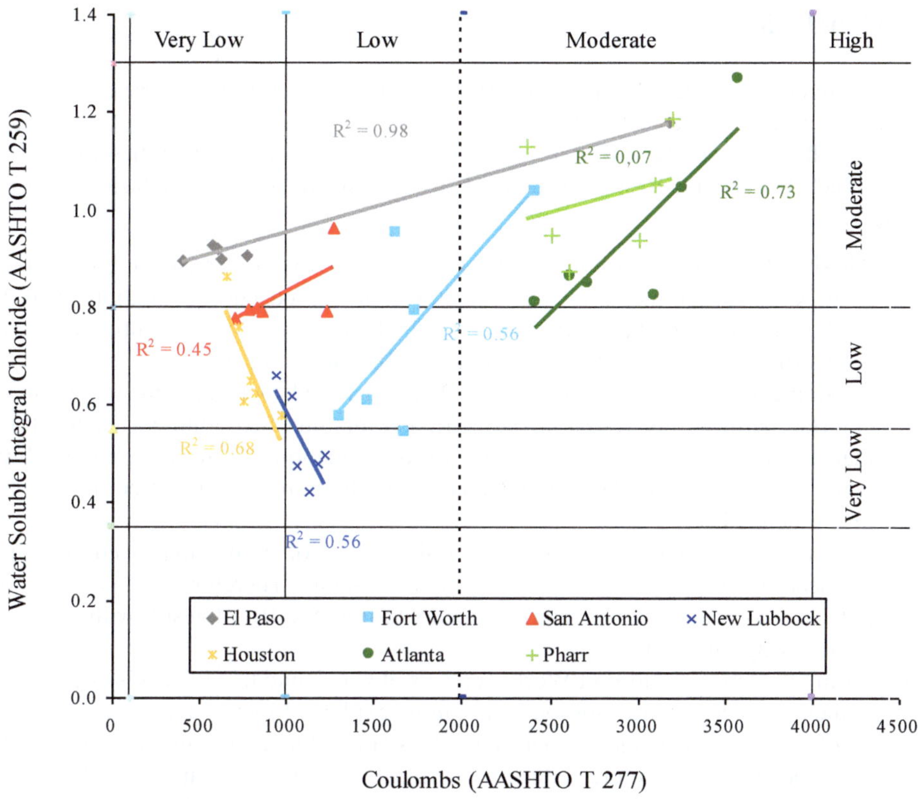

Fig. 5. Correlation of AASHTO T277 & AASHTO T259/260 for all curing durations

It can be seen from Fig. 5 that the correlation of AASHTO T277 and AASHTO T259/260 for Atlanta, Fort Worth and San Antonio are relatively good (R^2 = 0.73, 0.56 and 0.45 respectively). However, very poor correlation is shown for Pharr mixture. On the other side, New Lubbock and Houston mixtures show an inverse relationship. This is due to the inconsistent results obtained for the RCPT in these two sites as discussed in the previous section.

The following remarks can be depicted from Figs. 4 and 5.

(a) From the regression analysis of these three sets of data, it can be seen that a poor correlation exists in the 700–1300 coulombs range. This point seems to agree well with Whiting's original research (Fig. 4a).

(b) The original correlation between AASHTO T259/260 and T277 in the FHWA-81 report could not be used to accurately predict the permeability of concrete with a water cementitious range of 0.4–0.5 (Figs. 4b and 5). However, the RCPT can serve as a general tool to compare the relative permeability of similar mixtures design with similar constituents and admixtures.

7 Conclusions

To investigate the effect of curing on chloride ion permeability of concrete, the rapid chloride permeability test (AASHTO T277) was chosen. Seven concrete mix designs used in different districts within Texas were tested in laboratory facilities located at Texas Tech University. Based on the analysis conducted, several conclusions can be made:

(a) All sites (except New Lubbock and Houston) shows similar characteristics patterns. Chloride permeability decreased with increased curing duration. This is attributed to the fact that curing enhances the hydration process, resulting in dense pore structures that lead to a reduction in concrete permeability.
(b) Experimental results show that 2 days of curing were beneficial to the reduction of chloride permeability results for the El Paso mixture. San Antonio results show that 4 days of curing were necessary to the reduction of chloride permeability, while the Fort Worth, Atlanta and Pharr sites show that 8 days of curing can be recommended.
(c) Results from El Paso site show that only 2-days of curing improved concrete permeability from "moderate" to "very low" status. Therefore, the presence of 50% slag as cementitious material appears to improve RCPT result within a very short curing duration.
(d) From San Antonio and Fort Worth analysis, it was found that fly ash is effective in reducing concrete permeability. This can be explained by the pozzolanic reaction that leads to the formation of additional C-S-H that block the capillary void inside the concrete and therefore reduce its permeability. However, it is found that the effect of fly ash in reducing concrete permeability is limited when siliceous river gravel is used (Atlanta mixture).
(e) The Pharr and Atlanta sites, which contained siliceous gravel as coarse aggregate, displayed moderate permeability values and highest permeability among all the sites. This could be due to the development of a weak aggregate-cement interface transition zone in concrete mixes containing siliceous gravel. This point is confirmed by getting moderate water soluble integral chloride and coulomb values (Table 4).
(f) A comparison has been made between the RCPT (AASHTO T277) and the ponding test (AASHTO T259/260) based on the water soluble integral chloride content for all curing durations and on the percent weight of concrete. From this comparison, it can be concluded that there exists a correlation between AASHTO T277 and AASHTO T259/260, although the correlation is not perfect. This discrepancy arose from the fact that the integral chloride content resulted in a higher classification. It is also shown that the correlation using the percent weight of concrete (R^2 = 0.62) is higher than the correlation using the water soluble integral chloride (R^2 = 0.50).

(g) From the comparison of results of these two tests, it was found that the classification of El Paso, New Lubbock, San Antonio, and Houston mixtures based on these two tests showed different trends, while trends were the same for Pharr, Atlanta, and Fort Worth mixes, which leads to the conclusion that RCPT may not accurately represent the true long-term chloride penetration.

Acknowledgements. The writers would like to thank the Texas Department of Transportation for its support of this research. The technical assistance of Michel Laverton is also acknowledged.

References

1. Mindess, S., Young, J.F., Darwin, D.: Concrete, 2nd edn. Prentice Hall, Upper Saddle River (2006)
2. Mehta, P.K.: Sulfate resistance of blended Portland cements containing pozzolanic and granulated blastfurnace slag. In: Proceedings of 5th International Symposium on Concrete Technology, Mexico, pp. 1495–1508 (1981)
3. Young, J.F., Mindess, S., Gray, R.J., Bentur, A.: The Science and Technology of Civil Engineering Materials. Prentice-Hall, Upper Saddle River (1998)
4. Neville, A.M.: Properties of Concrete, 4th edn. Wiley, New York (1996)
5. Dhir, R.K., Hewlett, P.C., Chan, Y.N.: Near-surface characteristics and durability of concrete: assessment and development of in situ test methods. Mag. Concrete Res. **39**, 183–194 (1986)
6. Alhozaimy, A., Soroushian, P., Mirza, F.: Effects of curing conditions and age on chloride permeability of fly ash mortar. ACI Mater. J. **93**(1), 87–95 (1996)
7. Alsayed, S.H., Amjad, M.A.: Effect of curing conditions on strength, porosity, absorptivity, and shrinkage of concrete in hot and dry climate. Cem. Concrete Res. **24**(7), 1390–1395 (1994)
8. TxDOT: Standard Specifications for Construction and Maintenance of Highways, Streets, and Bridges, Adopted by the Texas Department of Transportation, June 2004
9. TxDOT: Special Provisions to Item 420 Concrete Structures, Texas Department of Transportation, Austin, Texas (1998)
10. AASHTO: Standard method of test for resistance of concrete to chloride ion penetration. AASHTO T259-02. AASHTO, Washington, D.C. (2012)
11. AASHTO: Standard method for sampling and testing for chloride ion in concrete and concrete raw materials. AASHTO T260-97. AASHTO, Washington, D.C. (2011)
12. ASTM: Standard test method for electrical indication of concrete's ability to resist chloride ion penetration. Annual Book of ASTM Standards, C 1202-10. 04.02, pp. 639–644 (2010)
13. Whiting, D.: Rapid Determination of the chloride permeability of concrete. Final Report No. FHWA/RD-81/119, Federal Highway Administration (1981)
14. Whiting, D.: Permeability of selected structures. Permeability of concrete, ACI SP-108, American Concrete Institute, Detroit, Michigan, pp. 195–222 (1988)
15. Whiting, D., Dzeidzic, W.: Resistance to chloride infiltration of superplasticizers concrete as compared with currently used concrete overlays systems. Final Report No. FHWA/OH-89/009, Construction Technology Laboratories (1989)

16. Hooton, D., Ramlochan, T., Thomas, M.: Pore solution analysis as a pore solution analysis as a tool for studying early tool for studying early–age hydration & predicting future hydration. International summit on Cement Hydration Kinetics, Quebec (2009)
17. Mehta, P., Monteiro, P.: Concrete: Microstructure, Properties, and Materials, 3rd edn. McGraw-Hill, New York (2006)
18. Liu, J., Mukhopadhyay, A., Celaya, M., Nazarian, S., Zollinger, D.: Best practices for the use of siliceous river gravel. In: Concrete Paving" Texas Transportation Institute, Report No. FHWA/TX-07/0-4826-1, Texas (2009)
19. Khan, J.: Influence of aggregate surface characteristic on pavement performance MS Thesis, Texas Tech University, USA (2003)
20. Yildirim, H., Oktar, O.N., Akkaya, Y., Sengul O., Tasdemir M.A.: Combined approach based on the mechanical behavior and microstructural examinations for the fracture of concrete. In: Proceedings of the International Association of Fracture Mechanics for Concrete and Concrete Structures, Seoul, Korea, pp. 283–290 (2010)
21. Ghanem, H.A., Phelan, R.S., Senadheera, S., Pruski, K.: Chloride ion transport in bridge deck concrete under different curing durations. ASCE J. Bridge Eng. **13**(3), 218–225 (2006)
22. AASHTO: Electrical indication of concrete's ability to resist chloride ion penetration. AASHTO T277, Washington, D.C (2015)
23. Stanish, K.D., Hooton, R.D., Thomas, M.D.A.: Testing the chloride penetration resistance of concrete: a literature review. In: Prediction of Chloride Penetration in Concrete, FHWA Contract DTFH61-97-R-00022 (1997)

The Effect of Underground Stories on the Dynamic Response of High-Rise Buildings

Lina Jaber[1(✉)], Yehya Temsah[1], Yasser El-Mossallamy[2], and Fadi Hage Chehade[3,4]

[1] Civil Engineering Department, Beirut Arab University, Beirut, Lebanon
{l.jaber,ytemsah}@bau.edu.lb
[2] Structural Engineering Department, Ain Shams University, Cairo, Egypt
yasser.elmossallamy@arcadis.com
[3] University Institute of Technology, Lebanese University, Saida, Lebanon
fadihagechehade@hotmail.com
[4] School of Science and Technology, Beirut, Lebanon

Abstract. The majority of high-rise buildings are designed with one or several underground stories in the urban area. However, many researchers don't incorporate it in the dynamic analysis assuming that the building is fixed at the ground level. This study aims to investigate the effect of the underground stories on the dynamic response of RC building. This is achieved by comparing the behavior of buildings without underground story with those with multi-level underground stories. In this scope, two dimensional finite element models are generated, where three buildings of 12, 16 and 24 stories resting on loose soil are modeled. The analysis is performed under the effect of a strong earthquake input motion by the use of PLAXIS software. The non-linear small strain hardening constitutive model is employed for an accurate simulation of the soil behavior. Based on the results, the embedded underground stories show a significant alteration of the structural dynamic response. This paper provides recommendations on the importance of including the underground stories during analysis.

Keywords: Soil-structure interaction · Underground stories · Earthquake Slenderness

1 Introduction

Most of the high-rise buildings are designed with one or several underground stories. Due to the lack of relevant code recommendations, neglecting or incorporating the embedded stories in the numerical model is still a controversial issue (G. Saad et al. 2012). When a building with multiple underground stories is subjected to lateral earthquake loads, it is common to crop the superstructure and analyze it as a fixed base building at the ground level. Then, the underground stories are designed based on the base shear and moment resulting from the superstructure. However, the presence of an embedded foundation changes the velocity of propagating wave and leads to reflection

H. Rodrigues and A. Elnashai (Eds.): GeoMEast 2018, SUCI, pp. 209–218, 2019.
https://doi.org/10.1007/978-3-030-01932-7_17

and refraction of the incident seismic wave. This deviation from free field motion, known as kinematic interaction in addition to the inertial interaction between the soil and foundation, leads to an alteration of the seismic structural response. Therefore, to study the impact of the underground stories' inclusion, the soil- structure interaction effect should be considered (Dutta and Roy 2002; Dutta et al. 2004; Shakib 2004; Naim et al. 2008; El Ganainy and El Naggar 2009; Raychowdhury 2010; Tabatabaiefar and Massumi 2010; Caglar et al. 2014). Naim et al. (2008) proved that fixing the structure at the base level, applying free-field motions as input at the base level and using horizontal foundation springs along basement walls with their end condition fixed to the free field ground motion provides poor results. El Ganainy and El Naggar (2009) investigated the seismic performance of low to medium rise moment-resisting frame steel buildings with multiple underground stories. It was found that soil–structure interaction (SSI) effects are significant for buildings with no underground stories, and gradually decrease with the increase of the number of underground stories. Caglar et al. (2014) demonstrated that when the height of the structure increased, the earthquake response of RC buildings without underground stories is detrimentally affected. Yet, when the same structure is built with underground story, the earthquake response became less damaging.

The main objective of this study is to better comprehend the dynamic behavior of high-rise concrete buildings with multiple underground stories. It aims to investigate the consequences of neglecting the underground stories in the analysis. This is realized by comparing the behavior of buildings having an incrementally increasing number of underground stories with those without underground stories. The number of stories of the building above the ground level is also considered. Two dimensional finite element models are generated by the use of PLAXIS software (Version 2015). Recommendations on the incorporation of the underground stories are provided at the end of the study.

2 The Numerical Model Description

The two dimensional finite element models consist of buildings supported on a raft foundation that is lying on a 120 m thick and 720 m wide loose sandy soil layer (Fig. 1). Three building structures of twelve, sixteen and twenty-four stories of 3.0 m story height are employed in the analysis (H = 36, 48 and 72 m). The buildings are composed of concrete frames of three bays equally spaced at 4.0 m (Fig. 2). The structural slenderness of the considered buildings that is their height to base width ratio is therefore 3, 4 and 6 respectively. A strong earthquake input motion (Parkfield, California 2004) of magnitude 6, peak ground acceleration 0.3 g, and 50 s duration is applied at the bedrock level beneath the soil. (Fig. 3). The analysis is performed in the time domain based on Newmark's (1959) implicit time integration equations.

2.1 Damping

The soil unbounded medium is idealized as a finite domain by assigning viscous boundaries of Lysmer type at its truncated boundaries (Lysmer and Kuhlemeyer 1969). They were placed at the vertical and bottom boundaries of the soil domain to reduce the

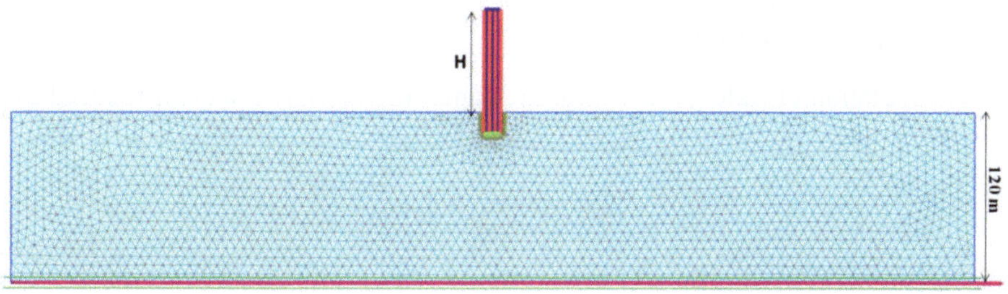

Fig. 1. Two dimensional finite element soil-structure interaction model

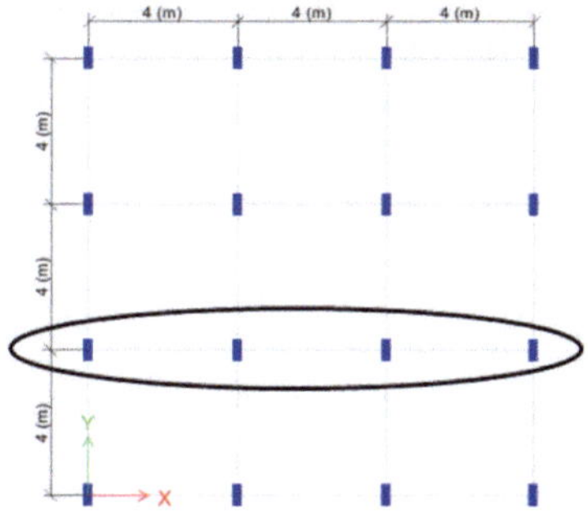

Fig. 2. Plan view of an intermediate frame columns layout

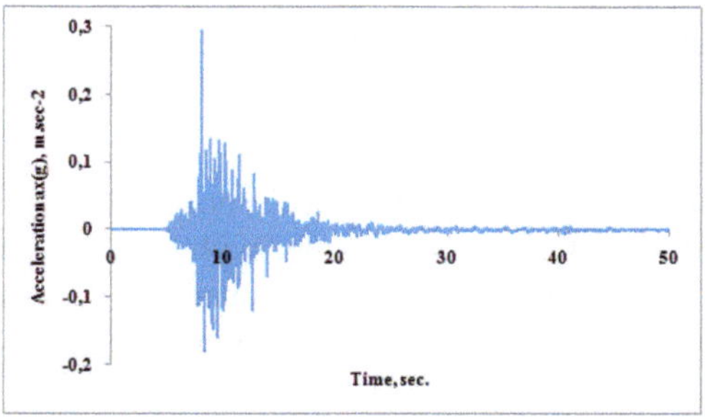

Fig. 3. Accelerogram record for Parkfield input motion

reflection of the radiation waves. These viscous boundaries were modeled according to Rayleigh damping approach in the form of:

$$[C] = \alpha_R[M] + \beta_R[K] \tag{1}$$

Where

[C]: damping matrix
[M]: mass matrix

[K]: stiffness matrix
α_R and β_R: the Rayleigh coefficients

In the formulation of Raleigh damping, the damping matrix [C] in (1) is reduced to the form

$$2\zeta_n \omega_{ni} = \alpha_R + \beta_R \omega_{ni}^2 \quad (2)$$

Where

ζ_n: the damping ratio
ω_{ni}: the soil circular natural frequency

This formula is implemented in the software, α_R and β_R are directly calculated for each soil type whenever the double frequencies of the soil bed and the constant damping ratio are defined, according to:

$$\omega_1 = \pi Vs/2D \quad (3)$$

$$\omega_2 = n\,\omega_1 \quad (4)$$

$$n = \omega_s/\omega_1 \quad (5)$$

Where

Vs: shear wave velocity
D: depth of the soil domain
ω_s: fundamental frequency of the seismic input motion
n: odd integer multiplier greater than the ratio between the fundamental frequency of the seismic input motion ω_s and the first natural frequency of the soil ω_1 (Hudson et al. 1994).

The double frequencies are calculated based on a 2% damping ratio. The shear wave velocity Vs is taken 100 m/s conforming with the 2000 International Building Code (IBC) requirements. The fundamental frequency of the seismic input motion ω_s = 5.29 Hz is determined from the earthquake Fourier spectrum (Fig. 4).

2.2 Soil Model

The small strain hardening constitutive model, proven effective in capturing the real soil behavior in dynamic analysis is employed (Gajo and Wood 1999). Hence, the hysteretic damping in addition to the Raleigh damping simulate the soil damping characteristics. Empirical equations for sandy soil, vidated by Brinkgreve et al. (2010), are used to determine the different soil parameters in the absence of experimental data (Table 1). An interface is placed around tnderground stories to ensure a full interaction between these concrete elements and the surrounding soil.

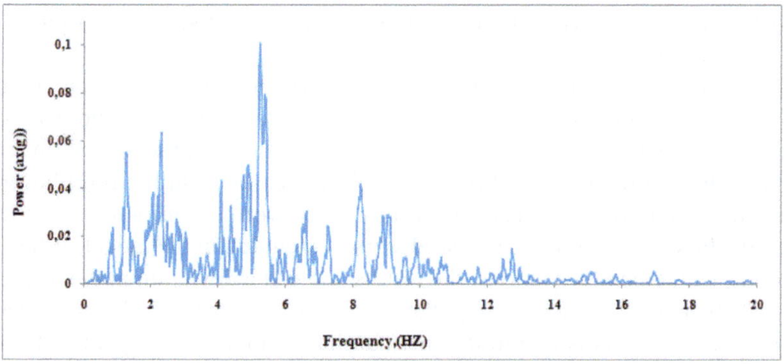

Fig. 4. Fourier Response Spectrum for Parkfield input motion

Table 1. Mechanical properties of HS small soil.

Parameter soil type: Loose sand	Symbol	Magnitude	Unit
Total unit weight	γ	17.5	KN/m^3
Small strain stiffness	G_0^{ref}	70.2	MPa
Shear strain at $0.7G_0$	$\gamma_{0.7}$	$1.85.10^{-4}$	-
Poisson's ratio	ν_{ur}	0.2	-
Triaxial compression stiffness	E_{50}^{ref}	9	MPa
Primary oedometer stiffness	E_{oed}^{ref}	9	MPa
Unloading/reloading stiffness	E_{ur}^{ref}	27	MPa
Reference pressure	p^{ref}	100	KPa
Rate of stress- dependency	m	0.65	-
Failure ratio	R_f	0.9	-
Stress ratio in primary compr.	K_0^{nc}	0.53	-

2.3 Mesh Element Size

The Finite element method is employed for analysis with element mesh size not exceeding one - eighth to one - fifth the ratio of shear wave velocity and the highest frequency of the earthquake input motion: $\Delta h \leq 1/8 - 1/5\ (Vs/f_{max})$ (Kuhlemeyer and Lysmer 1973). From Fig. 4, f_{max} is approximately equal to 20 Hz, requiring a mesh element size smaller than 1.0 m. A mesh size of 0.65 m is adopted in this study.

3 Results and Discussions

For a better understanding of the influence of the underground stories, the soil horizontal displacements at different depths in the domain are evaluated. The roof's lateral displacements, the base shear forces and the maximum moments are determined as well. A comparison between the responses is performed when the buildings are provided with zero, one, three and five underground stories symbolized by B0, B1, B3 and

B5 respectively. The behavior of the buildings with different slenderness ratios (S3, S4 and S6) is analyzed.

The graphical representation of the peak horizontal displacements in the soil domain from the rock layers (−120 m) to the surface (Figs. 5 and 6) displayed matching values when no embedded masses exist. In the absence of underground stories (B0), the soil displacement progressively increases from the rigid bottom, then shows a sudden amplification near the surface. It is worth mentioning that the higher building (S6) failed when no underground story is included in the model. When the soil is interacting with embedded stories, the results showed a deviation in the displacement curves at 3, 9 and 15 m designating the levels of one, three and five underground stories respectively. This is due to the fact that the vertically propagating shear waves emanating from the earthquake input load are getting hampered by the rigidity of the underground stories. The presence of underground stories has significantly reduced the displacement at the soil surface. For the buildings provided with one (B1) and three underground stories (B3), the higher building “S6” displayed the largest displacement at the soil surface, yet this value decreased as the building gets shorter (“S4” then “S3” respectively). On the contrary, when the building is modeled with five underground stories (B5), the displacement diminishes with the increase in building height (Fig. 7). Similar behavior is observed for the peak displacement at the building roof (Fig. 8(a)). For S3 and S4, the increase in number of underground stories augmented the building flexibility and consequently increased the displacement. Analogous performance is monitored for the higher building with one and three underground stories, whereas for the five underground stories, the displacement lessened.

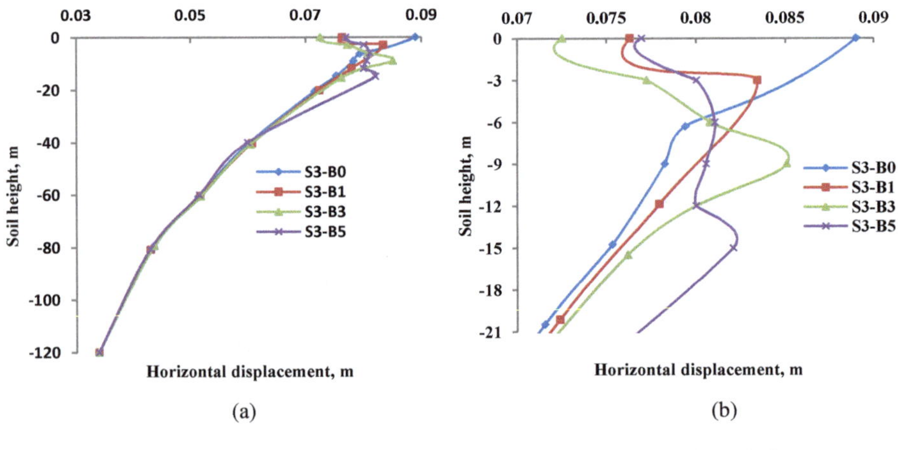

Fig. 5. Peak horizontal displacement on the subsurface for S3 building.

The buildings without underground stories exhibit the larger values for base shear forces and moments. These values diminished gradually while incorporating the embedded stories (Fig. 8(b) and (c)) except for building “S6” with 5 underground stories. The rate of reduction is calculated for both entities for B1, B2 and B3 with

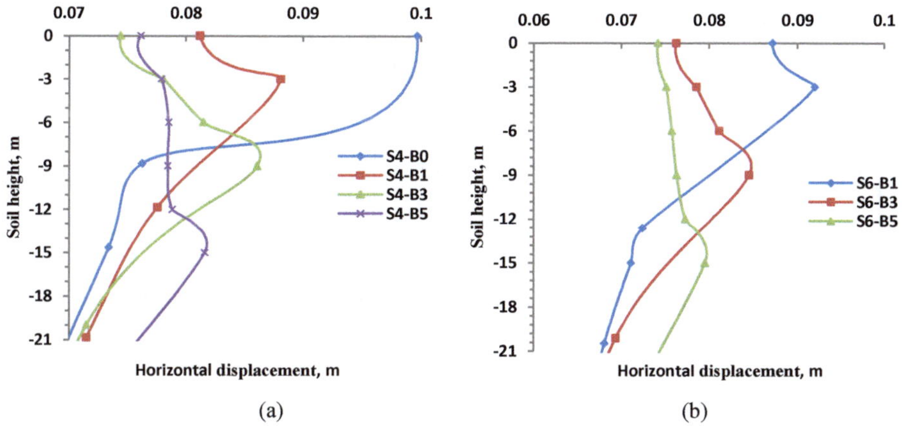

Fig. 6. Peak horizontal displacement on the subsurface for (a) S4 (b) S6 buildings.

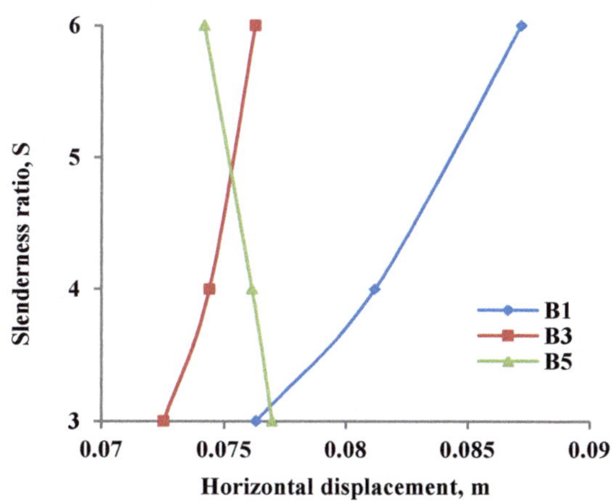

Fig. 7. Horizontal displacement at the soil surface function of the building slenderness for one, three and five underground stories.

respect to B0. The high values displayed in Table 2. Illustrated the major benefit that an underground story can offer. No significant changes in base shear forces are revealed after the third underground story's integration, while just one underground story was enough to produce a large drop in moment.

To summarize, the underground stories had shown a significant influence on the dynamic response of buildings. When embedded into the soil, these rigid masses hindered the shear wave propagation and therefore decreased the lateral displacement at the soil surface. The underground stories reduced the base shear force and moment and consequently provided stability for the buildings against the earthquake excitation.

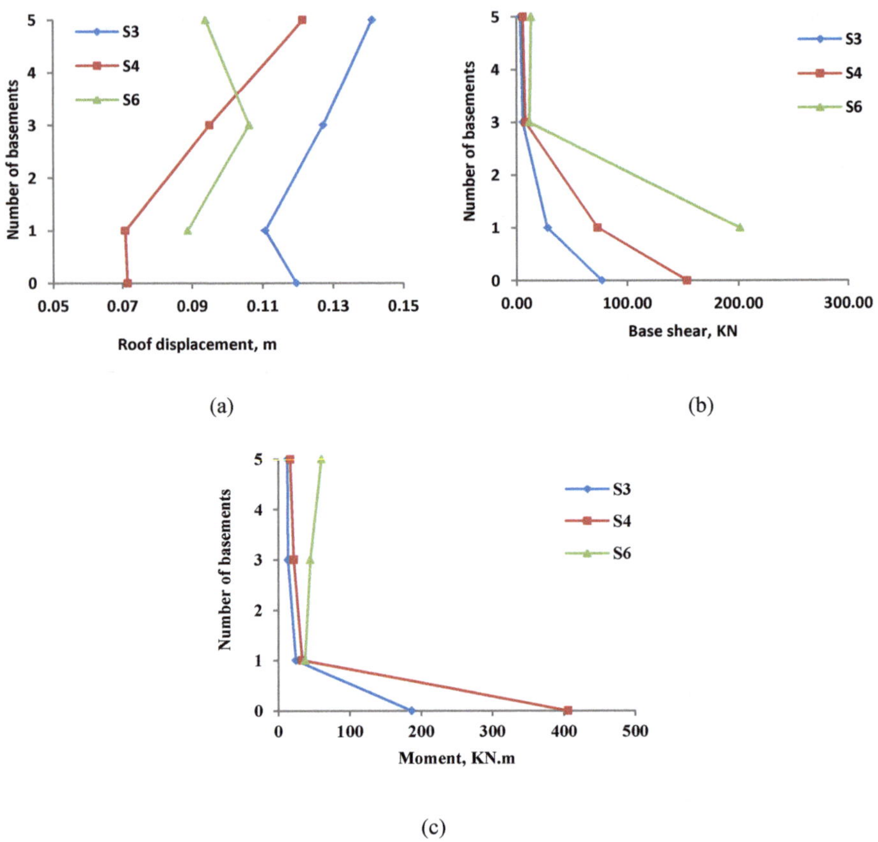

Fig. 8. (a) Roof displacement (b) Base shear (c) Maximum moment for S3, S4 and S6 function of the number of underground stories

The higher building that collapsed when no underground story is incorporated, resisted while designed with several underground stories. This is in accordance with Caglar's findings. In his study, it was pointed out that the seismic response of RC buildings without basement is adversely affected in absence of underground stories. It also proved that the presence of basement decreased the structural damage.

The integration of underground stories increased the building flexibility except for the higher building with 5 underground stories. Excluding the underground stories from the numerical model may be conservative or detrimental. The structural seismic performance is influenced by the interaction between the soil and the structure's number of stories above and below the ground. The results give clear evidence that the soil-structure interaction has to be considered. This outcome reinforced the importance of accounting for SSI as stated earlier by the researchers.

Table 2. Base shear forces and moments and their rate of reduction function of the number of underground stories and building heights

	Base shear (KN)	Reduction%	Moment (KN.m)	Reduction%
S3B0	76.98	-	186	-
S3B1	28	63.63	24.1	87.04
S3B3	5.66	92.65	13	93.01
S3B5	3.33	95.68	12.5	93.28
S4B0	153.6	-	405.5	-
S4B1	73.3	52.28	33	91.86
S4B3	8.07	94.75	21.26	94.76
S4B5	5.77	96.25	16.61	95.9
S6B0	-	-	-	-
S6B1	201.29	-	37.2	-
S6B3	11.81	-	44.95	-
S6B5	13.27	-	60.82	-

4 Conclusions

Designing a building with underground stories under the effect of earthquake load has always been a controversial issue. Whether these underground stories must be included or not in the numerical model necessitates a profound understanding of their influence on the structural dynamic response. The purpose of this study is to investigate the real impact of the incorporation of one and several underground stories on the behavior of three buildings of different heights on loose soil. The results demonstrated that:

- When the number of underground stories increased, the displacement at the building base, the base shear forces and maximum moments decreased simultaneously. The rate of reduction of the base shear force and moment exceeded 90 percent when the buildings had three underground stories in comparison with structures without underground story.
- When the higher building was designed without underground story, it collapsed. The integration of embedded stories improved its stability and consequently reduced the probable damages.
- The underground stories augmented the flexibility of the structure. If the underground stories were disregarded, this could have overestimated the lateral stiffness of the building and accordingly reduced the structural displacement.
- For the higher structure with five underground stories, the displacements at the building roof started to decrease and the base shear and the moment increased. This was not observed for the other buildings with the same number of underground stories. This proves that the effect of underground stories is related to the number of stories above the base, thus the soil-structure interaction effect must be considered.

Based on the above findings, it is recommended to include the underground stories in the dynamic analysis. Neglecting their presence might be conservative or detrimental since the structural seismic response is the result of an interaction between the number of underground stories and the structure height.

References

Brinkgreve, R.B.J., Engin, E., Engin, H.K.: Validation of empirical formulas to derive model parameters for sands. Numer. Methods Geotech. Eng. (2010). https://doi.org/10.1201/b10551-25

Caglar, N., Garip, Z.S., Atasoy, S.: The effect of basement storey on the response of RC buildings settled on soft soil. In: 2nd International Symposium on Innovative Technologies in Engineering and Science (2014)

Dutta, S., Roy, R.: A critical review on idealization and modeling for interaction among soil-foundation-structure system. Comput. Struct. (2002). https://doi.org/10.1016/S0045-7949(02)00115-3

Dutta, S., Bhattacharya, K., Roy, R.: Response of low-rise buildings under seismic ground excitation incorporating soil–structure interaction. Soil Dyn. Earthq. Eng. (2004). https://doi.org/10.1016/j.soildyn.2004.07.001

El Ganainy, H., El Naggar, M.H.: Seismic performance of three-dimensional frame structures with underground stories. Soil Dyn. Earthq. Eng. (2009). https://doi.org/10.1016/j.soildyn.2009.02.003

Gajo, A., Wood, D.M.: Severn-trent sand: a kinematic hardening constitutive model for sands: the q-p formulation. Géotechnique (1999). https://doi.org/10.1680/geot.1999.49.5.595

Hudson, M., Idriss, I.M., Bekaie, M.: QUAD4M: a computer program to evaluate the seismic response of soil structures using finite element procedures incorporating a compliant base. Center for Geotechnical Modeling, Department of Civil and Environmental Engineering, University of California, Davis (1994)

Kuhlemeyer, R.L., Lysmer, J.: Finite element method accuracy for wave propagation problems. J. Soil Dyn. Div. **99**, 421 (1973)

Lysmer, J., Kuhlemeyer, R.L.: Finite dynamic model for infinite media. J. Eng. Mech. Div. **95**, 859 (1969)

Naim, F., Tileyliouglu, S., Alimoradi, A., J. Stewart.: Impact of foundation modeling on the accuracy of response history analysis of a tall building. In: SMIP08 Seminar Proceedings (2008)

Newmark, N.M.: A method of computation for structural dynamics. J. Eng. Mech. Div. **85**, 67 (1959)

Raychowdhury, P.: Seismic response of low-rise steel moment-resisting frame (SMRF) buildings incorporating nonlinear soil–structure interaction (SSI). Soil Dyn. Earthq. Eng. (2010). https://doi.org/10.1016/j.engstruct2010.12.017

Saad, G., Saddik, F., Najjar, S.: Impact of soil structure interaction on the seismic design of reinforced concrete buildings with underground stories. In: Lisboa Conference American University of Beirut, Lebanon (2012)

Shakib, H.: Evaluation of dynamic eccentricity by considering soil-structure interaction: a proposal for seismic design codes. Soil Dyn. Earthq. Eng. (2004). https://doi.org/10.1016/j.soildyn.2004.01.003

Tabatabaiefar, R., Massumi, A.: A Simplified method to determine seismic responses of reinforced concrete moment resisting building frames under influence of soil-structure interaction. Soil Dyn. Earthq. Eng. (2010). https://doi.org/10.1016/j.soildyn.2010.05.008

Engineering Properties of Self-compacting Concrete Containing Class C Fly Ash and Processed Slag Sand

H. P. Vageesh[1(✉)], C. P. Ramesh[2], T. Raghavendra[1], B. C. Udayashankar[1], and A. Shashishankar[3]

[1] Department of Civil Engineering, VTU, R. V. College of Engineering, Bengaluru, Karnataka, India
{vageeshhp, raghavendrat, udayashankar}@rvce.edu.in
[2] Department of Civil Engineering, SET- JAIN University, Bengaluru, Karnataka, India
cp.ramesh@jainuniversity.ac.in
[3] Department of Civil Engineering, VTU, AMC Engineering College, Bengaluru, Karnataka, India

Abstract. Self-compacting concrete (SCC) is a concrete having special features, and is the most revolutionary development in concrete construction for several decades; since every country now faces a problem of skilled labour in construction industry. SCC provides rapid rate of concrete placement and hence faster construction times; it is more durable and also develops high early age strength. The exploitation of natural aggregates is harming environment and thereby necessitating sustainable construction industry. Several alternates to natural aggregates are being investigated to overcome this challenge, and few of them include manufactured aggregates, quarry waste, crushed sandstone aggregates, materials recycled from construction and demolition waste, copper slag, crushed sand from different mineralogical sources, marble and granite waste aggregates, aggregates etc. Research studies demonstrate that these waste or by-products may be used as partial replacement of river sand in concrete. Fly ash - a coal industry waste possessing Pozzolanic properties is highly recommended as cement replacement material for green initiatives. This paper illustrates the laboratory investigation results on SCC incorporating, European Federation of National Associations Representing for Concrete (EFNARC) guidelines, by complete replacement of fine aggregates with Processed Slag Sand (PSS) and partial replacement of cement with Class C fly ash. The results in terms of flow, strength and durability; for this novel SCC when compared with control concrete was found to be highly satisfactory and thereby encourage today's Engineers to adopt SCC technology with this novel mix combination for sustainable development.

Keywords: SCC · PSS sand · Class C fly ash

H. Rodrigues and A. Elnashai (Eds.): GeoMEast 2018, SUCI, pp. 219–229, 2019.
https://doi.org/10.1007/978-3-030-01932-7_18

1 Introduction

Self-compacting concrete (SCC) is an innovative concrete that does not require vibration for placing and compaction. It is able to flow under its own weight. Compaction of Concrete particularly in places with heavy reinforcement, beam-Column junctions has become a major issue. To overcome this difficulty Self Compacting Concrete was developed in 1980's at Japan.

Szweda et al. (2017) "modified SCC by different volumes of granulated imperial smelting process (ISP) slag; this was used as a replacement for natural sand. Earlier, mechanical properties of such SCCs mixes are already carried out, but their durability and possible life span of structures were studied further. They investigated on chloride ions infiltration into SCCs. The tests were planned in such a way that the achieved results would allow to use a theoretical diffusion model proposed by Szweda and Zybura (2013). The model was verified by a comparison with the results acquired in a real diffusion. Finally, the forecast durability of concrete structures made out of tested SCCs was calculated. The foreseen life span of the structures was differentiated due to thickness of concrete coating of rebars. They proved that replacing sand by ISP slag, even for small ratios, in terms of durability point of view it can become environmental impact study".

da Silva and de Brito (2016) investigated on the durability of self-compacting concrete (SCC) produced with binary and ternary mixes using fly ash (FA) and limestone filler (LF) as partial replacement of cement. Degradation mechanisms were studied and characterized by means of SEM image analysis, chloride migration, electrical resistivity, and carbonation tests were conducted. It was noted that the properties studied are strongly affected by the type and content of the additions. Also, the use of ternary mixes proved to be extremely favourable, confirming the expected beneficial effect of the synergy between Lime stone filler and Fly ash.

Omar Mohamed Oraibi et al. (2015) investigated the effect of fly ash on the mechanical properties of self-compacting concrete, modulus of elasticity and to determine the optimum fly ash content as cement replacement in improving mechanical properties at the age of 28 days. The results were compared to other standards with respect to compressive strength and tensile strength. Mechanical properties of self-compacting concrete modified with 0%, 15% and 35% Fly ash as cement replacement were considered. They concluded that Slump Flow values increased as the amount of Fly Ash increased within the allowable ranges of 500–700 mm in accordance to ASTM standard. The compressive strength and modulus of elasticity values decreased as the amount of fly ash increased from 0 to 35% whereas the value for improved mechanical property of flexural strength was found at 15% Fly ash content.

Craeye et al. (2014) collected a dataset consisting of results from more than 250 research papers on SCC during the last two decades. Here, an in depth analyses is provided on the modulus of elasticity and the tensile strength of SCC. The SCC results collected within the database are compared with those predicted from the formulations and existing models developed and validated for vibrated concrete (e.g. Eurocode 2 and the Model Code (MC 90 and/or MC 2010)). The influences of different mix-design parameters (aggregate type, paste volume, etc.) on the E-modulus and the tensile strength were analysed, and a more detailed view on the performance of SCC, and the different models, is obtained. They reported that modulus of elasticity of SCC was similar as that of Vibrated concrete.

Mahmoud et al. (2013) carried out a research to explore the use of supplementary cementitious materials (SCM's), and recycled materials on the performance of SCC concrete mixtures. "Twelve mixes were prepared with different combinations of fly ash, slag, and recycled asphalt pavement (RAP). Fresh and hardened concrete properties were measured, as they expected the inclusion of the SCM's, and recycled materials affected both fresh and hardened concrete properties. By the analysis of experimental data indicated that inclusion of RAP not only reduces the ultimate strength, but it also affected the compressive strength development rate. The addition of RAP to mixes showed a consistent effect, with a drop in strength after 3, 14, and 28 days as the RAP content increased from 0 to 50%. However, most of the mixes satisfied SCC fresh properties requirements, including mixes with up to 50% RAP. Besides, some mixes satisfied compressive strength requirement for pavements and bridges, those mixes included relatively high percentages of SCMs and RAP".

Uysal and Yilmaz (2011) reported the benefits of limestone powder (LP), basalt powder (BP) and marble powder (MP) as partial replacement of Portland cement. Also LP, BP and MP were used in the production of self-compacting concrete (SCC). The water to binder ratio was maintained at 0.33 for all mixtures. They inspected properties like workability, air content, compressive strength, ultrasonic pulse velocity, and static and dynamic elastic moduli. They concluded that it is possible to utilize waste LP, BP and MP as mineral admixtures in producing SCC, and the elastic and dynamic modulus of SCC mixes were decreased due to the incorporation of mineral admixtures. The incorporation of waste mineral admixtures improved the economic feasibility of SCC production on unit strength.

Joorabchian (2010) "investigated the effects of aggressive sulphuric acid attack on the concrete mixtures prepared with metakaolin (MK) and limestone filler (LF) at various replacement levels. In addition, rapid chloride permeability (RCPT), water sorptivity, water porosity and rapid freezing and thawing tests were also performed on the concrete samples. Three sulphuric acid solutions with concentrations of 3%, 5% and 7% were used for inspecting the resistance of concrete specimens for a total exposure period of eight weeks. The performance of the degraded specimens was evaluated by measuring the weight loss, change in strength and visual assessment. The results of the weight loss and visual assessment reveal that the increase in amount of MK would enhance the performance of concrete, while inclusion of LF into MK concretes has shown positive results in terms of resistance against sulphuric acid attack. The mixtures with ternary binders of OPC, MK and LF experienced the lowest strength loss after exposure to high concentrations of sulphuric acid. Measuring the change of weight was found to be a better way to evaluate the resistance of concrete specimens immersed in sulphuric acid solutions as the results of the load bearing capacity can be affected by several parameters such as the variable geometry of degraded specimens. The rate of the water absorption of concrete with MK and LF was lower than the reference mixture with only Portland cement. The RCPT results also showed that the reference mixture had considerably higher permeability than other concrete mixtures. The results of the water porosity on degraded samples indicate that the inner parts of the concrete (not in contact with sulphuric acid) have remained sound and maintained their original pore structure after different exposure periods".

The Purpose of the proposed research is to investigate the fresh and hardened properties of novel SCC mixes, by the complete replacement of fine aggregates with Processed Slag Sand (PSS) and partial replacement of cement with Class C fly ash in varying percentages, on compressive strength, split tensile strength and flexure at 7, 28 & 56 days age. The stress strain characteristics, effect of water-powder ratio on properties of SCC mixes and durability aspects were also analysed. As on date there is no IS code for SCC and hence, the Japanese method of mix design for SCC mixes by incorporating guidelines given by EFNARC (European Federation of Natural Association Representing for Concrete) are followed. Finally suitable concrete mix is proposed based on the workability property in its fresh state, compressive strength and durability requirements.

1.1 Materials and Methodology

The chemical and physical properties of the SCC ingredients were investigated. Bureau of Indian Standards (IS) procedures were followed for determining the properties of the ingredients in this investigation. Ordinary Portland Cement 53 grade was used as per IS-8112 (1989) and specific gravity of cement was found to be 3.15. Class C Fly ash a mineral admixture procured from Power Plant is replaced in terms of percentage to cement, the specific gravity of fly ash was 2.12. Super-plasticizer commercially termed as Master Glenium Sky 8233 of specific gravity 1.08, Crushed granite stones of 12.5 mm down size are used as coarse aggregate. As per IS: 2386 (Part III)-1963, the specific gravity in oven dry condition and water absorption of the coarse aggregate are 2.67 and 0.8% respectively. The Bulk density of the coarse aggregate as per IS: 2386 (Part III) 1963 is1620 kg/m^3. Processed slag sand is used as fine aggregate. As per IS: 2386 (Part III)-1963, the specific gravity in oven dry condition and water absorption of the sand are 2.53 and 0.3% respectively. Water of pH 7 is used.

1.2 Japanese Method of Mix Design

The Japanese method mix design procedure as indicated in literature is adopted. (Vengala and Ranganath 2004)

The mix design steps are as follows

(1) Decide the air content. This would depend on air entraining requirements, if any. If no air entrainment is specified one may assume about 2% of air content.
(2) The coarse aggregate content in concrete is tentatively fixed at 50 to 60% of volume of concrete. This is based on its bulk density. An example that follows will illustrate as to how the coarse aggregate content is determined. Actually, researchers at the University of Tokyo employed the term maximum possible packed density in a container; but for practical purposes it can be taken as bulk density as determined by standard tests. The lower the maximum size of aggregate the higher would be the proportion of coarse aggregate – but within the range specified. Likewise for rounded aggregate a higher proportion is used compared

with crushed (angular) aggregates. The suggested range is based on extensive laboratory trials, in particular on different types of obstacles employed in the U-box test.

(3) The fine aggregate content is also tentatively fixed at 40–50% of the mortar volume of concrete. Mortar is defined as a mixture consisting of cement paste, filler and fine aggregate.
(4) The type of filler (pozzolanic or non-pozzolanic) and its proportion to cement are then decided based on experience. In order to arrive at the appropriate values of powder ratio defined as the ratio of water to powder consisting of cement and filler, tests are conducted on mortar (cement, filler, sand and water) for different water powder ratios.
(5) Super plasticizer is then added to the mortar at different dosages and the flow cone and v-funnel are performed. From trials we got it 0.6% by weight of powder (cement + filler).
(6) Once the water powder ratio and super plasticizer dosage have been determined, coarse aggregate is added to the mortar and trial mixes are made of the concrete.
(7) U tube box test is carried out to ascertain the self-compatibility of the concrete. Specimens are also cast for strength assessment. According to Okamura and Ouchi strength rarely poses a problem since the water-powder ratio is low in SCC, unless the filler is totally non-pozzolanic and is present in too large quantity.
(8) In addition, slump flow and V-funnel tests may also be performed to ensure non segregated spread and flow. The advantages of Japanese method is that for a given set of materials, in particular the cement, super plasticizer and filler, the test performed to assess the water powder ratio and the super plasticizer dosage need not be repeated at different sites or at different occasions even if aggregates vary marginally. It may be enough if U box test is performed to ensure self compactibility. Such a situation is suited to Ready mix concrete.

It is considered to be the very simple and easy method and therefore is utilized for designing SCC in this experimental work, to prepare trial mixes for SCC and to satisfy the flow parameters like slump flow, V-Funnel, U-Box and J- Ring only super plasticizer dosage was varied and ultimate dosage was fixed using Marsh cone apparatus, by keeping all other parameters like water-powder ratio, fine aggregates, coarse aggregates were kept constant with complete replacement of fine aggregates by Processed Slag Sand (PSS) and partial replacement of cement with Class C fly ash in varying percentage 30% by the weight of cement i.e. (70:30) which is indicated as *SCC 73* whereas *SCC CM* for controlled mix, several trial mixes were carried out to arrive at desired mix ratio. Each and every SCC mix was verified to achieve workability property then cylindrical specimens of size 150 mm diameter and 300 mm height were cast to evaluate compressive strength, splitting tensile strength and Young's modulus. 500 mm × 100 mm × 100 mm beams were cast to evaluate flexure strength. To evaluate durability properties such as acid attack test and alkaline attack test; 150 mm × 150 mm cubes were cast.

Based on the EFNARC guidelines, tests such as slump flow, V-funnel, L-box and U-box tests were carried out to evaluate the fresh properties. EFNARC specifications for SCC workability tests requirements in fresh state for both the mixes *SCC 73* and *SCC CM* are given in (Table 1).

Table 1. Acceptance criteria for both the mixes (SCC 73 and SCC CM) as per EFNARC

Sl. no	Tests	Specification as per EFNARC	SSC CM	SSC 73	Units
1	Slump flow	600–800 mm	725	650	mm
2	T50 slump	2–5 s	3	3	sec
3	V Funnel	6–12 s	10	8	sec
4	L box	(H2/H) 1 0.8–1	0.8	0.8	Ratio
5	U Box	(H2–H1) 0–30 mm	18	0	mm

2 Results and Discussions

2.1 Compressive Strength

Compressive Strength indicates the resistance to permanent deformation of normal and modified self-compacting mixtures. As mentioned above, cylinders were cast and tested to determine compressive strength of SCC. Compressive strength was evaluated at 7, 28 and 56 days age. It has been noted that at the age of 7 days the *SCC 73* mixes obtained lesser strength of about 40% when compared with *SCC CM* and later due to pozzolanic action of fly ash the compressive strengths at 28 and 56 days was improved which is nearly equal to *SCC CM* mixes (Fig. 1).

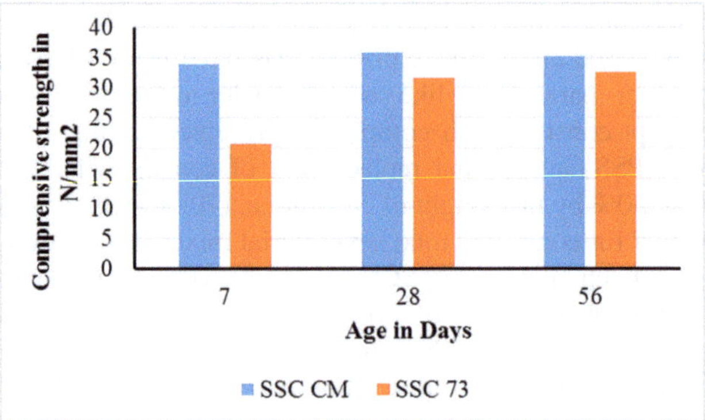

Fig. 1. Comparison of Compressive strength of SCC CM and SCC 73 at 7, 28 and 56 days

2.2 Tensile Strength

The split tensile strength value is affected by percentage of fly ash. As observed in Fig. 2, it is clear that there is a little difference on split tensile strength values among both self-compacting concrete mixtures. *SCC CM* mixes has comparatively more split tensile strength value of about +13.43% and +11% when compared with *SCC 73*, at the age of 28 and 56 days respectively.

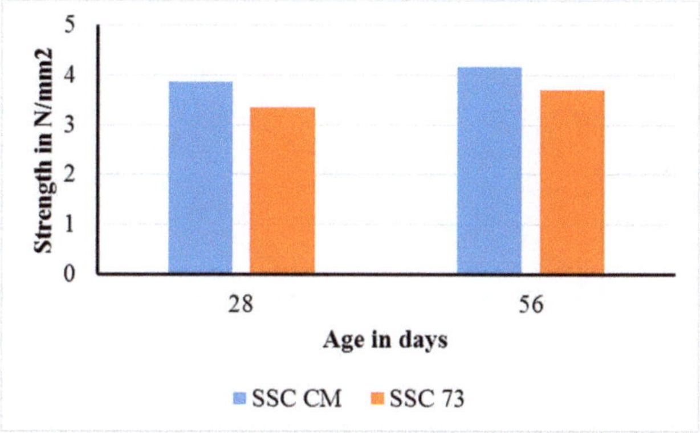

Fig. 2. Comparison of Split tensile strength of SCC CM and SCC 73 at 28 and 56 days

2.3 Flexural Strength

It may be observed in Fig. 3, that there is a little difference in Flexural strength test results among both the self-compacting concrete mixtures. *SCC CM* mixes has flexural strength value of about +20% and +15.77% when compared with *SCC 73* at the age of 28 and 56 days respectively.

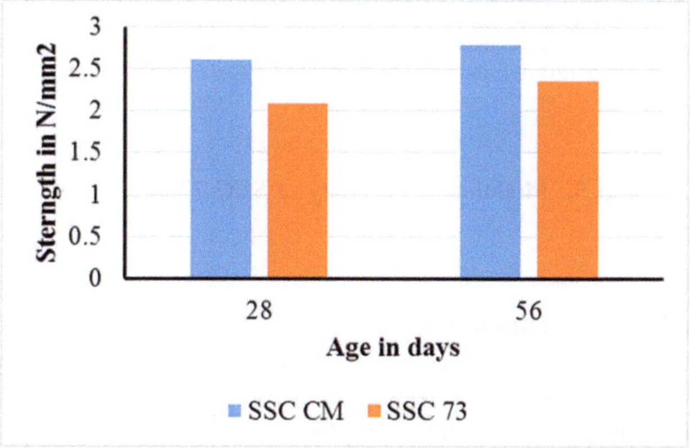

Fig. 3. Comparison of Flexure strength of SCC CM and SCC 73 at 28 and 56 days

2.4 The Modulus of Elasticity (MOE) (Young's Modulus)

The concrete cylinders were subjected to a slowly increasing longitudinal compressive stress and readings were recorded and MOE were computed. The highest value of modulus of elasticity was found for *SCC CM* when compared to *SCC 73* as shown in (Figs. 4 and 5).

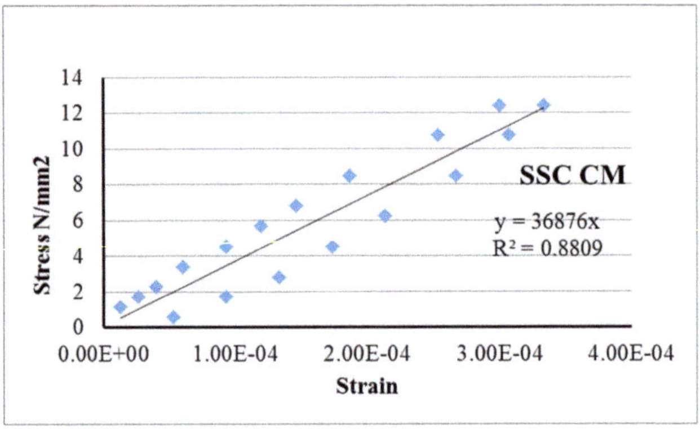

Fig. 4. Modulus of Elasticity of SCC CM at 28 days

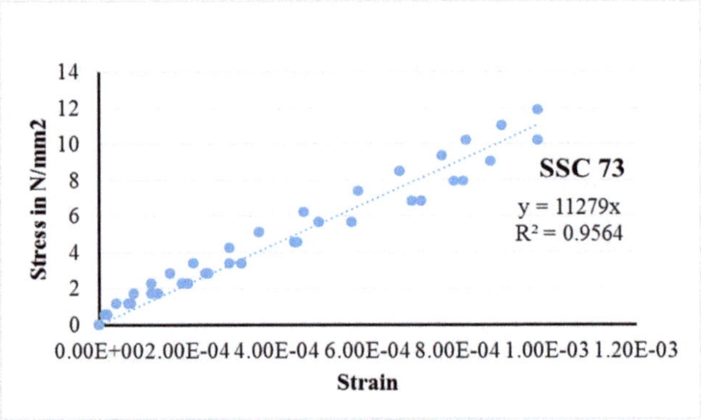

Fig. 5. Modulus of Elasticity of SCC 73 at 28 days

2.5 Durability

2.5.1 Acid and Alkali Attack Tests

The performance of the concrete specimens which were immersed separately into the 10% sulphuric acid solutions and 10% of 1 N NaOH solution were evaluated. The weight and dimension of the concrete samples was monitored weekly through-out the entire testing period was recorded (refer Tables 2 and 3); the compressive strength of

Table 2. Comparison of change in weight and dimensions of 10% Acid Cured SCC Specimen at the age of 56 days

SCC mix	Original weight for 10% acid in kg	Change in weight for 10% acid in kg	Original dimension for 10% acid in mm	Change in dimension for 10% acid in mm
SSC CM	8.18	6.64	150 × 150	141 × 142
SSC 73	8.22	7.58	150 × 150	146 × 145

Table 3. Comparison of change in weight and dimensions of 1 N 10% Alkaline Cured SCC Specimen at the age of 56 days

SCC mix	Original weight for 1 N 10% Alkaline IN kg	Change in weight for 1 N 10% Alkaline in kg	Original dimension for 1 N 10% Alkaline in mm	Change in dimension for 1 N 10% Alkaline in mm
SSC CM	8.25	8.32	150 × 150	150 × 150
SSC 73	8.41	8.46	150 × 150	150 × 150

Table 4. Compressive strength comparison of acid and alkaline cured SCC specimen at the age of 56 days

SCC mix	10% Acid cured strength (MPa)	1 Normal 10% Alkali cured strength (MPa)
SSC CM	9.57	35.88
SSC 73	28.87	37.03

specimens was evaluated and compared (Table 4) at the age of 28 days in acid/alkali exposure (refer Fig. 6). Acid exposure resulted in huge difference with respect to weight loss and compressive strength, of about 18% and 66% for *SCC CM* in comparison with *SCC 73*. Whereas, the alkaline exposure resulted in no major weight loss and compressive strength for both the mixes and were relatively same.

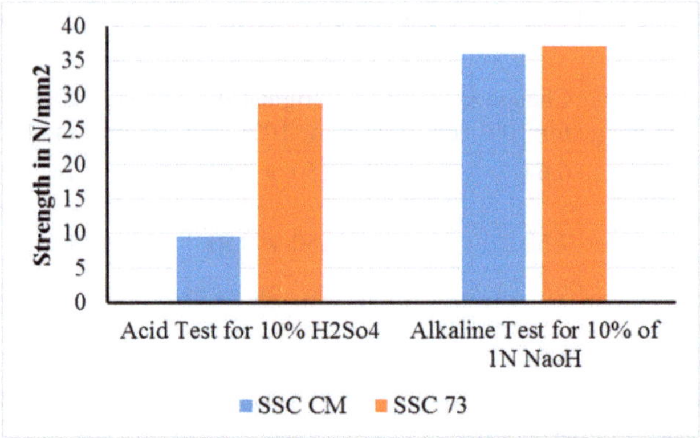

Fig. 6. Comparison of Compressive strength of Acid and Alkaline Cured SCC Specimens at the age of 56 days

3 Conclusions

From the detailed experimental exploration it has been observed that *SCC 73* mixes exhibited relatively equivalent strengths in terms of compression, tension and flexure, and also due to presence of filler material smaller values of modulus of elasticity were reported for *SCC CM* mixes. Due to presence of relatively higher volume of class C fly ash, the compressive strength of *SCC 73* mixes under 10% acid solution exposure have performed better compared to other mixes and hence more durable. 10% 1 N alkaline solution exposure resulted in satisfactory results similar to that of *SCC CM* Mixes. The concluding remarks are as follows.

(1) At the early ages, *SCC 73* mix gained very less compressive strength compared to *SCC CM* and at later ages such as 28 and 56 days, due to the pozzolonic action of Class C fly ash, the compressive strength of *SCC 73* was increased and found to be more or less equal to *SCC CM.*
(2) Both *SCC 73* and *SCC CM*, resulted in almost same values of splitting tensile strength and flexural strength at 28 days and 56 days, respectively.
(3) The modulus of elasticity of *SCC 73* mixes were relatively less in comparison with *SCC CM* mixes due to the incorporation of mineral admixtures, and the same was observed in literature (Mucteba Uysal and Kemalettin Yilmaz)
(4) High dimension loss and weight loss of *SCC CM* mix was observed when compared with *SCC 73* mixes, due to action of sulphuric acid and same was reported in literature (Seyed Mahmoodreza Joorabchian). Whereas, in case of alkaline exposure, the weight of specimens were found to be marginally increased and no remarkable changes in dimension were observed.
(5) *SCC 73* mixes resulted in comparatively better and acceptable compressive strength results when compared with *SCC CM* under acid and alkaline environment. Hence *SCC 73* may be suggested for the structures which require high durability aspects such as those exposed to marine environments.

(6) Both the SCC mixes satisfied the flow requirements such as V-Funnel, U-Box, and L-Box which are specified by EFNARC. Adoption of *SCC 73* will eventually reduce depletion of the natural resources, make structures durable and also sustainable by usage of PSS and Class C fly ash as replacements in SCC mixes, and hence reduce the adverse environmental effects.

References

Craeye, B., et al.: Modulus of elasticity and tensile strength of self-compacting concrete: survey of experimental data and structural design codes. Cem. Concr. Compos. **54**, 53–61 (2014)

Mahmoud, E., et al.: Self-consolidating concrete incorporating high volume of fly ash, slag, and recycled asphalt pavement. Int. J. Concr. Struct. Mater. **7**(2), 155–163 (2013)

Vengala, J., Ranganath, R.V.: Mixture proportioning procedure for self-compacting concrete. Indian Concr. J. **78**, 13–21 (2004)

Joorabchian, S.M.: Durability of concrete exposed to sulfuric acid attack. Theses and dissertations. Presented to Ryerson University, Toronto, Ontario, Canada (2010)

Omar Mohamed Oraibi, K.: Mechanical Properties Of Self Compacting Concrete Using Fly Ash. Ph.D Diss. Universitas Muhammadiyah Surakarta (2015)

Uysal, M., Yilmaz, K.: Effect of mineral admixtures on properties of self-compacting concrete. Cem. Concr. Compos. **33**(7), 771–776 (2011)

da Silva, P.R., de Brito, J.: Durability performance of self-compacting concrete (SCC) with binary and ternary mixes of fly ash and limestone filler. Mater. Struct. **49**(7), 2749–2766 (2016)

Szweda, Z., et al.: A study on replacement of sand by granulated ISP slag in SCC as a factor formatting its durability against chloride ions. J. Clean. Prod. **156**, 569–576 (2017)

Szweda, Z., Zybura, A.: Wyznaczanie współczynnika dyfuzji chlorków w betonie na podstawie pomiarów w polu elektrycznym. Roczniki Inżynierii Budowlanej **13**, 51–56 (2013)

Across Wind Load Analysis Using CFD for Sustainable Design of Tall Structures

K. Shruti[1], P. N. Rao[1(✉)], and G. R. Sabareesh[2]

[1] Department of Civil Engineering, BITS Pilani, Hyderabad Campus, Hyderabad, India
shruti.konka@gmail.com, pnrao.bits@gmail.com

[2] Department of Mechanical Engineering, BITS Pilani, Hyderabad Campus, Hyderabad, India
sabareesh@hyderabad.bits-pilani.ac.in

Abstract. With the advent of tall and complex shaped structures combined with urbanization, the structural designers are posed with a greater challenge to design the structures sustainable to rapid environmental changes. During the design of tall structures, wind loads acting on them are a major factor that needs to be considered. As a pre-requisite, the designer should have the information regarding wind environment of the region, wind forces on the structures and the response of the structure under these forces.

For tall and slender structures, across wind effects are more predominant than along wind effects. Recommendations in building codes for across wind effects are very limited. Several experimental studies were made to determine the across wind effects on structures. As wind tunnel tests are costlier and time consuming, an alternative computational fluid dynamics (CFD) approach has been emerged out to study these across wind effects.

This paper discusses on the across wind effects on tall structures like rectangular buildings and cooling towers in presence of other interfering structure using CFD analysis in ANSYS fluent. Experimental data of Kim et al. (2015) has been adopted to model the geometry of the structure and to simulate the wind environment around the structure in ANSYS fluent. LES (Large eddy simulations) turbulence model is considered to simulate the desired flow parameters around the structure. This 3D CFD model is validated with the experimental result of Kim et al. (2015). Further, the same scheme has been extended to cooling towers. Across wind effects acting on rectangular buildings and cooling towers are evaluated for the two cases: a. Stand alone case, b. interfering case with the same terrain category. In this manner, the response of the tall structures to across wind loading is studied using computational approach.

1 Introduction

Tall Buildings and other slender structures like towers, chimneys, street lighting poles and cable bridges are increasing day by day with the growing need for urbanization. Wind loads are more predominant for tall buildings and other slender structures. These structures are prone to damage under wind action, if wind loads on these structures are

H. Rodrigues and A. Elnashai (Eds.): GeoMEast 2018, SUCI, pp. 230–243, 2019.
https://doi.org/10.1007/978-3-030-01932-7_19

not properly assessed. This demands the sustainable wind resistant design of structures. Wind loading codes provide the guidelines for structural engineers to design isolated structures resistant to wind. These codes generally discuss the wind loads in along wind direction but the available recommendations for across wind loads (acting perpendicular to the direction of wind) are very limited in the codes.

The unstable nature of wind which causes turbulence in wind environment can be categorized into three cases: 1. Turbulence in approaching wind flow which is called as natural turbulence or gustiness of wind. 2. Turbulence in surrounding wind flow generated by the structure itself due to vortex shedding (flow separation) phenomenon. 3. Turbulence generated by the oscillations of the highly flexible structure (aero elastic model) which is known as aerodynamic damping. Net pressure fluctuations acting along the approaching wind flow direction produces along wind load while across wind load is produced by the pressure fluctuations caused by alternate shedding of vortices due to incident flow separation from the edges of the structure (vortex shedding phenomenon).The frequency of vortex shedding f_v is defined by Eq. 1. Non-uniform pressure distribution around the structure which causes twisting motion of structure generates torsion load. (Ref: Fig. 1)

$$f_v = \frac{S_t U}{D} \tag{1}$$

Where St = Strouhal Number, U = Approach Flow Velocity and D = Characteristic Length of body projected on a plane normal to mean flow velocity.

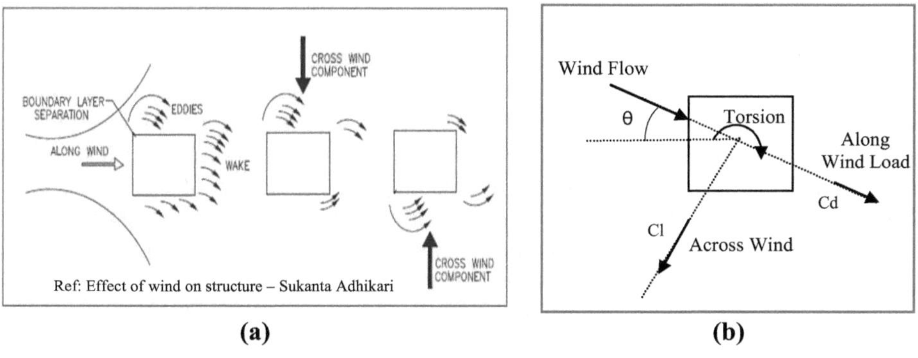

Fig. 1. Turbulence nature of wind: (a) vortex Shedding Phenomenon generating across wind load, (b) wind load directions

Across wind loading on structures can be analyzed by various methods like theoretical, numerical and experimental methods along with CFD approach. The present paper discuses on across wind load analysis of buildings and cooling towers through computational approach (CFD).

1.1 Past Research on Across Wind Effects

Majority of International codes and standards have adopted the concept of Gust Load factor (GLF) to account for the effect of gustiness of the wind on structures. The basic principle of this concept is that, the equivalent-static wind load (ESWL) used for design is equal to the mean wind force multiplied by the GLF. The 3D GLF concept is new framework developed by Kareem and Zhou (2003) for the formulation of wind load effects in the across wind and torsional directions. This is an advanced concept, which draws upon a database of aerodynamic wind loads on typical tall buildings, a mode shape correction procedure and a more realistic formulation of the ESWL and their effects. The researchers (Kareem and Zhou 2003; Zhou et al. 1999, 2002) have explained in detail the status of GLF and its advancements for along and across wind responses with numerical examples and empirical formulae. The three mechanisms of across-wind loading and their higher time derivatives were clearly described in Mendis et al. (2007). Patidar et al. (2014) reviewed and discussed on across wind analysis of slender structures such as chimney. This review states that, across wind analysis is required only if the critical wind speeds for any mode of oscillation is less than the mean design wind speed. Revuz (2011) worked on numerical simulation of the wind flow around a tall building and its resulting dynamic response. It is stated that commercial CFD can be suitably adapted to model the wind flow around high rise buildings as well as to obtain the dynamic response in across wind direction. The approach of using empirical expressions for determining across wind responses is discussed by Cenek and Wood (1990), Kijewski and Kareem (2001), Gu and Quan (2004) and Gu (2009). Tanaka et al. (2012) performed series of wind tunnel experiments to determine aerodynamic forces and wind pressures acting on square-plan tall building models with various configurations. Aerodynamic characteristics of various building models are studied and discussed through the experimental results. Vyavahare et al. (2012) developed a neural network model to predict the across wind response for a building with a given height to breadth to depth (h:b:d) ratio by computing the force spectrum coefficients. The same method was applied to determine the across wind response of tall buildings for all other h:b:d ratios overcoming the limitations of the draft code IS 875 Part-3(2011) All International wind loading codes and standards utilize the traditional displacement-based gust loading factor for assessing the dynamic along-wind loads and their effects on tall structures but incorporate different provisions for across wind loads (Kareem 2009). As per new code of IS 875: Part-III (2011), across-wind response has zero mean value and involves a different mechanism of excitation (vortex shedding). It is more structure specific, involving calculation models that are based on spectra generated from wind-tunnel studies. Computations of forces (across wind) due to incident turbulence and across wind displacement are beyond the scope of the code. Ranka and Shingade (2016) carried out wind load analysis on four different plan shapes (square, rectangular, T-shape and L-shape) of building with same plan area and height using CFD in ANSYS FLUENT 16.0 and compared with analytical method of IS 875 (part 3) – 1987. Yan et al. (2017) performed CFD analysis with FEM (Finite Element Method) to evaluate wind loads and wind-induced vibration response of typical high-rise building in urban area.

Across wind load response of regular plan shaped structures formulated in the latest draft Indian Wind code (2011) is for limited H:B:D and isolated structures only. Hence, two square plan shaped buildings (1:1:4) and two cooling towers are considered separately in the present research for across wind load analysis. The analysis has been carried out to investigate the across wind responses of these structures for standalone and interfering cases using CFD.

2 Computational Model Description

2.1 Model Geometry

Two regular square plan shaped buildings with B = D = 0.07 m and H = 0.28 m (1:400 scaled model) are arranged inside a fluid domain of dimensions $X_D \times Y_D \times Z_D = 2.52\ \text{m} \times 1.8\ \text{m} \times 2.2\ \text{m}$ (Fig. 2) to model the turbulence nature of wind flow around the structure. All the above geometrical parameters including position of structure inside the fluid domain and wind flow parameters like velocity, turbulence intensity etc., have been taken from the experimental data of Kim et al. (2015). In the first step, two rectangular buildings in tandem directions are considered and wind flow around the buildings is simulated by performing Large Eddy Simulation (LES). The pressure coefficients on principal building are evaluated and validated with the wind tunnel experimental results of Kim et al. (2015). In the next step, the same scheme is applied for the simulation of wind environment around two cooling towers arranged in tandem direction. The geometry of cooling tower is adopted from the wind tunnel test data of Goudarzi and Sabbagh-Yazdi (2008). The spacing between two structures is kept as 1.5B for both buildings and cooling towers.

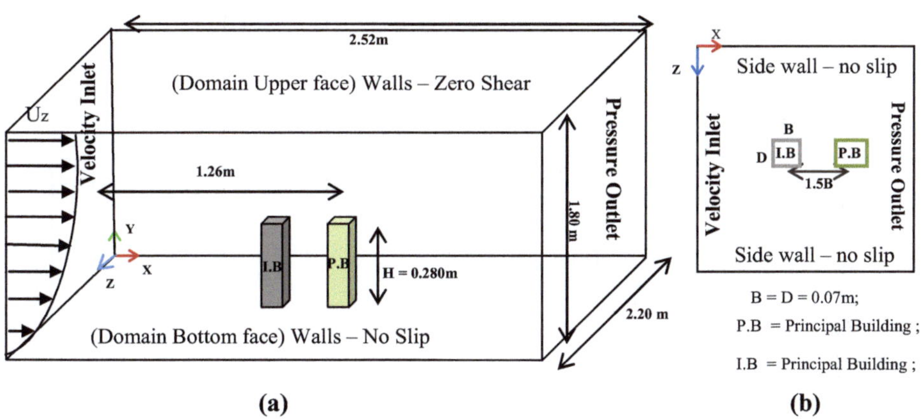

Fig. 2. Typical geometry of computational model with boundary conditions: (a) isometric view (b) top view

2.2 Boundary Conditions

Velocity inlet is imposed as boundary condition at the inlet of the fluid domain. Turbulent atmospheric boundary layer is simulated by adopting standard power law wind profile Eq. 2 at the velocity inlet. The imposed velocity profile shown in Figs. 1 and 2 is obtained using the Eq. 3 (derived from Eq. 2) in UDF. Incident turbulence intensity I profile calculated from Eq. 4 (Ref: AIJ code) is simulated with input parameters of k and ε at inlet boundary condition using Kinetic energy k Eq. 5 and energy dissipation rate ε Eq. 6 from Blockena et al. (2007). The adopted wind velocity profile and longitudinal turbulent intensity profile are based on the measured incident wind profiles of TPU aerodynamic data base and Kim et al. (2015). Pressure outlet boundary condition is imposed at domain outlet. Top surface as zero shear wall boundary condition and bottom surface of the domain as no slip wall boundary condition with spire roughness height as Ks = 0.00015 m (as per experimental data) have been adopted. All other walls of buildings and side walls of domain are specified as no slip wall boundary condition. All the above imposed boundary conditions are shown in Fig. 3.

$$U_Z = U_G\left(\frac{Z}{Z_G}\right)^{\alpha} \tag{2}$$

$$U_Z = 9.393\left(\frac{Z}{0.482}\right)^{\frac{1}{4}} \tag{3}$$

$$I = 0.1\left(\frac{Z}{Z_G}\right)^{-\alpha-0.05} \tag{4}$$

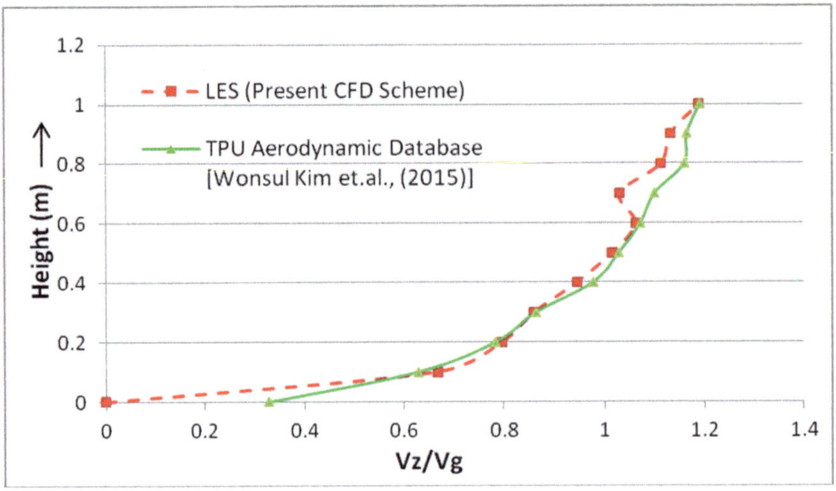

Fig. 3. Simulated wind velocity profile at the inlet

Where, $C_\mu = 0.09$; K = 0.41; $y_o = 7.6585 \times 10^{-6}$ m from Eq. 7;

$$k = \frac{3}{2}\left(I.U_{avg}\right)^2 \tag{5}$$

$$\varepsilon = \frac{\left(U^*_{ABL}\right)^3}{K(y + y_o)} \tag{6}$$

$$U^* = C_\mu^{1/4} k_p^{1/2} \tag{7}$$

$$k_s = \left(\frac{9.793}{C_{ks}}\right) y_o \tag{8}$$

2.3 Computational Grid Details

The fluid volume in the domain is divided into finite number of volumes and each fluid volume is modeled using finite volume method by creating a hexagonal structured grid as shown in Fig. 4. The minimum grid size is 2.2×10^{-2} m with 51,72656 cells in the domain. For simulation of atmospheric boundary layer, the mesh at the bottom portion of the geometric model is more refined up to 7 layers (slightly above building Height) with growth rate kept as 1.2.

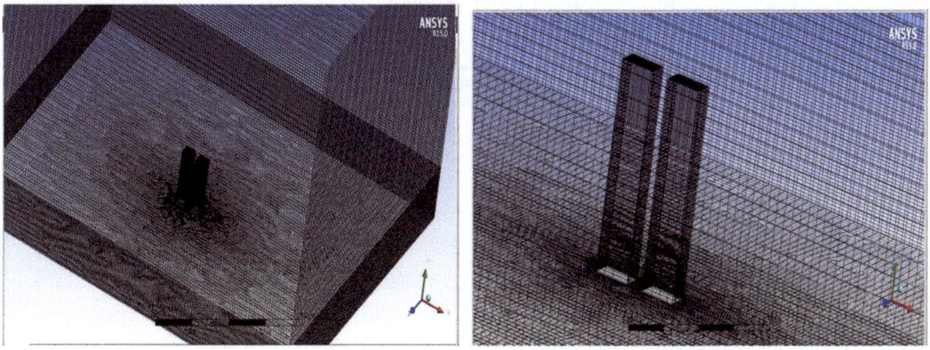

(a) (b)

Fig. 4. Computational grid; (a) domain-entire grid distribution; (b) refined grid at the buildings

2.4 Numerical Scheme for Computing the Wind Flow

The Atmospheric boundary layer with wind speed profile and turbulence characteristics were simulated using LES in ANSYS Fluent for urban terrain category with an exposure factor of $\alpha = 1/4$ (as per AIJ codes). The mean wind velocity of 8.2 m/s and turbulent Intensity I of 20% at the building height was considered in the simulation.

The fluid volume inside the domain is modeled as per above mentioned flow parameters using LES Smagorinsky-Lilly model with SGS (Sub grid scale) to model eddies smaller than the minimum grid size while simulating the larger eddies. In this model, governing equation consists of filtered forms of continuity equation and Reynolds Averaged Navier Stokes (RANS) equation, which are solved using the applied boundary conditions. Simulation is initialized at the inlet. SIMPLE Pressure velocity coupling algorithm is used as solver. Steady state simulations were performed initially using realizable k-ε model with standard wall functions to obtain the converged solution. Further, after obtaining the converged solution, transient simulations were turned on with 0.002 s as time step size and 150 number of time steps. All the simulations are performed with 0.0001 as convergence criteria. Second order upwind scheme with bounded second order implicit transient formulation is opted in this scheme.

3 Results and Discussions

3.1 Validation of Numerical Scheme

The two building computational scheme used in the present study is validated with the wind tunnel experimental results (Kim et al. 2015). The computed mean pressure coefficients along the left corner, centre and right corner of each face of principal building for two building model are quantitavely compared with the experimental results as shown in Figs. 5, 6 and 7, respectively. The results computed were found to be in fair agreement with the experimental values.

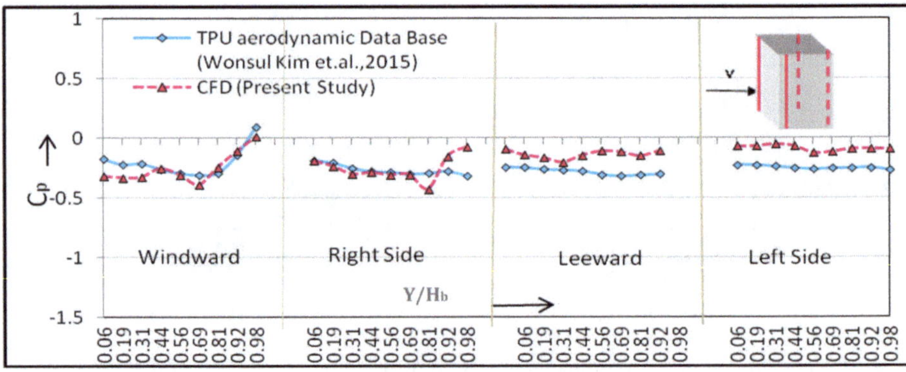

Fig. 5. Mean pressure coefficients on left corner of each face of principal building

3.2 Across Wind Load Analysis

Wind loads acting on structures can be computed either by pressure coefficient method or force coefficient method. In the present work, across wind load analysis has been performed on buildings and cooling towers by investigating the wind force

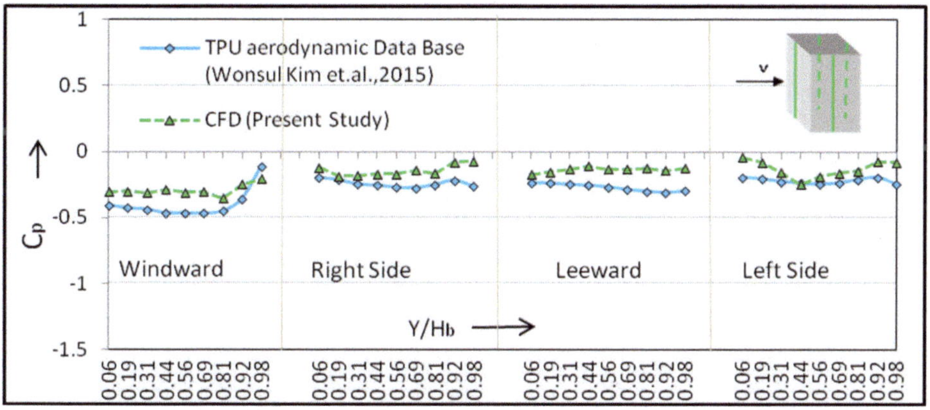

Fig. 6. Mean pressure coefficients on centre line of each face of principal building

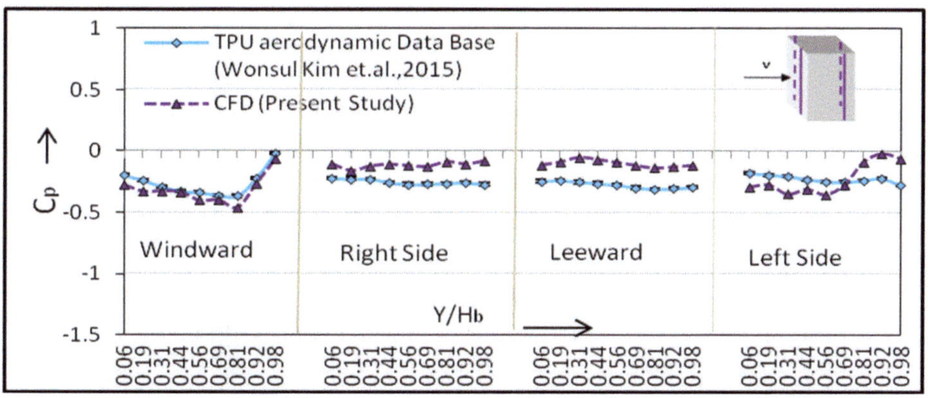

Fig. 7. Mean pressure coefficients on right corner of each face of principal building

characteristics like lift force coefficient and power spectral density for different wind angles. And the influence of these parameters on across wind loading has been studied. Further, the effect of adjacent structure on across wind load is also discussed in the present work through mean pressure coefficients.

3.2.1 Power Spectral Density Curves for Buildings and Cooling Towers

Power spectral density curves of lift force coefficients C_L obtained for buildings and cooling towers from the CFD analysis are plotted for different wind angles as shown in the Figs. 8, 9, 10 and 11.

From the Figs. 8 and 9 it is observed that the standalone case of building is exhibiting higher peaks of spectrum curve than two buildings tandem case and the Strouhal number is in the range of 1×10^{-4} to 1×10^{-3}. 60° wind angle is giving higher peak values compared to other wind angles for both standalone and two buildings cases while 90° wind angle case is showing different curve pattern from other

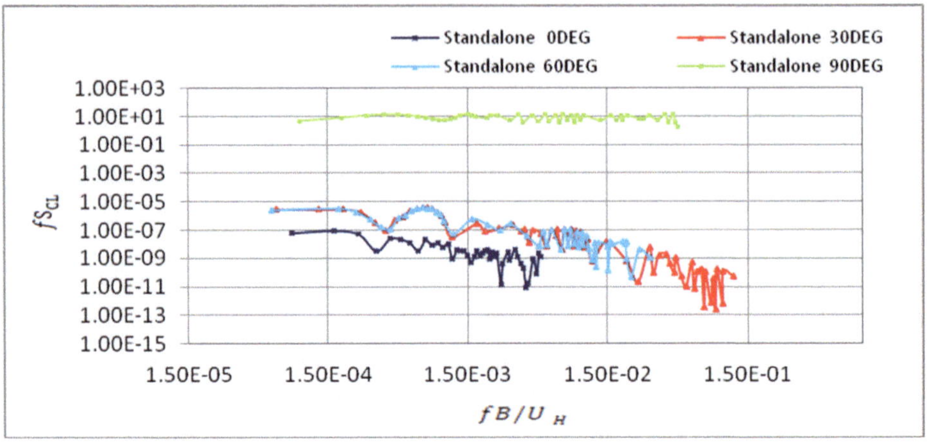

Fig. 8. Power spectral density of C_L for standalone building case

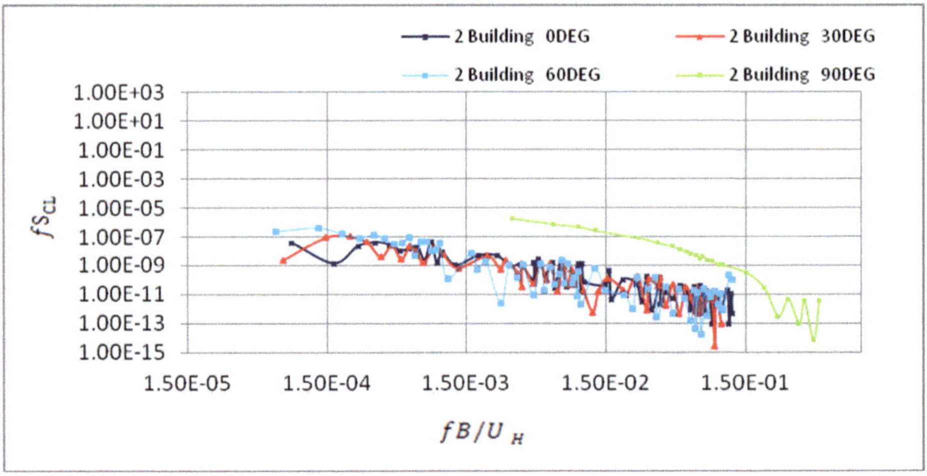

Fig. 9. Power spectral density of C_L for two building case

wind angles cases. Broader Peak curves of spectrum are observed when Strouhal number $<1.5 \times 10^{-3}$ for standalone building case and Strouhal number $<1 \times 10^{-2}$ for two building tandem case.

Comparatively, higher peaks of spectrum curve are observed for two cooling towers case than standalone case cooling tower and the Strouhal number is in the range of 1×10^{-4} to 1×10^{-2} (Ref: Figs. 10 and 11). 90° wind angle case is showing different pattern of spectrum curve from other wind angles cases.

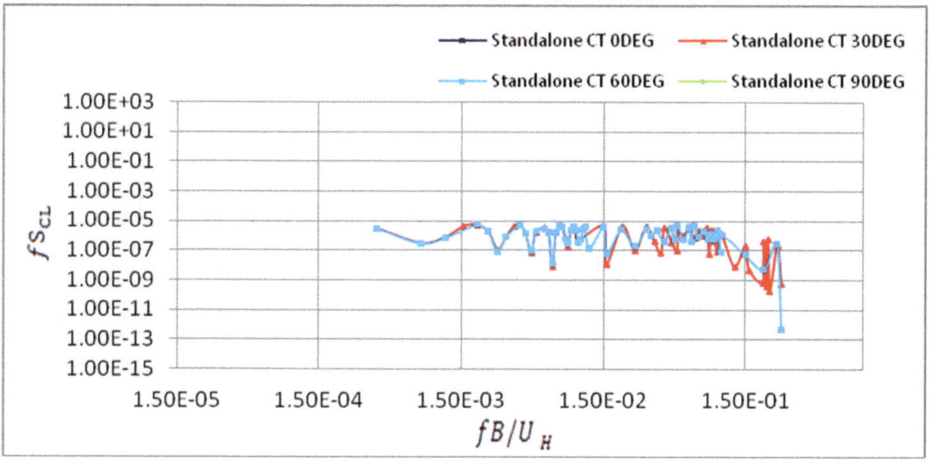

Fig. 10. Power spectral density of C_L for standalone cooling tower case

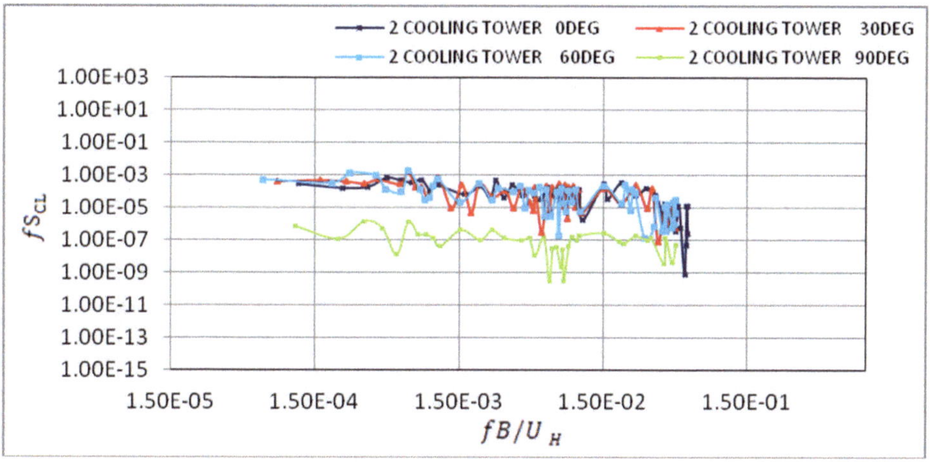

Fig. 11. Power spectral density of C_L for two cooling tower case

3.2.2 Lift Force Coefficient Variation on Buildings and Cooling Towers

The results of Lift force coefficient C_L variation on buildings and cooling towers are discussed to study across wind load effects on these structures.

From the Figs. 12 and 13, lift force coefficient C_L variation is very less between two buildings and its isolated case and its value is decreasing with increase in wind angles. C_L values are negative for all wind angles except for 0° wind angle. Lower C_L values are observed on cooling towers when compared with buildings for all wind angles. At 0° wind angle, the C_L is almost same for both buildings and cooling towers.

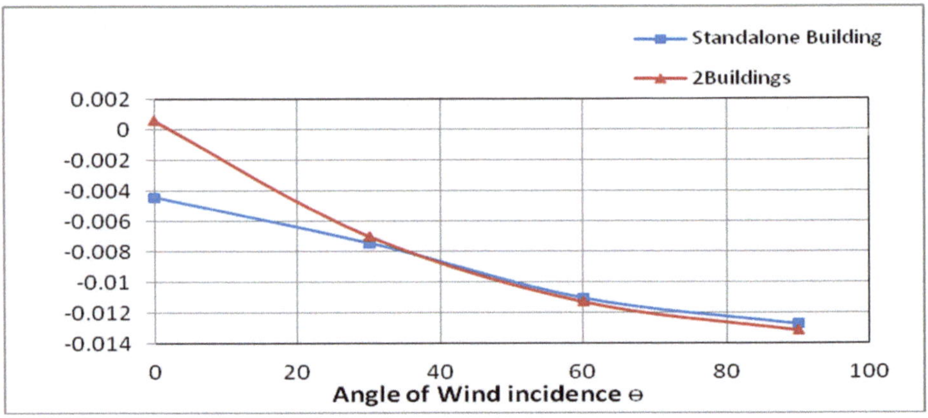

Fig. 12. Lift force coefficient variation C_L for buildings

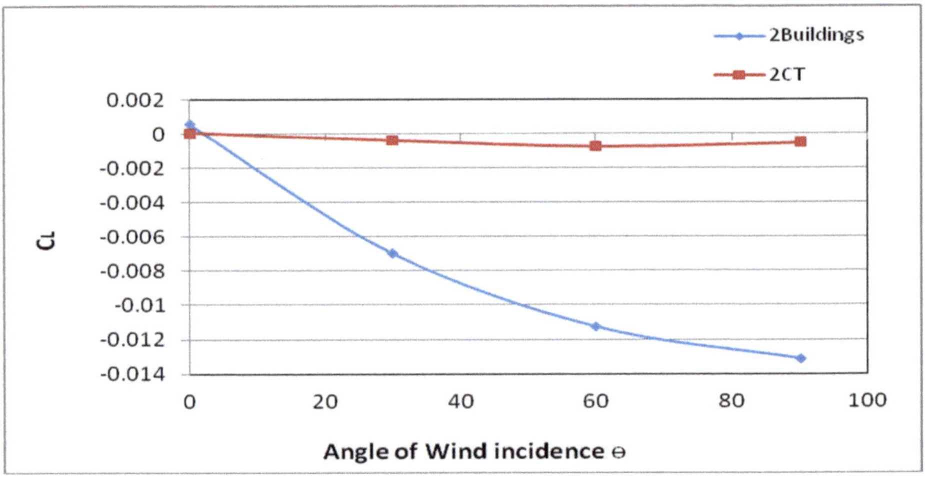

Fig. 13. Lift force coefficient C_L comparison for buildings and cooling towers

3.2.3 Mean Pressure Coefficients for Buildings and Cooling Towers

The results of mean pressure coefficients on each face of principal building for different wind angles are shown in Fig. 14 for two building case and in Fig. 15 for standalone case.

From the Figs. 14 and 15, the mean pressure coefficient is negative on side and leeward faces of principal building, as the lift force is acting more on these side faces and the same effect is been carried to leeward face creating a wake region on leeward side. But in presence of adjacent building, the wake region is created on windward face of principal building too and hence negative pressure coefficients are observed on windward face for two buildings case. As the wind angle is approaching towards 90°, the windward face and right side face of principal building exhibits positive pressure coefficient values.

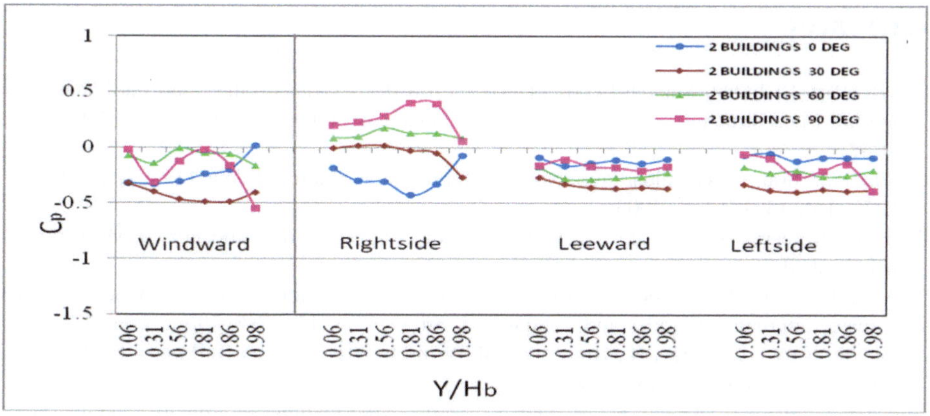

Fig. 14. Mean pressure coefficient variation for different wind angles on two building case

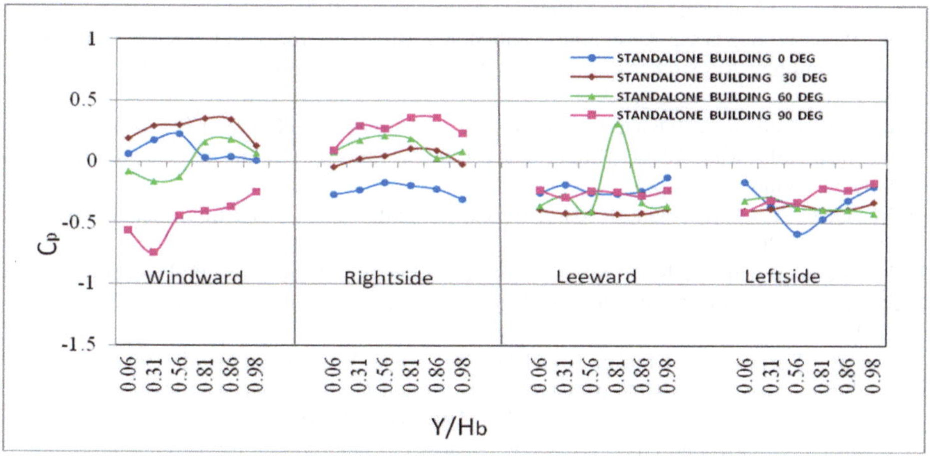

Fig. 15. Mean pressure coefficient variation for different wind angles on standalone case

4 Conclusions

1. The power spectral density of lift coefficient is decreasing with increase of shedding frequency for standalone and two buildings cases.
2. Less variation is observed in power spectrum density of lift force with sharp peaks for standalone and two cooling towers cases as shedding of vortices are more periodic.
3. High spectrum energy is dominant in cooling towers than buildings. Across wind effects are more for isolated rectangular buildings than isolated cooling towers for 0–90° wind angles.
4. Across wind effects are more for two cooling towers than two tandem buildings for wind incidence angles 0–90°.

References

IS: 875 (Part 3): Indian Standard code of practice for design loads (other than Earthquake) for buildings and structures, Part 3 wind loads, Bureau of Indian Standards, New Delhi, India (1987)

Kareem, A., Zhou, Y.: Gust loading factor—past, present and future. J. Wind Eng. Ind. Aerodyn. **91**(12–15), 1301–1328 (2003)

Zhou, Y., Kareem, A., Gu, M.: Gust loading factors for design applications (1999). https://nathaz.nd.edu

Zhou, Y., Kijewski, T., Kareem, A.: Along-wind load effects on tall buildings: comparative study of major international codes and standards. J. Struct. Eng. **128**, 788–796 (2002)

Mendis, P., Ngo, T., Haritos, N., Hira, A., et al.: Wind Loading on Tall Buildings, EJSE Special Issue: Loading on Structures (2007)

Patidar, B., Patil, A.R., Thiele, K., Mandal, S.: Across-wind loading for structures: an overview. In: Proceedings of 7th National Conference on Wind Engineering (NCWE 2014), Thapar University Patiala, 21–22 November 2014

Revuz, J.: Numerical simulation of the wind flow around a tall building and its dynamic response to wind excitation [PhD Thesis], Univ. of Nottingham (2011)

Cenek, P.D., Wood, G.H.: Designing Multi-storey Buildings for Wind Effects, Branz study report: Building Research Association of New Zealand, Report No. 25 (1990). ISSN 0113 3675

Kijewski, T., Kareem, A.: Dynamic Wind Effects: A Comparative Study of Provisions in Codes and Standards with Wind Tunnel Data (2001). https://www3.nd.edu/~nathaz/journals/Dynamic_Wind_Effects.pdf

Gu, M., Quan, Y.: Across-wind loads of typical tall buildings. J. Wind Eng. Ind. Aerodyn. **92**, 1147–1165 (2004)

Gu, M.: Study on wind loads and responses of tall buildings and structures. In: Proceedings of the Seventh Asia-Pacific Conference on Wind Engineering, Taipei, Taiwan, 8–12 November 2009

Tanaka, H., Tamura, Y., Ohtake, K., Nakai, M., Kim, Y.C.: Experimental investigation of aerodynamic forces and wind pressures acting on tall buildings with various unconventional configurations. J. Wind Eng. Ind. Aerodyn. **107–108**, 179–191 (2012)

Vyavahare, A.Y., Godbole, P.N., Nikose, T.: Analysis of tall building for across wind response. Int. J. Civ. Struct. Eng. **2**(3), 979–986 (2012)

IS: 875 (Part 3): Wind Loads on Buildings and Structures—Proposed draft & Commentary, Document No.: IITK – GSDMA-Wind 02-V 5.0 (2011)

Kareem, A.: United States Report for APEC-WW (2009). https://nathaz.nd.edu

Blockena, B., Carmelieta, J., Stathopoulos, T.: CFD evaluation of wind speed conditions in passages between parallel buildings- effect of wall-function roughness modifications for the atmospheric boundary layer flow (2007)

Kim, W., Tamura, Y., Yoshida, A.: Interference effects on aerodynamic wind forces between two buildings. J. Wind Eng. Ind. Aerodyn. **147**, 186–201 (2015)

Ranka, S., Shingade, V.S.: Analytical and numerical analysis of wind load on tall buildings. Int. J. Sci. Eng. Technol. Res. (IJSETR), **5**(11), November 2016. ISSN: 2278-7798

Yan, Q., Li, D., Liu, K., Bowen, Y.: CFD/FEM based analysis framework for wind effects on tall buildings in urban areas, Second International Conference on Industrial Aerodynamics (ICIA 2017) (2017). ISBN: 978-1-60595-481-3

Goudarzi, M.-A., Sabbagh-Yazdi, S.-R.: Modeling wind ribs effects for numerical simulation external pressure load on a cooling tower of KAZERUN power plant-IRAN. Wind Struct. **11** (6), 479–496 (2008)

Utilization of Ground Penetrating Radar (GPR) in the Non-destructive Assessment of FRP Laminate-Concrete Bond Strength

Mina Riad(✉), Nur Yazdani, Yazan Almomani, and Enrique Gonzalez

Civil Engineering, University of Texas at Arlington, Arlington, TX 76013, USA
mina.riad@mavs.uta.edu, Yazdani@uta.edu

Abstract. Repair and rehabilitation methods for civil infrastructure have become a topic of great interest to engineers, and fiber reinforced polymer (FRP) laminate is one of the most popular and practical solutions for strengthening and retrofitting of concrete structures. Several past studies indicated that significant increase in strength and stiffness can be achieved by using this technology. The quality of the FRP-concrete bond is critical in transferring stresses through the interface, and the in-situ evaluation of the bond strength is still a challenging issue.

To quantitatively evaluate the bond strength, non-destructive technique (NDT) using ground penetrating radar (GPR) was utilized on 32 laboratory concrete beams strengthened with carbon FRP (CFRP) laminates. Various parameters that may affect the bond strength, such as surface roughness, voids, epoxy type and thickness, and FRP type, were considered. The experiment was conducted with the objective of non-destructively detecting the parameters. The associated bond capacity of each sample was then found through a three-point bending test and also correlated through finite element modeling. Quantitative relationships between each of the parameters in the study and the associated bond strengths were then developed. The results of this study will be very useful in estimating the in-service bond conditions of applied FRP laminate, thereby estimating the expected strength contribution of the laminate in the overall flexural strength of structural members.

1 Introduction

External strengthening with fiber reinforced polymer (FRP) laminate composites have emerged as a popular alternative for deteriorating and/or under-designed concrete infrastructure (Meier 1995). The advantages are high strength to weight ratio (Bakis et al. 2002), fast and easy application, confinement of concrete elements and corrosion resistance. The application of the external very thin laminate has negligible effect on the dimensions of the member under repair. Therefore, the technique is widely used for structures with usage change, construction defects, code changes, seismic/blast retrofit, and deterioration.

H. Rodrigues and A. Elnashai (Eds.): GeoMEast 2018, SUCI, pp. 244–267, 2019.
https://doi.org/10.1007/978-3-030-01932-7_20

Although externally bonded FRP performed well in practice, premature debonding failure was observed and identified in several past studies as one of the associated problems. For the repair to be effective, the bonded layer must work together with the original structure. However, under variable conditions, interfacial debonding of the bonded layer from the concrete surface may occur. Debonding may initiate at the end of the bonded layer, where a high strain gradient exists (Täljsten 1997). When the workmanship is inadequate, imperfect bonding may lead to the formation of debonded zones under the FRP layer. Debonding may also initiate at the bottom of a flexural crack or flexural shear crack, leading to the formation of a debonded region away from the edge of the FRP layer (Leung 2001). If the debonding is left unrepaired, its continued growth may lead to failure of the repair. The detection of the interfacial debonding in its early stage is, therefore, very important in monitoring the health of the strengthened structures. Visual inspection and acoustic sounding (hammer tapping) are commonly used to detect delaminations. However, these current practices are unable to assess the depth and width of debonded areas and the level of adhesion between the FRP and the concrete (Taillade et al. 2012).

Various past studies investigated the effectiveness of nondestructive techniques (NDT) to assess the quality of the FRP-concrete bond (Karbhari et al. 2005). Based on the impact-echo technology (Maerz and Galecki 2008), the ultrasound technology (Mirmiran and Wei 2001; Ribolla et al. 2016; Schabowicz 2014), the infrared detection technology (Galietti et al. 2007; Starnes et al. 2003), and the Microwave and radar technology (Akuthota et al. 2004; Yu and Büyüköztürk 2008). Most of these NDE methods involve techniques adequate for limited characterization (location and size) of defects like pores, delamination or debonding within the adhesive layer. There is so far, no NDE technique that is able to quantify the strength of the bond in structure applications (Ehrhart et al. 2010). This lack remains a major issue set against a wider application of the FRP adhesive bonding technology.

This paper aims to develop reliable and convenient procedures to be followed to quantitatively evaluate the FRP-concrete bond strength. The proposed concept consists of using the ground penetrating radar technique to obtain NDE test parameters (such as amplitude of the reflected wave) from test specimens and then to correlate the parameters with the actual bond strengths measured when these same specimens are destructively examined. Various parameters affecting the bond strength, such as surface roughness, voids, epoxy type and thickness, and FRP type, were considered.

2 Experimental Design and Specimens Preparation

The first step in the design is to determine the number of required specimens that incorporates all the parameters considered in this study. Since that the interaction between the parameters could not be ignored, the simple approach of changing only one parameter at a time is not suitable for this test. The design of the experiment is necessary for this test for a sound statistical approach.

Due to limitations on space, cost and time, the "Full Factorial" design with 128 specimens was not possible. Therefore, the “Fractional factorial resolution 4” method was selected (NIST 2017). The use of the Fractional Factorial method resolution 4 for 6 parameters each having two levels (High and low) reduces the required specimens to only 16 by confounding the effect of the higher-order interactions. For this test, replicability is important to reduce any noise and reduce the chance of false conclusions. Hence, two replicates of the same specimen were used, resulting in a total of 32 specimens.

Two FRP types with different thickness were selected. The adhesive layer is considered to have a considerable impact on the bond strength. Two Epoxies with low and high viscosity were selected. Also, an additional parameter investigated the possibility of applying “Sikadur 31 to the concrete surface in addition to the “Sikadur 300” or the “Sikadur 330” to form the second layer. This type of epoxy is used to seal the cracks at the concrete surface.

The friction between the concrete surface and the FRP layer is one of the forces contributing to the bond strength. The surface roughness of the concrete surface is considered to increase the friction and consequently increase the bond strength. Two types of surface roughness were considered in this research. The two types selected are classified based on the method used to achieve the surface roughness, the first type using sandblasting, in this method the surface profile of the beam is prepared to a minimum of concrete surface profile (CSP) 3 as defined by the ICRI-surface-profile chips. The second method using hand grinding, in this method the surface profile of the beam is prepared to a maximum of concrete surface profile (CSP) 2.

The last two parameters considered in this research, study quality of the FRP application in the field as of the presence of any bubbles or voids under the FRP surface and the effect of these imperfections on the total bond strength. The two parameters consider the presence of voids of different sizes. Table 1 summarize the properties of the parameters selected in this study.

After assigning each of the before mentioned parameters to its corresponding Fractional Factorial level, the final experimental matrix is presented in Table 2. The final step in the experimental design is randomization. The randomization is important to remove any potential biases or judgments.

The width of the beams was chosen as 20.3 cm. The depth and the length of the beams were selected following the ACI 440 (ACI committee 440 2017) guidelines to make sure that the failure occurs in the bond. The design was based on concrete material with a compressive strength (f’c) of 24.1 MPa. The schematic of the beam is illustrated in Fig. 1.

All specimens were cast horizontally, compacted, and trowel finished. Ready-mix concrete was used to make sure that all beams have the same properties. All the 32 beam specimens were cast using normal weight concrete with a water-cement ratio of 0.40 and a maximum aggregate size of 19 mm with a target 28-day compressive strength (f’c) of 24.1 MPa. Some predetermined voids were inserted at the surface of the concrete using foam cubes with the exact dimensions of the required voids. After curing process the foam pieces were easily removed. The specimens were cured for 28 days by spraying water. After curing the surface of each beam was prepared corresponding to the required type (CSP 3 or CSP 2).

Table 1. Parameters in study

	Type I	Type II
1st parameter "FRP type"	"Sika Wrap Hex 117C" Cured laminate Properties: • Tensile Strength: 724 MPa • Modulus of Elasticity: 56,500 MPa • Elongation at Break: 1.0% • Thickness: 0.51 mm	"Sika Wrap Hex 103C" Cured laminate Properties: • Tensile Strength: 1110 MPa • Modulus of Elasticity: 71,700 MPa • Elongation at Break: 1.45% • Thickness: 1 mm
2nd parameter "Number of Epoxy Layers"	"Sikadur 300 or 330 only"	"Sikadur 31" in addition to the "Sikadur 300 or 330" •Tensile Strength: 22.7 MPa • Elongation at Break: 0.9% • Flexural Strength: 42.0 MPa • Flexural Modulus: 11,52 Mpa
3rd parameter "Epoxy type"	"Sikadur 300" •Tensile Strength: 55 MPa • Elongation at Break: 3% • Flexural Strength: 79 MPa • Flexural Modulus: 3,450 MPa	"Sikadur 330" • Tensile Strength: 33.8 MPa • Elongation at Break: 1.2% • Flexural Strength: 60.6 MPa • Flexural Modulus: 3,489 MPa
4th parameter "Surface Roughness"	(CSP) 3	(CSP) 2
5th parameter "small voids"	–	20 mm × 20 mm
6th parameter "large voids"	–	40 mm × 40 mm

Table 2. Experimental matrix

DOE #	FRP type	Epoxy Nb. of layers	Epoxy type	Surface roughness	Voids 1 (mm)	Voids 2 (mm)	No.
1	Type I	Type I	Type I	Sandblasting	–	–	2
2	Type II	Type I	Type I	Sandblasting	2 × 20	–	2
3	Type I	Type II	Type I	Sandblasting	–	–	2
4	Type II	Type II	Type I	Sandblasting	–	2 × 40	2
5	Type I	Type I	Type II	Sandblasting	–	–	2
6	Type II	Type I	Type II	Sandblasting	–	2 × 40	2
7	Type I	Type II	Type II	Sandblasting	–	–	2
8	Type II	Type II	Type II	Sandblasting	2 × 20	–	2
9	Type I	Type I	Type I	Hand grinding	–	2 × 40	2
10	Type II	Type I	Type I	Hand grinding	–	–	2
11	Type I	Type II	Type I	Hand grinding	2 × 20	–	2
12	Type II	Type II	Type I	Hand grinding	–	–	2
13	Type I	Type I	Type II	Hand grinding	2 × 20	–	2
14	Type II	Type I	Type II	Hand grinding	–	–	2
15	Type I	Type II	Type II	Hand grinding	–	2 × 40	2
16	Type II	Type II	Type II	Hand grinding	–	–	2

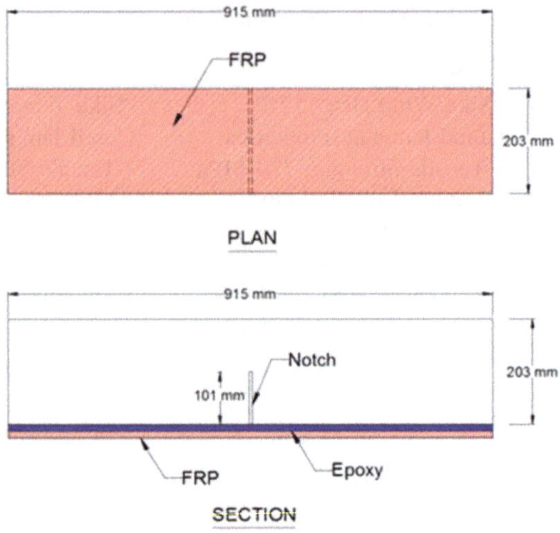

Fig. 1. Schematic of the beam specimen

The FRP was then applied at an ambient temperature of 15°C and in absence of direct light to prevent any undesirable problems with the epoxy. The temperature was cooling down to prevent the phenomena of outgassing of the concrete which might increase the voids under the FRP. Before applying the epoxy to the concrete surface, the surface was cleaned and dry. The final step in the FRP application was laying the FRP sheets on top of the concrete surface and making sure that the interface is free of any voids or imperfections. After attaching the FRP, the specimens were left for 3 days to cure. Figure 2 show the sample preparation.

Fig. 2. Samples preparation

3 Non-destructive Testing Procedures

In order to evaluate the quality of the bond, several unknown values that contributes to the overall strength are required in order to evaluate the bond (FRP type, number of epoxy layers used, Epoxy type, surface roughness and voids). As the FRP layer is exposed, the FRP type could be determined visually or by simple examination. Also, the presence of voids and delaminations could be determined using a visual examination of the GPR scans as presented below. However, the Epoxy types, number of epoxy layers used and surface roughness effects are combined together and cannot be easily separated. Therefore, three different equations are needed to determine the value of each of these parameters separately.

3.1 Ground Penetrating Radar (GPR) Test 1 and Test 2

The first two equations are prepared using the ground penetrating radar device. A GPR system acquired from Geophysical Survey System Inc. (GSSI) is used with a 2600 MHz GPR antenna. The 2600 MHz antenna is able to scan concrete up to a depth of 10 in with high resolution. Ground Penetrating Radar (GPR) uses high-frequency radio waves that propagate into the material under investigation. The transmitter sends a signal into the material. The transmitting signal gets reflected from the different surfaces and the objects throughout the depth and the reflected signal is received by the receiver antenna. The received data is then processed and displayed. The output of the GPR scan is presented in form of A-scan or B-scan. The GPR device provides information about the amplitude of the reflected wave through the depth of the specimen. The value of the amplitude is used in this test to compare the effect each parameter. Figure 3 presents the details of the GPR handle and a typical B scan.

Test 1, uses the data from a line scan performed directly on top of the FRP surface. The reflected wave considered in test 1 is within the direct coupling. While test2 rely on attaching an additional material with a known dielectric constant. The thickness of the additional material taken more than the thickness of the direct coupling zone.

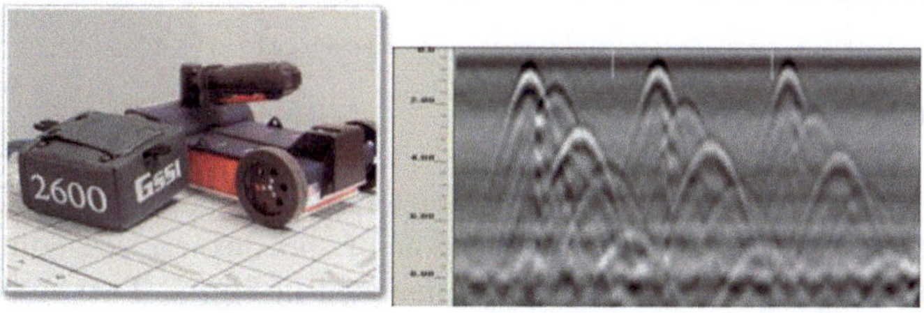

Fig. 3. GPR device

3.2 Schmidt Hammer Test

To develop the third equation the method of Schmidt hammer was used. Schmidt hammer defines the ratio between the rebound velocity and the impact velocity as the quotient value "Q". The Q value is used in this test to compare the surface hardness of each specimen. The surface hardness of each specimen to be correlated with the overall bond strength.

The first step was using the Schmidt hammer device to record the Q value at 18 preselected test points. In order to compare the surface hardness of the FRP surface with the surface hardness of the concrete surface of each beam as illustrated in Fig. 4, additional test points were used to record the Q value from the concrete surface of the beam.

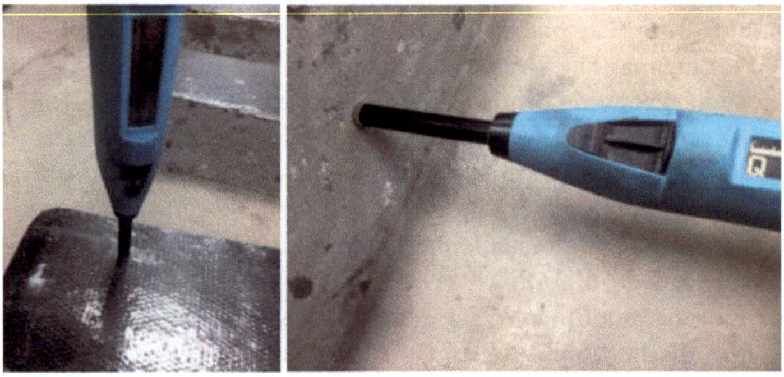

Fig. 4. Schmidt hammer testing at the FRP surface

3.3 Ground Penetrating Radar Test for Void Detection

In this test, a simple approach was undertaken to identify and locate any voids and delamination that might be hidden under the FRP surface. The method depends on visually analyzing the B-scan of each beam to identify the presence of hidden voids.

4 Destructive Testing

In order to evaluate the effect of the parameters on the bond strength, three points bending test was performed as an indication of the flexural strength of the beam strengthened with FRP. The test setup is illustrated in Fig. 5. Two strain gages were attached to each beam. The first strain gage is attached to the FRP layer at mid-span to capture strain corresponding to the maximum tensile stress at the bottom of the beam section. The second strain gage is attached to the top of the concrete beam to record the strain corresponding to the maximum compressive strain at the top of the beam. Two displacement transducers (LVDT) were used to record the displacement at the mid-span during the bending test. The LVDT were placed at both sides at the mid-span of the beam

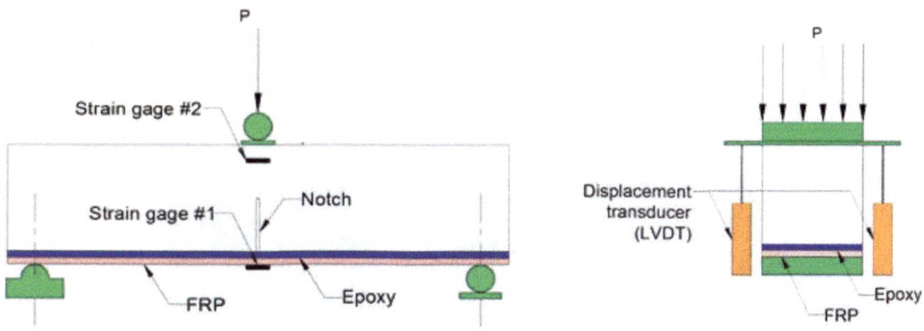

Fig. 5. Bending test setup

After preparing the beams for the test, a 1780 KN compression test machine was used. The beam was simply supported. A plate was placed on top of the beam at the mid-span to distribute the load uniformly at the total width of the beam. A loading rate of 0.2 N/min was selected.

5 Numerical Modelling

The results obtained from the experimental analysis are used as inputs to construct a series of calibrated finite element models using the finite element software package ABAQUS (ABAQUS Manual 2010). The finite element models are required to accurately reflect the effect of the parameters and their interactions on the bond strength with no aliases.

The first material considered in this FE model is the concrete. In this study, the concrete compressive and tensile properties were modeled as concrete damaged plasticity. ABAQUS uses the plasticity model proposed by Ref. (Lubliner et al. 1989). The concrete damage plasticity model requires the values of elastic modulus, Poisson's ratio, the plastic damage parameters and description of compressive and tensile behavior. The dilation angle and Poisson's ratio (ν) were chosen to be 35° and 0.15°, respectively.

For FRP, linear elastic behavior up to failure was assumed. Elastic modulus, Poisson's ratio, and tensile strength were needed for the simulation. The material type was assumed to be lamina with Poisson's ratio as 0.222.

A cohesive surface model was evaluated for describing the concrete-FRP interface. The cohesive model available in ABAQUS is a recommended choice for representing the interface behavior (Obaidat 2011). The definition of the model is characterized by the parameters, initial stiffness, shear strength, fracture energy and curve shape of the bond slip model. In order to find the values of initial stiffness, shear strength and fracture energy that gave the best fit, the equations from Ref. (Obaidat 2011) were used.

The concrete was modeled using C3D8R (eight-node solid) element. The FRP sheets were modeled using S4R (shell, 4-node) element. Preliminary results obtained with a rather coarse mesh showed that it was fairly difficult to obtain convergence and the results were not acceptable when compared to the bending test. The results obtained from a fine mesh were more accurate. An even finer mesh gave almost the same result as the previous mesh but more time was needed for computations. Therefore, a moderately fine mesh (25 mm × 25 mm) was chosen in this study (Fig. 6).

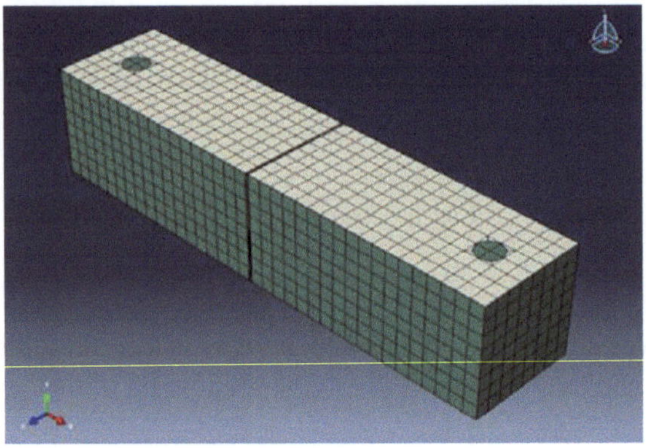

Fig. 6. Mesh used in ABAQUS model

Finally, the boundary conditions and loads were defined to complete the inputs of the ABAQUS model. The boundary conditions for the simulated beam were a hinge on one side and a roller for the other side to simulate the condition of the experimental test. The load was applied at mid-span corresponding to the experimental situation.

In order to determine the materials properties of the concrete used in the test, twelve concrete cylinders were taken as per ASTM C39 (ASTM C293 2016) during the concrete casting. After 28 days, six cylinders were tested according to the ASTM C39 to find the compressive strength of the concrete. Another six cylinders were tested simultaneously with the bending test to capture the material properties of the concrete used. The Concrete compressive strength (f'c) calculated based on these values is 22.75 MPa (E = 27,753 MPa).

A total of six FRP coupons were tested to determine the tensile properties of the cured FRP laminate according to the ASTM D3039 (ASTM D3039 2014). Figure 7 illustrates the coupon test performed to determine the material properties of the FRP. The results of this test were used as inputs of the material properties of the FRP in ABAQUS.

As the contact properties differ from one specimen to the other corresponding to the parameters incorporated in each specimen, it was necessary to determine the contribution of each parameter separately on the contact properties. Each parameter was assigned separate properties in the model inputs. The surface roughness effect was included in the friction coefficient properties of the contact. The Epoxy type and thickness effect were included in the damage properties of the contact. On the other hand, the voids effect was simulated by creating the voids in the model. Finally, the FRP type was included in the material properties of the FRP.

Fig. 7. ASTM FRP coupon test

6 Results and Discussions

6.1 Ground Penetrating Radar Results Test 1

A sample of the B-scan of the GPR test for two different beams is presented in Fig. 8. The results show that the parameters in the study seem to have an effect on the amplitude values from the GPR scan.

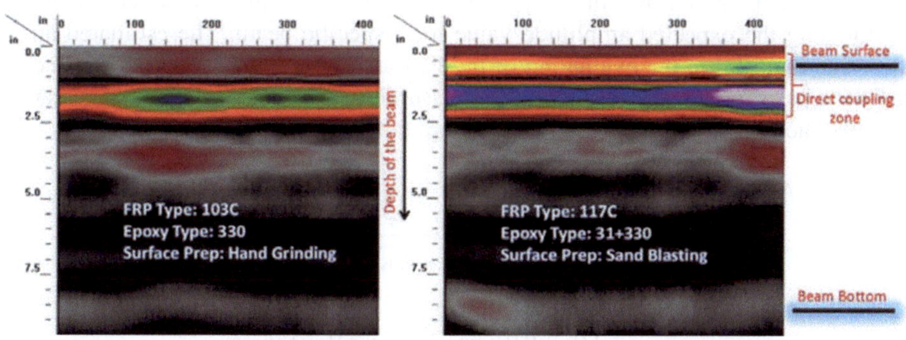

Fig. 8. A sample of the B-scan of two different beams

To further analyze the data and determine the effect of each parameter on the Amplitude, The value of the reflected amplitude from the surface of each beam was selected. The selected amplitude value for all the beams were analyzed using the Minitab software package for statistical analysis. Figure 9 examine the assumption that the residuals are approximately normally distributed, are independent and have equal variances. The four charts show a large spread of the data and that there is no clear pattern to the residuals. The residuals do appear to have approximately, a normal distribution. Minitab software summarizes the effect of each of the main parameters and their interactions on the bond strength as shown in Fig. 10.

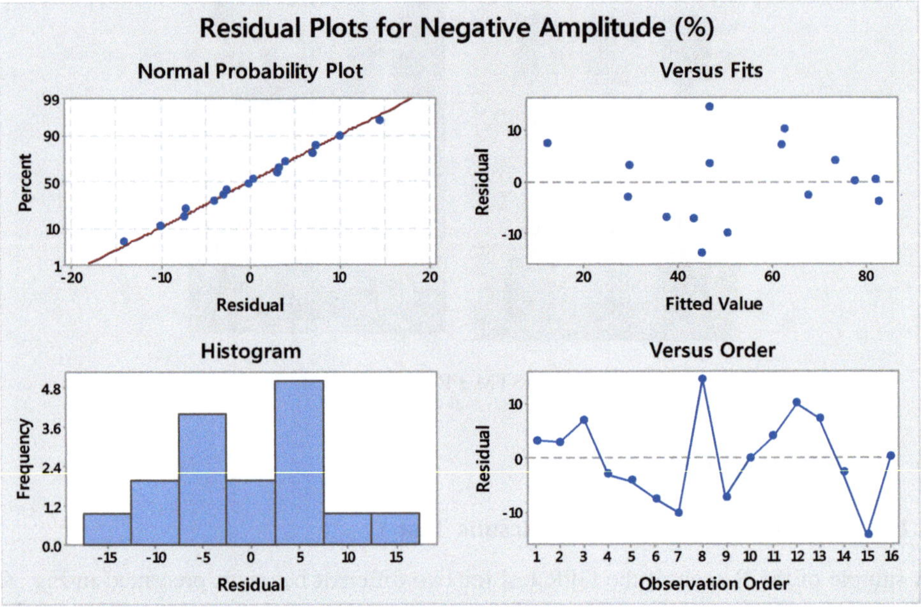

Fig. 9. Residual Plots for the amplitude (%)

The model prepared by Minitab present the regression equation that could be used in estimating the expected amplitude value for a specific beam. The estimated amplitude value is useful in the evaluation of the FRP application. The amplitude value can provide a base for quality inspection and assurance. The model has an R square of 87.64%. The equation for estimating the amplitude value is presented in Eq. 1.

$$\begin{aligned} \text{Negative Amplitude } (\%) = {} & 53.04 - 11.05[\text{FRP type}] + 2.91[\text{Nb of Epoxy Layers}] \\ & + 1.85[\text{Epoxy type}] - 2.28[\text{Surface Roughness}] \\ & - 12.03[\text{FRP type} * \text{Nb of Epoxy Layers}] \\ & - 6.54[\text{FRP type} * \text{Epoxy type}] \\ & - 2.98[\text{FRP type} * \text{Surface Roughness}] \\ & - 7.31[\text{Nb of Epoxy Layers} * \text{Epoxy type}] \\ & + 0.17[\text{Nb of Epoxy Layers} * \text{Surface Roughness}] \\ & - 3.44[\text{Epoxy type} * \text{Surface Roughness}] \end{aligned} \quad (1)$$

Where, FRP type is −1 for FRP type I and +1 for FRP type II. Similarly, for each parameter in the previous equation, a value of −1 is assigned to the type I and a value of +1 is assigned to type II.

6.2 Ground Penetrating Radar Test 2

Similarly, Minitab software presents the data for the GPR test 2. The effect of each of parameters and their interaction in a combined figure as illustrated in Fig. 11

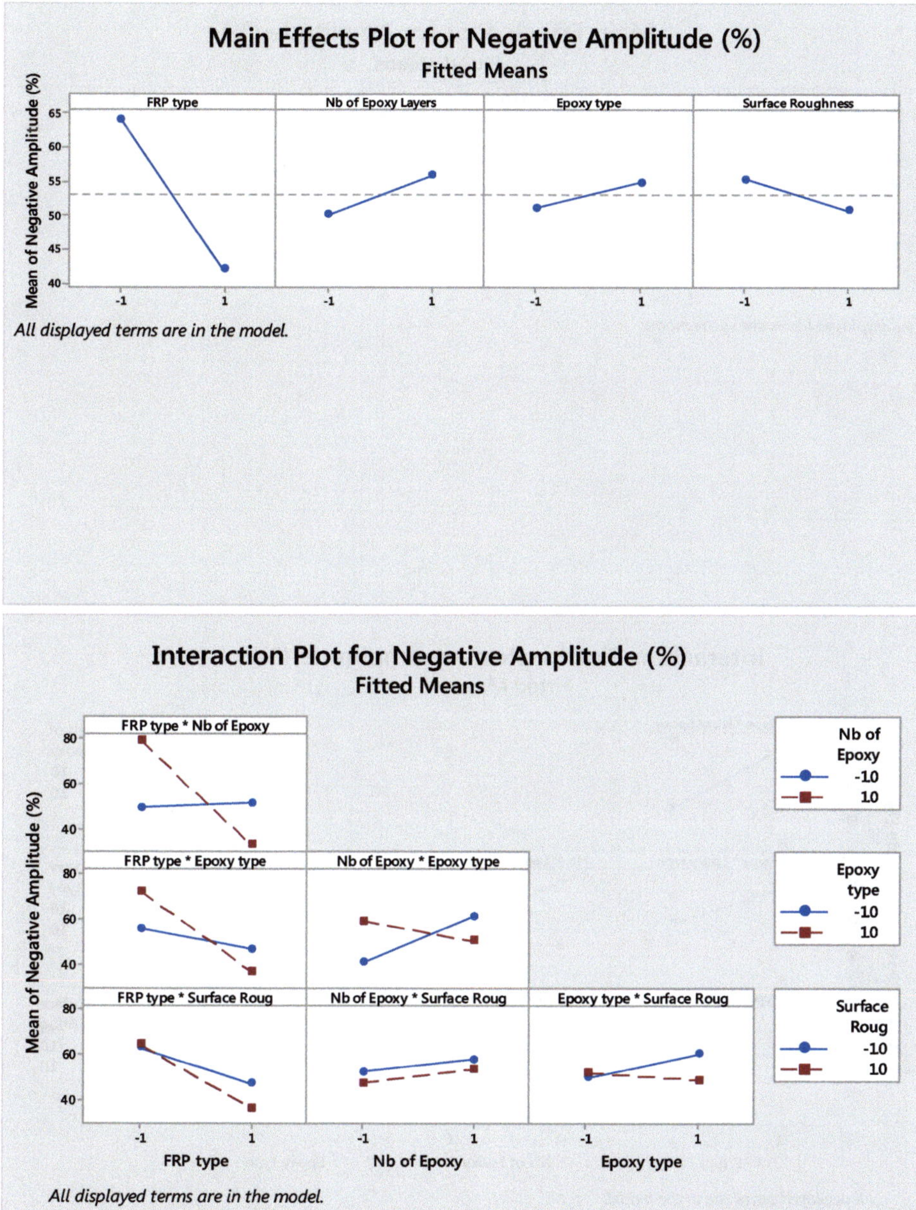

Fig. 10. Effect of the parameters on the bond strength using ultrasound test 1

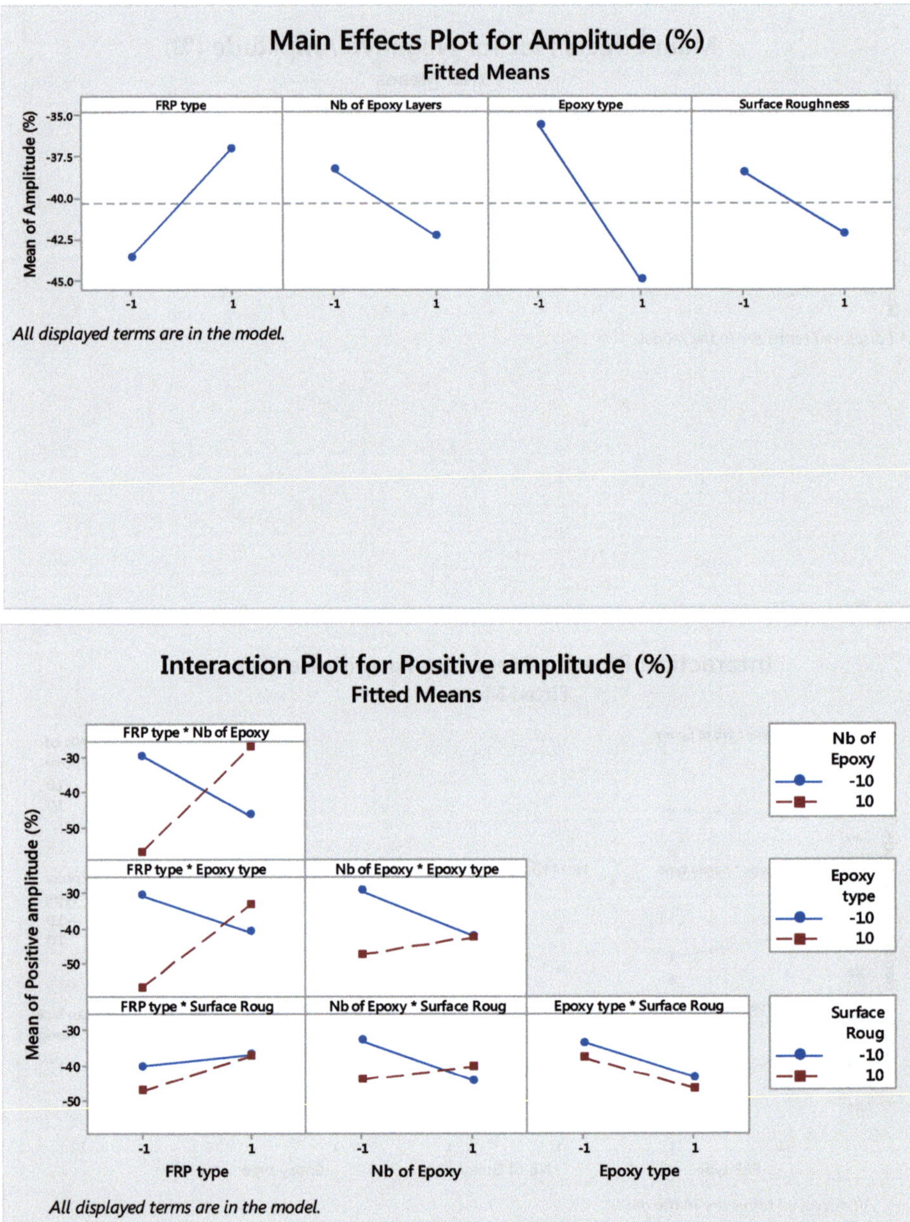

Fig. 11. Effect of the parameters on the bond strength using ultrasound test 2

Finally, the regression equation developed is presented. This additional equation can be used along with the first equation to evaluate the bond strength. The second equation for estimating the amplitude value is presented in Eq. 2.

$$
\begin{aligned}
\text{Negative Amplitude}\,(\%) = {} & -40.30 + 3.31[\text{FRP type}] - 2.00[\text{Number of Epoxy Layers}] \\
& - 4.67[\text{Epoxy type}] - 1.87[\text{Surface Roughness}] \\
& + 11.70[\text{FRP type} * \text{Nb of Epoxy Layers}] \\
& + 8.51[\text{FRP type} * \text{Epoxy type}] \\
& + 1.60[\text{FRP type} * \text{Surface Roughness}] \\
& + 4.37[\text{Number of Epoxy Layers} * \text{Epoxy type}] \\
& + 3.75[\text{Number of Epoxy Layers} * \text{Surface Roughness}] \\
& + 0.30[\text{Epoxy type} * \text{Surface Roughness}]
\end{aligned} \tag{2}
$$

6.3 Schmidt Hammer Results

Similar to the GPR test 1 and 2, Minitab software was used. The inputs were selected to be the percentage change of Q value of the FRP surface relative to the Q value of the concrete surface. Following are some of the results of the design of experiment using Minitab software as shown in Figs. 12 and 13.

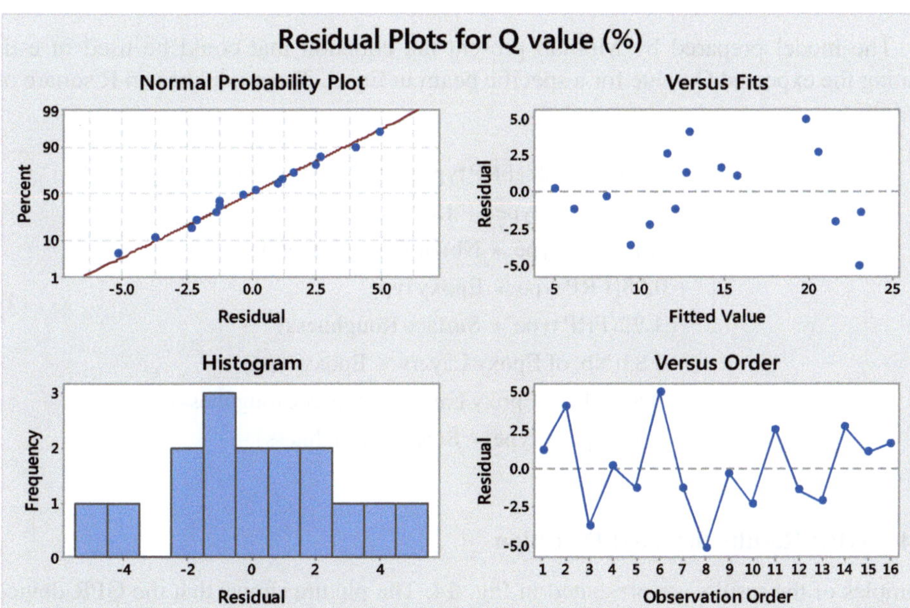

Fig. 12. Residual Plots for Q value (%)

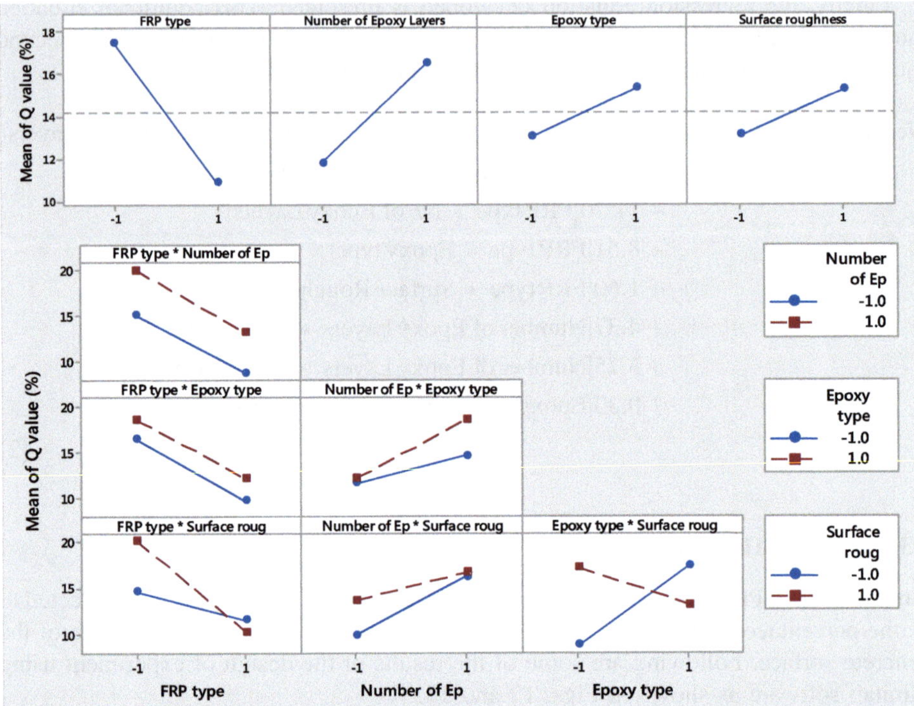

Fig. 13. Effect of the parameters

The model prepared by Minitab present the equation that could be used in estimating the expected Q value for a specific beam in Eq. 3. The model has an R square of 82.09%

$$\begin{aligned} \text{Q value } (\%) = {} & 14.16 - 3.28(\text{FRP type}) + 2.34(\text{Nb. of Epoxy Layers}) \\ & + 1.14(\text{Epoxy type}) + 1.07(\text{Surface Roughness}) \\ & - 0.11(\text{FRP type} * \text{Nb. of Epoxy Layers}) \\ & + 0.08(\text{FRP type} * \text{Epoxy type}) \\ & - 1.72(\text{FRP type} * \text{Surface Roughness}) \\ & + 0.83(\text{Nb. of Epoxy Layers} * \text{Epoxy type}) \\ & - 0.84(\text{Nb. of Epoxy Layers} * \text{Surface roughness}) \\ & - 3.16(\text{Epoxy type} * \text{Surface Roughness}) \end{aligned} \tag{3}$$

6.4 GPR Results for Void Detection

Samples of the results are presented in Fig. 14. The picture shows that the GPR device could be used for the detection of voids. The results show that if the FRP is not bonded to the concrete surface, the voids appear in the B-scan in the form of a hyperbola.

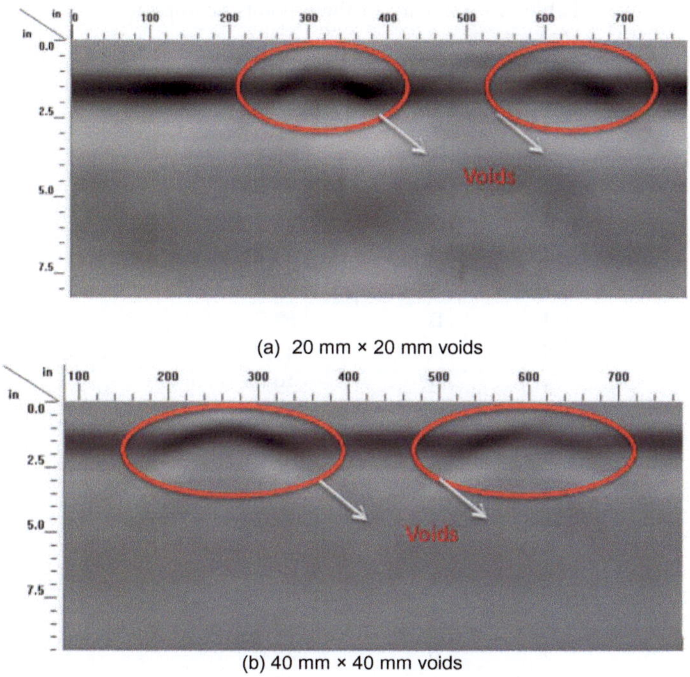

Fig. 14. GPR B-scan beams with voids

6.5 Destructive Testing Results

With the application of loads, the beam started to deflect to a point where flexure cracks started to appear at mid-span. These flexure cracks lead to debonding between the FRP and the concrete. The debonding progressed from the mid-span toward the edge of the beam with the increase of loading until failure. The failure happened when the FRP layer was completely debonded from the FRP. Table 3 presents the failure load from each beam. Figure 15 shows a selected sample from the three points bending.

The results from the bending test presented in the above section were used for calibration purposes of the finite element model. The Finite element model presented in the following section uses the results as input properties of the contact between the FRP and concrete.

6.6 Numerical Models Results

A series of iterations were performed to calibrate the models. For each iteration, the parameter values were selected to achieve a similar failure load in the model as per the corresponding experimental test. The load-displacement curve was plotted for each beam. Figure 16 illustrates the load-displacement curve of a beam after calibration the graph show an agreement between the performance of the experimental results and the ABAQUS model.

Table 3. Summary of three points bending test.

Beam type	Replicate no.	Failure load (KN)
1	A	5.3
1	B	34.7
2	A	56.1
2	B	50.3
3	A	50.3
3	B	43.3
4	A	66.8
4	B	67.7
5	A	42.7
5	B	39.9
6	A	52.6
6	B	61
7	A	52.1
7	B	47.2
8	A	24.6
8	B	55.7
9	A	23.7
9	B	29.6
10	A	45
10	B	39.7
11	A	25.9
11	B	39.3
12	A	55.7
12	B	63.7
13	A	27.2
13	B	46.8
14	A	5.3
14	B	51.7
15	A	55.7
15	B	46.3
16	A	44.5
16	B	61

After several iterations, the following values were selected to represent the bond properties. The surface roughness was assigned a value of 0.1 and 0.3 in the friction coefficient of the contact properties were assigned to simulate the hand-grinding and the sandblasting respectively. The number of layers of Epoxy used was assigned a value of 0.29–0.42 and 0.43–0.5 in the shear damage parameter of the contact corresponding to type I and type II respectively. The Epoxy type was assigned a value of

Fig. 15. Three points bending test failure

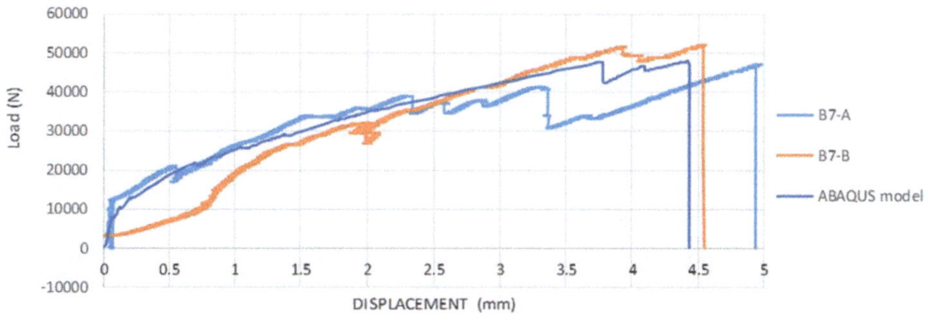

Fig. 16. Load-displacement curve of a beam

1.7–3.5 and 3–4 in the fracture energy of the damage parameter of the contact corresponding to type I and type II respectively. Table 4 presents the final results of the calibration process.

After completing the calibration process, a value was assigned to each parameter in the study and the full factorial method was replicated using the FEM models. Figure 17 illustrates the effect of the main parameters and their interactions on the bond strength.

Table 4. Results of the calibration process.

Beam no.	Failure Load experimental test (KN)	Failure Load ABAQUS model (KN)	Percentage change (%)
1	34.7	34.8	−0.3
2	53.2	52.7	1
3	46.8	42.8	8.6
4	67.2	56.5	16
5	41.3	38	8
6	56.8	50.2	11.7
7	49.7	48.2	3.1
8	55.7	58.5	−5.1
9	26.6	24.6	7.6
10	42.4	43.9	−3.6
11	39.3	42.2	−7.4
12	59.7	50.1	16.1
13	37	32.5	12.2
14	51.7	44.3	14.4
15	51	44	13.8
16	52.8	50.6	4.2

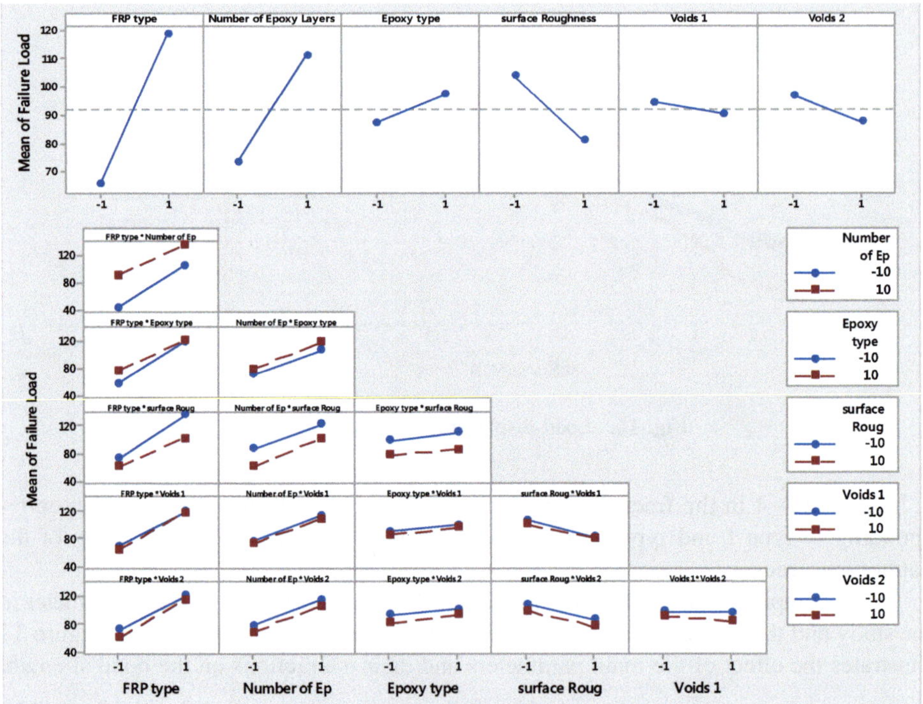

Fig. 17. Effect of the parameters on the bond strength

Finally, a comprehensive equation combining all the parameters in the study is presented in Eq. 4. The equation developed could be used to calculate the expected bond strength by using certain parameters for the FRP application.

$$
\begin{aligned}
\text{Failure load } (\%) = {} & 92.153 + 26.311[\text{FRP type}] \\
& + 18.975[\text{Number of Epoxy Layers}] + 5.054[\text{Epoxy type}] \\
& - 11.434[\text{surface Roughness}] - 2.028[\text{Voids 1}] \\
& - 4.752[\text{Voids 2}] - 4.494[\text{FRP type} * \text{Number of Epoxy Layers}] \\
& - 4.080[\text{FRP type} * \text{Epoxy type}] - 5.248[\text{FRP type} * \text{surface Roughness}] \\
& + 0.816[\text{FRP type} * \text{Voids 1}] + 1.187[\text{FRP type} * \text{Voids}] \\
& - 5.248[\text{FRP type} * \text{surface Roughness}] + 0.816[\text{FRP type} * \text{Voids 1}] \\
& + 1.187[\text{FRP type} * \text{Voids}]
\end{aligned} \tag{4}
$$

7 Proposed Procedures

In order to implement the findings of this study and to successfully evaluate actual structures strengthened with FRP, a proposed procedure in form of guidelines is presented. The method included in this study can be helpful to engineers to investigate the quality of a new FRP application or an in-situ condition assessment of FRP strengthening to determine the extent the extent of the adequate placement of the FRP and the long-term performance of the FRP wrapping.

A flowchart was prepared to summarize the steps that need to be followed for a complete evaluation of the bond strength is illustrated in Fig. 18. The first step is to collect the available information about the repair work. If the information is available, a quick check of the bond strength could be performed. If not, a detailed procedure needs to be followed.

In the case of a quick check, one NDE equation might be appropriate for the quick evaluation. The first step is to calculate the NDE parameter from the selected NDE equation using the repair information as inputs. Then, a complete scan of the area that requires evaluation need to be performed. After completing the scan, a comparison between the expected value calculated using the equation and the actual values recorded from the scan needs to be performed. If the values are similar, a conclusion about the quality of the bond can be drawn that the FRP application was properly executed and the bond strength is as expected. If the values are not similar, an additional step is required. In this step, the percentage of the area where the values are different need to be identified. If the area can be considered local, a quick solution could be to neglect the contribution of these areas on the total bond strength. To do so, these areas might be considered as voids and neglect the contribution of this zone. If the loss of strength is acceptable, the FRP bond could be considered as adequate. If the loss in bond strength exceeds the required values, a detailed procedure is necessary to judge the bond strength.

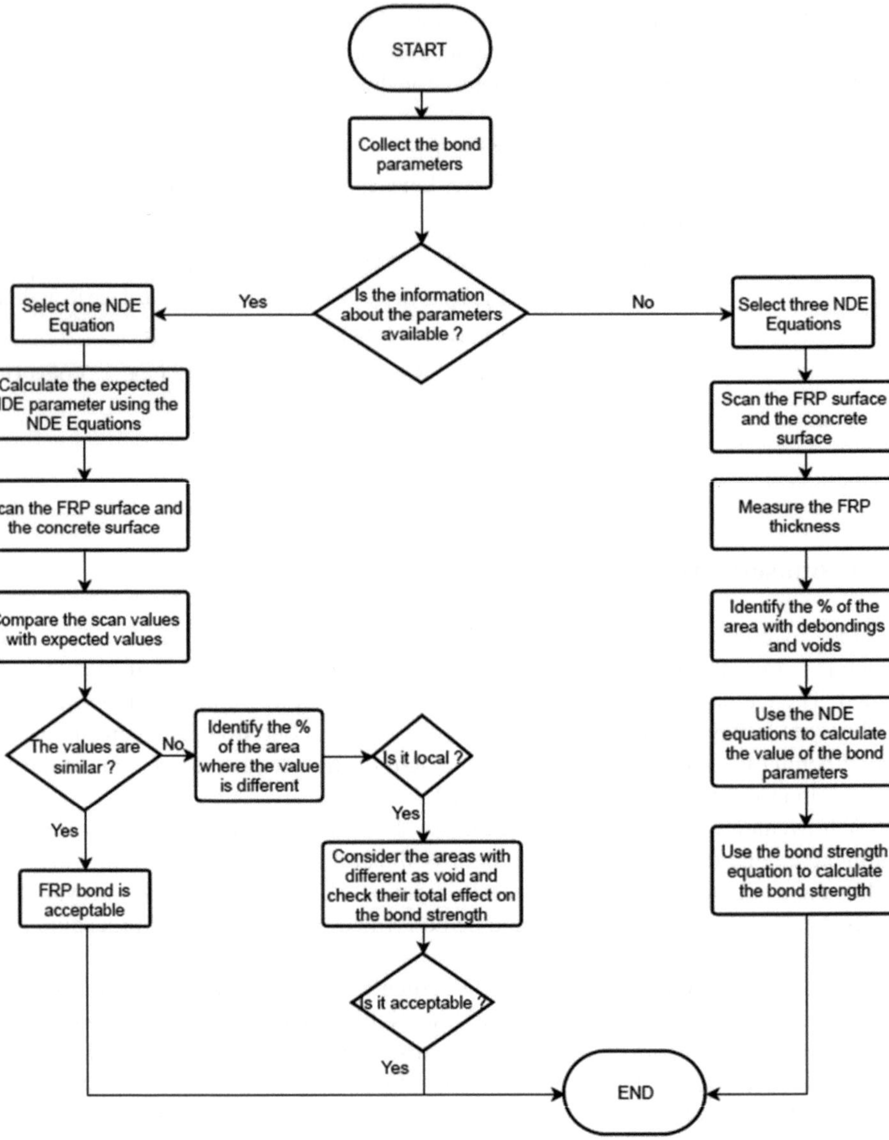

Fig. 18. Proposed procedures to evaluate the bond strength flowchart

In the cases where a detailed check is required, at least three equations are required in that case. The surface of the concrete and the surface of the FRP need to be scanned. After scanning, the areas that require evaluations, the FRP thickness needs to be measured and the areas with deboning and voids need to be identified. Using these data, the NDE equations could be used to calculate the state of each parameter. Finally, the bond strength could be calculated using the information about the bond parameters deducted from the NDE equations and adding the voids area. The bond strength could be calculated using the bond strength formula presented.

8 Conclusions

The following conclusions and recommendations may be made based on the results from this study

This study explored the ability of Schmidt hammer and Ground Penetrating Radar techniques on identifying several parameters that are considered to have an effect on the bond strength between the FRP and the concrete. The parameters considered in this study were FRP type, number of Epoxy layers used, Epoxy type, surface roughness and voids. In summary, the study demonstrated these NDE techniques were capable of identifying the parameters that affect the bond strength.

The addition of a layer on top of the scan surface with known dielectric constant and with thickness equal to the direct coupling zone proved to be beneficial to the quality of the scan. It increases the ability to detect the parameters and reduced any noise from the direct coupling zone.

Using the Schmidt hammer method, the ratio of Q value from the FRP surface compared to the Q value from the concrete surface is used to evaluate the quality of the bond. This method was capable of identifying all the parameters in the study. One equation was developed to represents the relationship between the Q value readings and the parameters that affect the bond strength. Additionally, this method proved to be successful in detecting any voids or delamination that might be hidden under the FRP surface. This method requires a point to point inspection and some voids may end up undetected.

Using the GPR test, the ratio of the amplitude of the radar wave reflected at the FRP surface relative to the amplitude of the wave reflected at the concrete surface is used to evaluate the quality of the bond. This method was capable of identifying all the parameters in the study. Two equations were developed to represents the relationship between the amplitude readings and the parameters that affect the bond strength. Additionally, the method proved to be successful in detecting any voids or delamination that might be hidden under the FRP surface.

The destructive testing was required to analyze the effect of each parameter on the bond strength. In order to complete this task, two destructive tests were performed. The first test was the pull-off test. This test investigates the strength of the bond in direct tension. The second destructive method was three points bending test. This test investigates the strength of the bond in bending.

The results from both destructive tests were used to calibrate several FEM models. The FEM models were required to evaluate the effect of all the parameters and the interaction between them. An equation was developed based on the results. The equation developed could be used to calculate the expected failure load based on the actual parameters used in the FRP application

In addition to the previous results, a series of FEM models were prepared to investigate the effect of voids on the bond strength. Each model included voids with specific size under the FRP layer to represent a percentage loss of the total area of the voids. The results showed a clear downtrend of the bond strength with the increase of voids size.

The procedures presented in this paper is recommended for in situ condition assessment of FRP strengthening to determine the extent the extent of the adequate placement of the FRP and the long-term performance of the FRP wrapping.

References

ABAQUS Manual, A.U.: Dassault Systems Simulia Corporation. Providence USA (2010)

ACI Committee 440: Guide for the Design and Construction of Externally Bonded FRP Systems for Strengthening Concrete Structures (ACI 440.2R-17), American Concrete Institute (2017)

Akuthota, B., Hughes, D., Zoughi, R., Myers, J., Nanni, A.: Near-field microwave detection of disbond in carbon fiber reinforced polymer composites used for strengthening cement-based structures and disbond repair verification. J. Mater. Civ. Eng. **16**(6), 540–546 (2004)

ASTM C293/C293M: Standard Test Method for Flexural Strength of Concrete (Using Simple Beam with Third-Point Loading). American Society for Testing and Materials, Philadelphia, PA (2016)

ASTM D3039/D3039M Standard Test Method for Tensile Properties of Polymer Matrix Composite Materials, West Conshohocke (2014)

Bakis, C.E., Bank, L.C., Brown, V., Cosenza, E., Davalos, J.F., Lesko, J.J., Triantafillou, T.C.: Fiber-reinforced polymer composites for construction—state-of-the-art review. J. Compos. Constr. **6**(2), 73–87 (2002)

Ehrhart, B., Valeske, B., Muller, C. E., Bockenheimer, C.: Methods for the quality assessment of adhesive bonded CFRP structures-a resumé. In: 2nd International Symposium on NDT in Aerospace (2010)

Galietti, U., Luprano, V., Nenna, S., Spagnolo, L., Tundo, A.: Non-destructive defect characterization of concrete structures reinforced by means of FRP. Infrared Phys. Technol. **49**(3), 218–223 (2007)

Karbhari, V. M., Kaiser, H., Navada, R., Ghosh, K., Lee, L.: Methods for detecting defects in composite rehabilitated concrete structures (No. FHWA-OR-RD-05-09) (2005)

Leung, C.K.: Delamination failure in concrete beams retrofitted with a bonded plate. J. Mater. Civ. Eng. **13**(2), 106–113 (2001)

Lubliner, J., Oliver, J., Oller, S., Onate, E.: A plastic-damage model for concrete. Int. J. Solids Struct. **25**(3), 299–326 (1989)

Maerz, N. H., Galecki, G.: Preservation of Missouri transportation infrastructures: non-destructive testing of FRP materials and installation, Gold Bridge: Validation of FRP Composite Technology Through Field Testing (No. UTC R96). Missouri S & T, Center for Infrastructure Engineering Studies (2008)

Meier, U.: Strengthening of structures using carbon fibre/epoxy composites. Constr. Build. Mater. **9**(6), 341–351 (1995)

Mirmiran, A., Wei, Y.: Damage assessment of FRP-encased concrete using ultrasonic pulse velocity. J. Eng. Mech. **127**(2), 126–135 (2001)

NIST/SEMATECH: e-Handbook of Statistical Methods http://www.itl.nist.gov/div898/handbook/. Accessed 18 Dec 2017)

Obaidat, Y.T.: Structural retrofitting of concrete beams using FRP-debonding issues. Department of Construction Sciences, Lund University (2011)

Ribolla, E.L.M., Hajidehi, M.R., Scimemi, G.F., Spada, A., Giambanco, G.: Assessment of bonding defects in FRP reinforced structures via ultrasonic technique. Challenge **2**(3), 139–146 (2016)

Schabowicz, K.: Ultrasonic tomography—the latest nondestructive technique for testing concrete members—description, test methodology, application example. Arch. Civ. Mech. Eng. **14**(2), 295–303 (2014)

Starnes, M.A., Carino, N.J., Kausel, E.A.: Preliminary thermography studies for quality control of concrete structures strengthened with fiber-reinforced polymer composites. J. Mater. Civ. Eng. **15**(3), 266–273 (2003)

Taillade, F., Quiertant, M., Benzarti, K., Dumoulin, J., Aubagnac, C.: Nondestructive evaluation of FRP strengthening systems bonded on RC structures using pulsed stimulated infrared thermography. In: Prakash, R.V. (ed.) Infrared Thermography. IntechOpen (2012). https://www.intechopen.com/books/infrared-thermography/pulsed-stimulated-infrared-thermography-applied-to-a-nondestructive-evaluation-of-frp-strengthening-

Täljsten, B.: Strengthening of beams by plate bonding. J. Mater. Civ. Eng. **9**(4), 206–212 (1997)

Yu, T.Y., Büyüköztürk, O.: A far-field airborne radar NDT technique for detecting debonding in GFRP–retrofitted concrete structures. NDT E Int. **41**(1), 10–24 (2008)

Indirect Tensile Stiffness and Permanent Deformation of Microwave Treated Cold Bituminous Emulsion Mixtures (CBEMS) Containing Cementitious Activated Waste Fly Ash

Hassan Al Nageim[1(✉)], Abbas Al-Hdabi[2], and Shakir Al Busultan[3]

[1] Department of Civil Engineering, Liverpool John Moores University, Liverpool, UK
h.k.alangeim@ljmu.ac.uk
[2] Civil Engineering Department, Faculty of Engineering, Kufa University, Kufa, Iraq
[3] Department of Civil Engineering, University of Kerbala, Kerbala, Iraq

Abstract. The mechanical properties in terms of indirect tensile stiffness modulus (ITSM) properties and permanent deformation of CBEMs containing new cementitious materials made from activated blend of waste fly ash and inert Limestone filler were investigated. CBEMs were prepared with these fillers, and then these mixtures were subjected to microwave treatment at four treatment time and high power level. The aim was to identify the influence of microwave treatment on the mechanical properties and identify the microwave suitable application time ranges.

The papers reports that microwave treatment has significantly improved the indirect tensile stiffness and permanent deformations resistance of all the mixes studied and shows that the mixtures containing the new developed cementitious filler provide high quality materials in terms of both stiffness and deformation resistance which are acceptable by road engineers. More importantly, the new fast curing cold mix asphalt is satisfying the current British standard.

Keywords: Cold mix asphalt · Bituminous emulsion · Treated fly ash
Indirect tensile stiffness · Creep of cold mix asphalt

1 Introduction

1.1 Cold Mix Asphalt (CMA)

Cold Mix Asphalt (CMA) are normally prepared by mixing aggregates, filler and bitumen emulsion at ambient temperature. The technology has been developed in the 1970s and since then it has been widely used in many countries such as France, and the USA Leech (1994).

H. Rodrigues and A. Elnashai (Eds.): GeoMEast 2018, SUCI, pp. 268–276, 2019.
https://doi.org/10.1007/978-3-030-01932-7_21

In the UK the SROH "Specification for Reinstatement of Opening in Highways" by the Highway Authorities and Utilities Committee (HAUC) allowed the use of cold laid asphalt mixtures for reinstatement works in low volume roads and footways (Highway Authorities and Utilities Committee 1992).

The main disadvantages of CMA can be summarised as:

(i) CMA full curing to yield its ultimate strength may take from 2 to 24 months depending on the mixture's ingredients and weather conditions. This is associated with high air voids for the compacted CMAs.
(ii) Partial coating percentage of the ingredients of CMA is another shortfall. This is due to the incompatibility between the aggregates and emulsion and the high contents of water in emulsions.
(iii) Additionally binder drainage during storage because of the low viscosity of the emulsion and
(iv) Binder stripping from the aggregate due to high water sensitivity and weak adhesion Thanaya et al. (2009) and Leech (1994).

Therefore and in order to improve the strength and durability properties of CMA a large range of products, preparing procedures and laying techniques are tried by researchers and CMA producers. For example Cold mix asphalt (CMA) mixture range from gap graded such as Hot Rolled Asphalt (HRA) or continuous dense graded (such as AC mixtures); it may include; virgin aggregates or Reclaimed Asphalt Pavement (RAP); cationic or anionic bitumen emulsion binder; normal emulsion or emulsion modified by polymer or solvent. Fibre and cement can also be used to enhance its properties. CMA may or may not be stockpiled before final laying, spreading and laying by (a) hand (in some cases or road-opening reinstatement) and (b) graders or asphalt pavers. CMAs might be used as structural layers in heavily trafficked base layers, low trafficked wearing courses, and as non-structural layers in surface treatment layers (National Cooperative Highway Research Program 1975).

Ordinary Portland cement (OPC), other type of cements and Supplementary Cementitious Materials (SCMs) has attracted the interest of many researchers and used as either additional fine materials or as replacement of the conventional limestone filler in CMA. The main benefits that can be reported when using OPC and SCMs in CMAs are:

- mechanical properties: enhancement of ultimate strength due to the production of cementitious and/or pozzolanic bonding properties,
- economic advantages: as the SCMs are mostly industrial waste or by-products,
- ecological advantage factor when SCMs are used, by reducing CO_2 emissions and landfilling's associated health and environmental problems Thanaya (2003), Thanaya (2007), Cliff et al. (2004), Al-Busaltan et al. (2012).

1.2 Microwave Technique for Heating Bituminous Materials

Microwaves are a kind of electromagnetic wave between infrared radiation and radio waves, with wavelengths between 0.001 and 1 m and 900 and 2450 MHz frequencies Meredith (1998). In terms of efficiency, Meredith (1998) reported that the overall efficiency of well-designed microwave heating systems is very high, 80% at 2450 MHz, while conventional heating methods operate at around 30% efficiency.

Microwaves can penetrate materials and deposit energy throughout the volume of the material as microwave energy is delivered directly to materials through molecular interaction with the electromagnetic field, as reported by Thostenson and Chou (1999). Thus, there is a possibility to achieve fast and uniform heating of whole materials by using microwave energy.

The microwave technology can process mineral materials and carbon-related material, as reported by previous studies Haque (1999) and Menedez et al. (2010). Also, using the microwave heating process for the pre-heating process in the bituminous materials' mixing process prevents the oxidation of bitumen during this process in addition to the previous advantages. Furthermore, Jeon et al. (2012) stated that there is a possibility that the microwave can break up the larger molecules in the bitumen and generate light oil fractions.

Nieftagodien (2013) conducted a study to investigate the suitability of using microwave application to heat-reclaimed bituminous mixtures and crushed aggregates as an energy-efficient technique to produce a Half-Warm Mix (HWM). The advantages when using this method of heating comprise energy saving due to its volumetric heating capability and fast heating, which in turn improve productivity when using suitable materials. The properties of Half-Warm Foamed (HWF) mixtures are a compromise between HMA and Foamed Bitumen Stabilised mixtures' (BSM-foam) properties. He produced a new HWA utilising microwave heating technique to improve the engineering characteristics of BSM-foam. He evaluated the HWF properties and BSM-foam by implementing Indirect Tensile Strength (ITS) and monotonic triaxial tests on specimens as well as by means of pavement analysis. He concluded that the compaction density increased for HWF mixtures and a huge reduction in water contents is observed after curing in comparison with BSM-foam mixtures.

In this study, the microwave technique has been used to produce a new Half-Warm Rolled Asphalt (HWRA) in the pre-heating process before compaction of the materials.

2 Materials and Sample Preparations

2.1 Materials- Cold Mix Asphalt (CMA) and Aggregates

A 55/14C gap graded surface course mixture gradation based on BS EN 13108-4 European Committee for Standardization (2006) is used for preparing the Cold mix asphalt used in this study. To produce HRA mixtures for comparison with CMA a 100/150 and 40/60 grade binders are used. Tables 1 and 2 shows the properties of the aggregates and binders respectively. All CMA mixtures produced in this research were prepared based on the mix design procedure adopted by the Asphalt Institute (1989). Whereas HMA is prepared according to BS EN 13108-4 European Committee for Standardization (2006) for HRA.

Table 1. Aggregates' physical properties

Material	Property	Value
Coarse aggregate	Bulk particle density, mg/m^3	2.79
	Apparent particle density, mg/m^3	2.83
	Water absorption, %	0.6
Fine aggregate	Bulk particle density, mg/m^3	2.68
	Apparent particle density, mg/m^3	2.72
	Water absorption, %	1.6
Mineral filler	Particle density, mg/m^3	2.71

Table 2. Properties of 100/150 and 40/60 bitumen binders

Bituminous binder 40/60		Bituminous binder 100/150	
Property	Value	Property	Value
Appearance	Black	Appearance	Black
Penetration at 25 °C	43	Penetration at 25 °C	122
Softening point, °C	54	Softening point, °C	43.6
Density at 25 °C	1.02	Density at 25 °C	1.05

2.2 Bitumen Emulsion

For manufacturing in the lab the CMA, a cationic slow setting bitumen emulsion [C56B7] was used to prepare all the new CMA with or without replacement of the conventional Limestone mineral filler. Table 3 shows the properties of the selected bitumen emulsion.

Table 3. Properties of (C56B7) bitumen emulsion

Bitumen emulsion (C56B7)	
Property	Value
Appearance	Black to dark brown liquid
Relative density at 15 °C, g/ml	1.05
Residue by distillation, %	56
Residual bitumen penetration, 1/10 mm	100

3 Results and Discussion

The results presented in this section shows the influence of using microwave technology in reducing the CMA void contents. CRA mixtures containing supplementary cementitious waste fly ash with high lime content as a replacement for the conventional mineral filler.

Fig. 1. Domistic microwave apparatus

In this investigation, the domestic microwave apparatus shown in Fig. 1 is used to heat CRA mixtures at the pre-compacted stage. The wavelength frequency was 2450 MHz and the output power was set as 300 W. All CMA mixtures produced in this research were prepared based on the mix design procedure adopted by the Asphalt Institute (1989). After conditioning mixtures in the microwave, the materials were moulded and compacted to indicate their mechanical properties in terms of ITSM. A Marshall Hummer is used to compact the specimens which received 75 blows at each phase/individual specimen. The normal curing process is adapted, i.e. leave the sample in the mould for 24 h to cool down and then cure it at 20 °C to the curing timer required before testing.

Figure 2 shows the air voids and ITSM after microwave heating of the specimens made from conventional CMA containing for; 1, 2, 3, 4, 5 and 6 min and curing the sample at 20 Co for one day. The figure also reports the summary of the volumetric properties, ITSM and the temperature which are observed within this investigation (Tables 4 and 5).

From Fig. 2, it is obvious that air voids decreased, from 10.1% to 5.3%, with increasing of microwave conditioning time from 0 to 6 min and at the same duration the temperature of the mixes increased from 20 °C to 100 °C; while stiffness modulus decreased slightly until 3 min of microwave conditioning, then the rate of decrease increased sharply between 3 and 4 min of microwave curing.

Finally, stiffness modulus values increased after 5 min of microwave curing when the temperature of the mixtures was close to 100 °C. The air voids decrease can be attributed to the loss of trapped water due to the increment in mixture temperature. Stiffness modulus behaviour is more complicated than that of air voids. But, generally, the early decrease in their values might be due to the negative effect of temperature increase on the progress of the hydration process of the secondary binder, which is generated from the hydration process between the TBF-2 filler and the trapped water; while the later increment of the stiffness modulus values, after 5 min of microwave

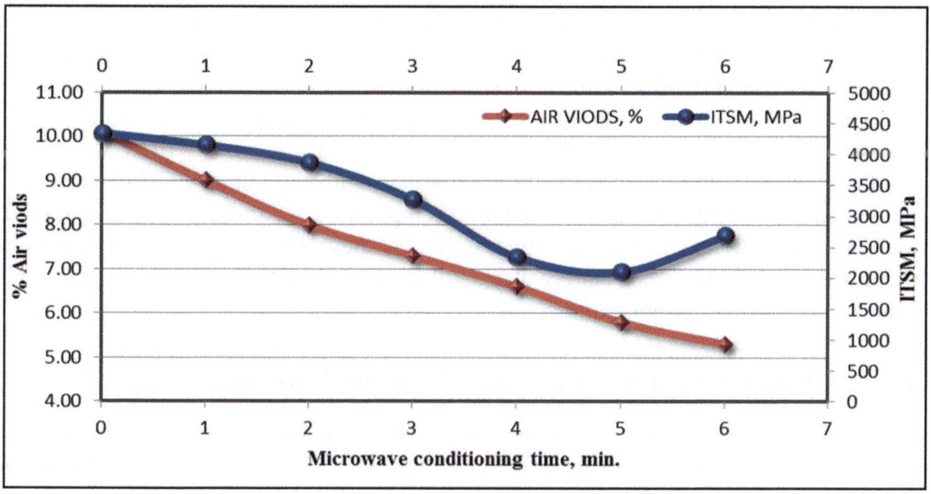

Fig. 2. Air voids and ITSM after microwave heating of the specimens

Table 4. Air voids and ITSM after microwave heating of the specimens

Microwave conditioning time, min.	Compaction temperature °C	Air voids, %	Dry density, mg/m^3	VMA, %	VFB, %	ITSM after 1 day, MPa
0	20	10.1	2.15	24	58	4340
1	33	9.20	2.18	23	60	4150
2	50	8.00	2.20	22	64	3869
3	69	7.30	2.21	22	66	3275
4	85	6.60	2.23	21	68	2343
5	91	5.80	2.25	20	71	2100
6	100	5.30	2.26	20	73	2688

Table 5. ITSM of other mixtures [MPa]

Curing days	ITSM of other mixtures [MPa]				
	CMA [LS]	CMA [OPC]	CMA [TSCM]	HMA [100/150 pen]	HMA [40/50 Pen]
1	284	3224	5539	1941	5707
28	519	6530	7107	1941	5707

curing, can be attributed to: (i) the densification of the mixture which leads to increase the interlock between the mixture particles and (ii) the high temperature, which is more than 90 °C, leads to decrease the viscosity of the bitumen (improve the consistency) of the primary binder, i.e. bitumen, and motivated the bonding reaction between the bitumen film and aggregate particles.

An optimisation study can be conducted according to the air voids, stiffness modulus and compaction temperature to indicate the more suitable microwave conditioning time for TBF-2 mixtures. Compaction of TBF-2 mixtures after 3 min of microwave conditioning is selected as the best one for several reasons, namely: (i) it decreases the air voids from 10.1% to 7.3% with a 28% decrease, (ii) the stiffness modulus value is slightly decreased in comparison with the control TBF-2 mixtures, then decreased sharply, and (iii) compaction temperature for these mixtures is 69 °C, which represents a very moderate temperature. The new optimised CRA mixtures are termed within this thesis as Half-Warm Rolled Asphalt (HWRA) because they are produced at moderate temperatures.

3.1 ITSM Results for Half-Warm Rolled Asphalt Mixtures

Figure 3 shows the influence of curing time at normal curing temperature, i.e. 20 °C, on stiffness modulus results for the HWRA mixtures. The ITSM test was conducted at 1/6, 1, 3, 7, 14 and 28 days after de-moulding. Generally, it can be stated that stiffness modulus of HWRA mixtures decreased for all the curing times in comparison with those for TBF-2 mixtures. Interestingly, these results are still higher than OPC mixtures for the same duration of curing and stiffness modulus at 4 h (2024 MPa), are more than those for the 100/150 HRA mixture (1941 MPa) and achieve the target stiffness modulus, i.e. 2000 MPa.

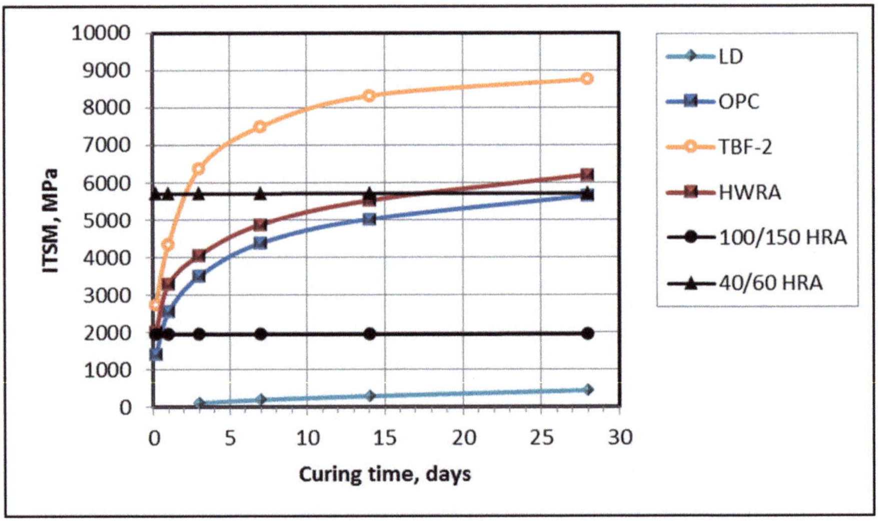

Fig. 3. Effect of curing time on ITSM of HWRA mixtures

3.2 Resistance to Permanent Deformation of Microwaved CBEMs

The Uniaxial Compressive Cyclic Test (UCCT) was conducted at 40 °C in accordance with the European Committee for Standardization (2005) to investigate the creep performance of Half-Warm Rolled Asphalt mixtures (HWRA) mixtures. Figure 4 below shows the creep strain vs. number of pulses relationship. It is obvious that HWRA behaves much better than the TBF-2, control CRA (LD) and conventional HRA mixtures. Again, these results identify the positive effect from (i) increasing particle interlock together with the (ii) enhanced bonding due to the hydration process in HWRA mixtures which is more than the bonding in the TBF-2 mixtures caused only by the generation of asphalt bonding between the particles at lab temperature and the bonding resulted from the hydration of SCM's which is a function of the duration of curing.

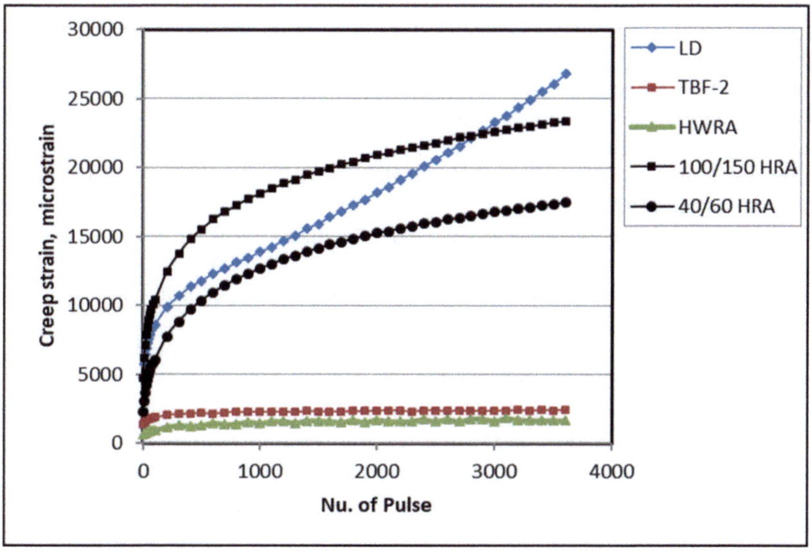

Fig. 4. Creep strain vs. number of pulses for HWRA mixtures

4 Conclusions

1. Microwave heating the CBEMs above laboratory temperature and up to 100 °C contributes significantly to reducing porosity with minimum acceptable effect on stiffness modulus and durability properties.
2. Microwave heating of samples increase both polarization and heating. Polarization improve bitumen aggregate adhesion as it acts as coalescence accelerator. Whereas heating reduces the bitumen viscosity and thus improves the compactability of the bituminous mixtures under the compaction efforts.
3. Heating also, beneficial for hydration process and shortening curing time to accomplish the desirable mechanical properties.
4. The research work produced a new HWRA with;

a. improved properties in terms of ultimate strength due to the production of enhanced cementitious bonding properties,
b. economic advantages: as the used SCMs are mostly industrial waste or by-products,
c. Ecological and health advantages as when SCMs are used, CO_2 emissions and landfilling's associated health and environmental problems will be reduced.

References

Leech, D.: Cold bituminous materials for use in the structural layers of roads, TRL Project report 75. Transportation Research Laboratory, UK (1994)

Highway Authorities and Utilities Committee: Specification for the Reinstatement of Opening in Highways, New Roads and Street Works, London, UK (1992)

Thanaya, I., et al.: A laboratory study on cold-mix, cold-lay emulsion mixtures. In: Proceedings of the Institution of Civil Engineers (2009)

National Cooperative Highway Research Program: Bituminous Emulsion for Highway Pavements. Synthesis of Highway, Practice No. 30, p. 76, Transportation Research Board, Washington, D.C. (1975)

Thanaya I.: Improving the performance of cold bituminous emulsion mixtures incorporating waste materials. Ph.D. thesis, The University of Leeds, UK (2003)

Thanaya, N.A.: Review and recommendation of cold asphalt emulsion mixtures (CAEMs) design. Civ. Eng. Dimens. **9**, 49–55 (2007)

Cliff, E., et al.: Properties of GGBS-bitumen emulsion systems with recycled aggregates. Road Mater. Pavement Des. **5**, 373–383 (2004)

Meredith, R.: Engineers' Handbook of Industrial Microwave Heating. The Institution of Electrical Engineers, London (1998)

Al-Busaltan, S., et al.: Green bituminous asphalt relevant for highway and airfield pavement. Constr. Build. Mater. J. **31**, 243–250 (2012)

Thostenson, E.T., Chou, T.W.: Microwave processing: fundamentals and applications. Compos. Part A: Appl. Sci. Manuf. **30**, 1055–1071 (1999)

Haque, K.: Microwave energy for mineral treatment processes-a brief review. Int. J. Miner. Process. **57**, 1–24 (1999)

Menedez, J., et al.: Microwave heating processes involving carbon materials. Fuel Process. Technol. J. **91**, 1–8 (2010)

Nieftagodien, R.: Suitability of microwave application to heat reclaimed asphalt and crushed aggregates as an energy efficient method in the production of half warm mix. M.Sc. thesis, University of Stellenbosch (2013)

European Committee for Standardization: BS EN 13108: Part 4. Bituminous Mixtures Materials Specification-Hot Rolled Asphalt. British Standard Institution, London, UK (2006)

Asphalt Institute: Asphalt cold-mix manual, MS-14. The Asphalt Institute, UK (1989)

Analytical Fragility Curves of Open Frame Platform Structures

Yasser S. Salem(✉), P. E. Aren Azizian, and Jin Sung Cho

Civil Engineering Department, Cal Poly Pomona, Pomona, CA, USA
{ysalem, aazizian, jinsungcho}@cpp.edu

Abstract. The seismic performance of open frame platform structures which is a complex structure commonly found in large oil refineries is evaluated in this study. Analytical Fragility curves are developed to evaluate the performance of each platform models at Immediate Occupancy damage state after an earthquake. The methodology that were used to develop the curves, are based on calculating axial deformation of the brace member using nonlinear time history analysis for different ground motions with different intensities and frequency contents. The fragility curves are developed based on the acceptance criteria for the brace member given from the FEMA 356. It was found that as the bracing member size increases within the same structural shape, the performance of the structure in reaching/exceeding immediate occupancy performance level decreases. Additionally, as the height of the structure increases the probability of reaching immediate occupancy increases. It was also found that certain structural members are not to be used as bracing members for these structures.

1 Introduction

Open frame platform structures are commonly found in large refineries where they are used to process heavy crude oil. The unit is a steel structure that consists of steel towers carrying maintenance platforms. The current practice is to estimate the seismic loads for these structures using parameters of similar building-type structures. These structures are substantially different than typical building structures where there are no floor or roof diaphragms with lumped mass. Therefore, building code design equations are not necessarily suitable to predict their performance during earthquakes. Many of these structures are constructed or planned to be constructed in an area of high seismic activities and a safe and economic design of these units is of a great value to the society (Salem and Nasr 2014). It is important to note that a failure in OFS structures in refineries pose a great danger to the surrounding society similarly to explosions and fires in the Chevron El Segundo Facility in Los Angles. Therefore, it is important to design such structures to not only remain their integrity during a seismic event, but also remain operational immediately after. As a result, computer analysis is performed to develop a better understanding of the structural failure modes and how to prevent significant failures of such structures during a seismic event. The objective is to identify bracing members that provide the most improvement to the building seismic response. In addition, it is important to address the economical effects of construction in comparison with optimal structural members.

H. Rodrigues and A. Elnashai (Eds.): GeoMEast 2018, SUCI, pp. 277–291, 2019.
https://doi.org/10.1007/978-3-030-01932-7_22

Figures 1 and 2 show examples of OFS structures showing the concentric braced frames used the lateral resisting system. OFS bracing systems can be made with various configurations such as single diagonals, X-bracing, K-Bracing, V-braced and Chevron (Inverted V) bracing as shown in the photos. Chevron bracing is most popular in such structural due to the openness of the frame which allows crew to move large equipment and material onto the decking. Hence, all computer models created for the analysis of this research use chevron bracing. In these examples, the lateral resisting system for the open steel frames and the steel tower are formed of ordinary concentric brace frames. Lateral stability of the platforms is provided by horizontal braced members below the grating elevation. Due to constructability consideration, [it appears that] all steel connections are bolted at the field using bearing type connection (Salem)."

Fig. 1. Typical OPFS structure

Fig. 2. OPFS tower

In this study, different cases with different parameters of OFS structure are investigated to study the behavior of the structure under the seismic demand. Analytical fragility curves will be developed to assess the OFS structure vulnerability risk.

2 Sample Buildings

In this study, representative samples of OPFS structures were considered. Figures 3, 4 and 5 show on of the sample structures, which include framing members that are consistent with the industry standard. All beams, columns, and platform framings are wide flange members. The brace members that were considered in this study are Hollow Steel Tube (HSS), Double Angle (2L), and Wide Flange (W-Shape) in order to observe the behavior changes of the structures based on the changes of the bracing members.

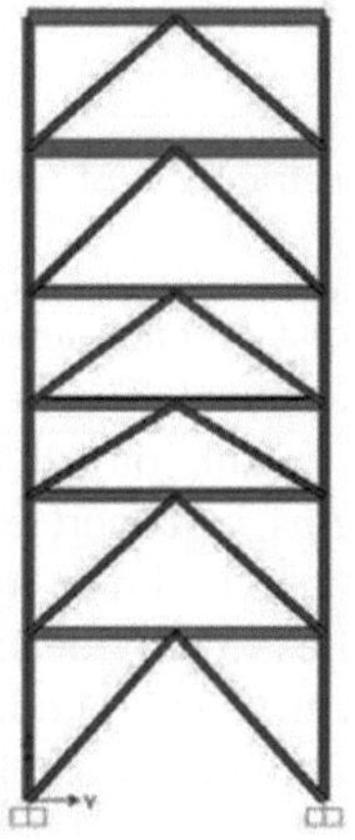

Fig. 3. Elevation view OPFS structure

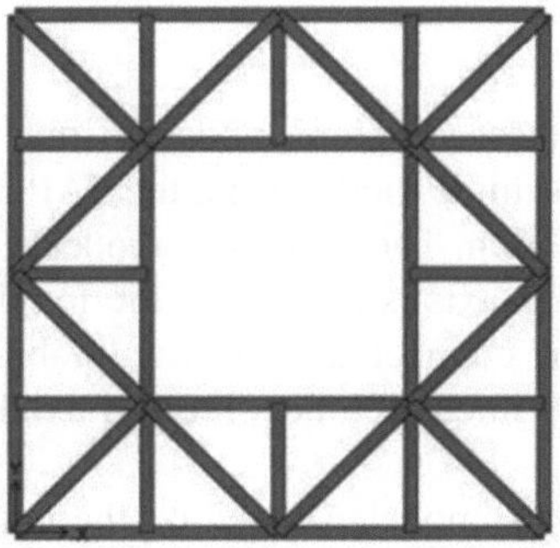

Fig. 4. Plan view of OPFS structure

For the OPFS, the longitudinal and transverse directions are typically constructed of Ordinary Concentrically Braced Frames (OCBF). OPFS vary in terms of capacity, shape, and member sizes, but their main function is to support the main tower structure for maintenance purposes. This study will focus on OPFS made of steel to carry an

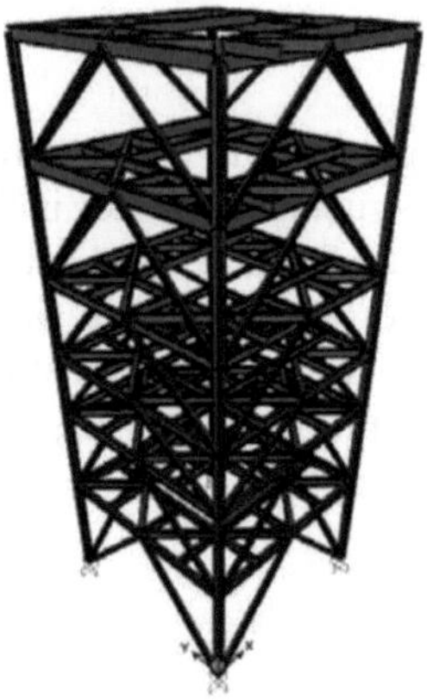

Fig. 5. Isometric view of OPFS structure

average distributed dead load of 15psf on each level. These loads are listed in Process Industry Practices (PIP) STC01015 manual. The loads for the supporting grating of the structure are divided per tributary width on all the beams per each floor. Unlike buildings, OPFS have no diaphragms or concrete slabs to allow load path between transverse bents. Horizontal light diagonal brace members in the forms of W. Shapes are usually installed in these structures to provide stability to horizontal surfaces that is usually covered by grating or checker plates. The horizontal bracing members transfer the lateral forces (wind and Seismic) to the vertical braced frames. Three different cases of OPFS structure are considered in this study. Each case has four different parameters for the OPFS. Tables 2 and 3 show the parameters of each different case used for the research.

3 Methodolgy

To simulate the response of these structures under seismic loads, a finite element model was developed for a typical sample model using the SAP 2000 © program. The base of the columns is a fixed connection. The numerical model will consider a fixed bent in its analysis. The column's strong axis is parallel to the longitudinal direction.

All floors of the model are connected by inverted V-bracing is placed on all sides of the structure. The struts and brace members rely on axial strength and together resist applied lateral loads.

The sample structures were modeled based on the nonlinear behavior of materials and its damping characteristics. Modal analysis was performed first to account for structural behavior accuracy and observe at least 95% participation factor of the structure. Modal output data for one of the sample structures is shown per Table 4. Pushover curves of the sample structures at each step of pushover analysis were produced for each case. Pushover curves indicated most critical hinging location of the braces. Therefore, the axial deformation of the most critical brace was considered for the remaining analysis and generation of the fragility curves. It is critical to note that

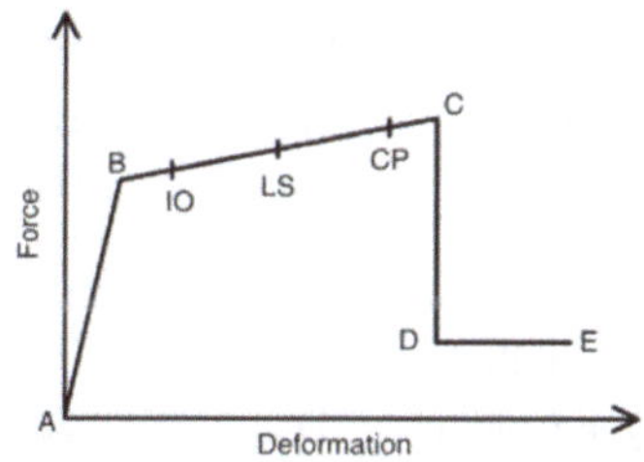

Fig. 6. Inelastic model used for beams and columns

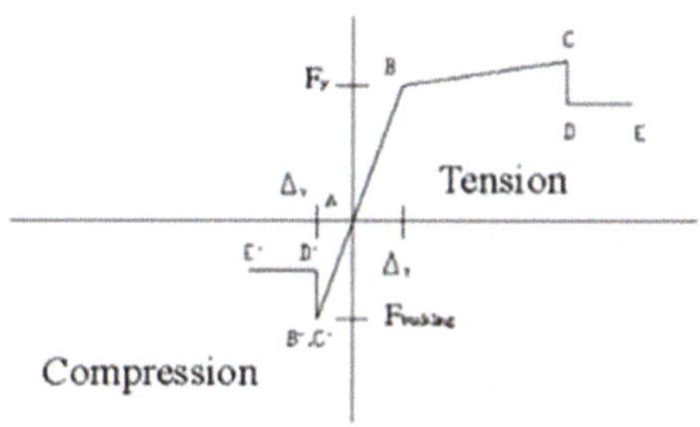

Fig. 7. Inelastic model used for bracing elements

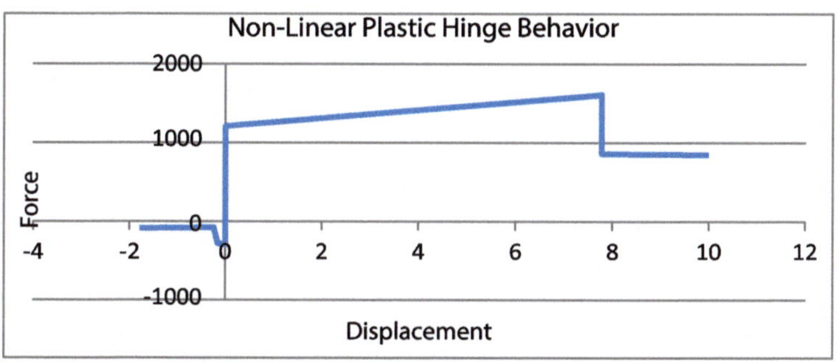

Fig. 8. Inelastic Model Used for W10X88

pushover analysis was performed after the dead load analysis. The dead load is to remain constant at all times while the structure is laterally pushed over an iterative process up to a preset displacement.

Time history analyses were carried out to simulate the response of structure based on 22 far-field records, which are selected from (PEER) ground motion database and ATC63 project. All selected ground motions are given in Table 1. Figure 9 shows the response spectra for the 22 ground motions that were used in this study. As it can see from this graph, the selected ground motions represent a wide variety of earthquake records that have different peak ground accelerations (PGA) and frequency contents. The selection criteria for the earthquake records are as follows:

Table 1. Selected ground motions (PEER strong motion database)

Earthquake	Mag. (Ms)	Station name	NEHRP class	Distance (Km)	PGA Max (g)	Fault type
1994 Northridge	6.7	Beverly Hills – 14145 Mulhol	D	17.2	0.4158	Blind thrust
1994 Northridge	6.7	Canyon Country – W Lost Cany	D	12.4	0.4100	Blind thrust
1999 Duzce, Turkey	7.1	Bolu	D	12.4	0.7276	Strike-slip
1999 Hector Mine	7.1	Hector	C	12	0.2656	Strike-slip
1979 Imperial Valley	6.5	Delta	D	22.5	0.2186	Strike-slip
1979 Imperial Valley	6.5	El Centro Array #11	D	13.5	0.3640	Strike-slip
1995 Kobe, Japan	6.9	Nishi-Akashi	C	25.2	0.5030	Strike-slip
1995 Kobe, Japan	6.9	Shin-Osaka	D	28.5	0.2432	Strike-slip
1999 Kocaeli, Turkey	7.5	Duzce	D	15.4	0.2159	Strike-slip
1999 Kocaeli, Turkey	7.5	Arcelik	C	13.5	0.1502	Strike-slip
1992 Landers	7.3	Yermo Fire Station	D	23.8	0.1687	Strike-slip
1992 Landers	7.3	Coolwater	D	20	0.2754	Strike-slip
1989 Loma Prieta	6.9	Capitola	D	35.5	0.4508	Strike-slip
1989 Loma Prieta	6.9	Gilroy Array #3	D	12.8	0.5365	Strike-slip
1990 Manjil, Iran	7.4	Abbar	C	13	0.5146	Strike-slip
1987 Superstition Hills	6.5	El Centro Imp. Co.	D	18.5	0.3579	Strike-slip
1987 Superstition Hills	6.5	Poe Road (temp)	D	11.7	0.4463	Strike-slip
1992 Cape Mendocino	7.0	Rio Dell Overpass	D	14.3	0.2439	Thrust

(*continued*)

Table 1. (*continued*)

Earthquake	Mag. (Ms)	Station name	NEHRP class	Distance (Km)	PGA Max (g)	Fault type
1999 Chi-Chi, Taiwan	7.6	CHY101	D	15.5	0.3100	Thrust
1999 Chi-Chi, Taiwan	7.6	TCU045	C	26.8	0.4014	Thrust
1971 San Fernando	6.6	LA - Hollywood Stor FF	D	25.9	0.2079	Thrust
1976 Friuli, Italy	6.5	Tolmezzo	C	15.8	0.3513	Thrust (part blind)

Table 2. Different cases considered in this study

Percentage	Total height (ft)	Total height (in)	Number of floors
125%	149.5	1794	7
100%	119.5	1434	6
80%	99.5	1194	5

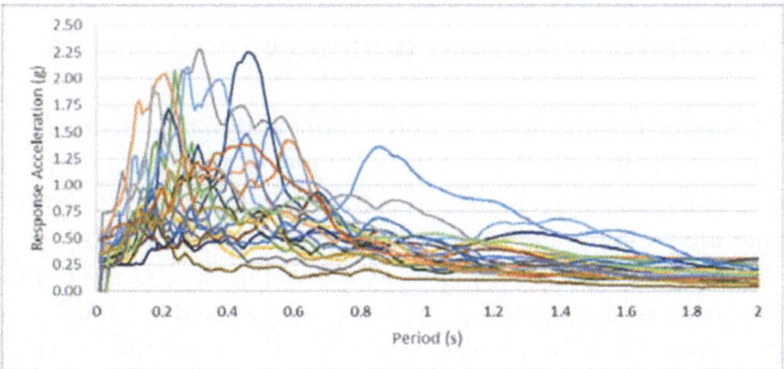

Fig. 9. Unscaled far-field spectra for ground motion records used in the non linear dynamic analysis

- PGA = 0.2 g to 0.8 g
- Magnitude > 6.5
- PGA > 15 cm/s
- Strike-Slip & Thrust
- Earthquakes from 1971–1999

The next step is to apply the correct value of the earthquake records to the structure. Far field records are provided in units of "g"; therefore, in order to import the proper value to SAP 2000, the records have to be individually scaled by a factor of 386.4, conversion to inches per second squared to be interpreted as actual ground motion values. It is important to note that in the development of fragility curves for some members, the 22 data is not sufficient to generate a smooth curve. Therefore, a selection of PGAs were selected and scaled to lowest and highest percentiles of PGAs for additional data points. This factor is calculated by dividing the desired PGA value by the PGA value of the ground motion selected, then multiplied by 386.4. Equation below shows the scaling factor calculation.

$$SF = \frac{\mathrm{X}}{\mathrm{PGA}} 386.4$$

Acceptance Criteria from the FEMA 356 lists different target performance after the earthquake for each of the structural members. Plastic rotation that occurs in the plastic hinge of the beam and column determines that target performance. For the axial compression member, braces, the axial deformation identifies the target performance. The FEMA 356 separates three target performances, which are Immediate Occupancy (IO), Life Safety (LS), and Collapse Prevention (CP). The IO level of performance classifies the structure to remain "safe to immediately occupy" with minor repairs. The LS targeted structure needs to remain stable and has "significant reserve capacity". Lastly CP describes the structure to remain standing, but "only barely" and any other damage or loss is acceptable. For the purpose of the research, IO is chosen to depict the behavior of the structure due to the necessity of such structures to society.

4 Analytical Model of Sample Buildings

To perform the nonlinear static analysis and nonlinear dynamic analysis plastic hinge is introduced to the structure members. Plastic hinge is assigned to each structure elements, such as beams, and braces, to model the material nonlinear properties for SAP 2000 to simulate the yielding and post-yielding behaviors. The plastic hinge properties are based on FEMA 356 standards. Figure 6 shows the steel moment hinge model for beams and columns. The parameters A, B, and C are defined in FEMA 356, where A is the Origin, B is yielding, C is ultimate capacity of the Pushover, D is the residual strength, and E is the failure point. Strain-hardening slope in between 3%–6% can be used. For steel brace elements, steel axial hinge model is used. The Fig. 7 represents the hinge for the steel brace member. Unlike the steel moment hinge, the brace hinge model is not symmetrical. The initial compression slope is same as the initial tension slope (FEMA 273). However, compression buckling will always occur in due to local buckling capacity of a member, which is much lower than the tension yield capacity. The values of C, D, E and C', D', E' the points called out is defined in Table 5–7 of FEMA 356. The parameters for this plastic hinge model can be determined by using the following equations:

$$P_c = 1.7 F_a A \tag{1}$$

Where, Pcr is the allowable axial stress obtained from AISC Table, 4-11 is the area of the brace cross section, and P_c represents the compressive axial force of the brace member. With P_c, deformation of the brace member can be used by using Eq. (2), where, L is the length of the brace member.

$$\Delta_c = \frac{P_c L}{AE} \tag{2}$$

For the tension side of the force-deformation relation model, the axial force and the deformation are defined per Eqs. (3) and (4), where the F_y is the yield stress of the steel. Figure 8 depicts the plastic hinge model of the brace frame used in the structure. W10X88 section is used for the brace member and the length of the brace member is 34 ft. The steel materials of F_y = 50 ksi is used.

$$P_y = F_y A \tag{3}$$

$$\Delta_T = \frac{P_y L}{AE} \tag{4}$$

5 Fragility Modeling

Fragility functions, a fundamental component in the process of assessing seismic risk, can be defined as the probability of exceeding a set of limit states, given a certain level of ground motion. Fragility curves show the probability of reaching or exceeding a specific limit state at a given level of ground shaking intensity. A limit state usually represents a damage condition, or a limitation of usage. Limit states can be defined in terms of inter-story drift ratio, strain or plastic rotation. Erberik and Elnashai (2003) explained different methodologies which may be utilized for derivation of fragility curves in their study. The first one is empirical fragility curves which are based on available damage data from previous earthquakes. The second one is judgmental fragility curves. This approach is based on expert opinions when there is not enough data to develop fragility curves. The third one is analytical fragility curves. The fragility curves which are constructed using this approach are based on engineering analysis. Probability is obtained using damage distributions simulated from analyses of structural models under increasing earthquake intensity. Hybrid methods attempt to compensate for the scarcity of observational data, subjectivity of judgmental data and modeling deficiencies of analytical procedures by combining data from different sources. In this study, analytical method is used to construct the fragility curves.

$$P[ds|Sa] = \phi\left(\frac{1}{\beta_{ds}} \cdot \ln\left(\frac{S_a}{S_{a,ds}}\right)\right) \tag{5}$$

Fragility curves express the probability of reaching or exceeding a specific limit state at a given level of ground shaking intensity. Lognormal distribution for damage probability is a common assumption in the literature (Karim and Yamazaki 2003). The probability of reaching or exceeding a limit state (LS) at a given earthquake intensity can be expressed as follows:

A log-normal distribution is used because all values are expected to be positive. The cumulative probability function is given by Eq. 7, where Φ is the standard log-normal cumulative distribution function; S_a is the spectral acceleration amplitude (PGA in this case); $S_{a,ds}$ is the median value of spectral acceleration (PGA) at which the building reaches the threshold of damage state, d_s; and β_{ds} is the normalized composite log-normal standard deviation respectively (FEMA 2012). Demand spectra and capacity curves are described probabilistically by median properties and variability parameter, βD and βC, respectively" (FEMA 2012).

$$S_{a,ds} = \delta_{R,S_{ds}} \alpha_2 \cdot h \tag{6}$$

It is noted that "Structural fragility is characterized in terms of spectral displacement and by equivalent-PGA fragility curves for buildings that are components of lifelines". FEMA also explains that damage state medians are developed using functions to describe not only the drift at the damage state, but also variability in the capacity curve, demand spectrum, and the threshold of damage itself. First, damage state drift ratios are converted to spectral acceleration using Eq. 8, where $\delta_{R,Sds}$ is the drift ratio at the threshold of structural damage state, ds; α_2 is the fraction of the building height at the location of pushover mode displacement; and h is the typical height, in inches, of the model building type of interest (FEMA 2012).

To account for variability, FEMA used Eq. 9 where βSds is the log-normal standard deviation that describes the total variability for structural damage state, ds; βC is the log-normal standard deviation parameter that describes the variability of the capacity spectrum; βD is the log-normal standard deviation parameter that describes the variability of the demand spectrum; βM(Sds) is the log-normal standard deviation parameter that describes the uncertainty in the estimate of the median value of the threshold of structural damage state, ds (FEMA 2012).

$$\beta_{S_{ds}} = \sqrt{\left(\mathrm{CONV}\left[\beta_C, \beta_D, S_{d,S_{ds}},\right]\right)^2 + \left(\beta_{M(S_{ds})}\right)^2} \tag{7}$$

"The variability of building response depends jointly on demand and capacity, since capacity curves are nonlinear. The function "CONV" implies a complex process of convolving probability distributions of the demand spectrum and the capacity curve, respectively. Demand spectra and capacity curves are described probabilistically by median properties and variability parameter, βD and βC, respectively" (FEMA 2012). FEMA uses these median values and variability to develop median values of drift,

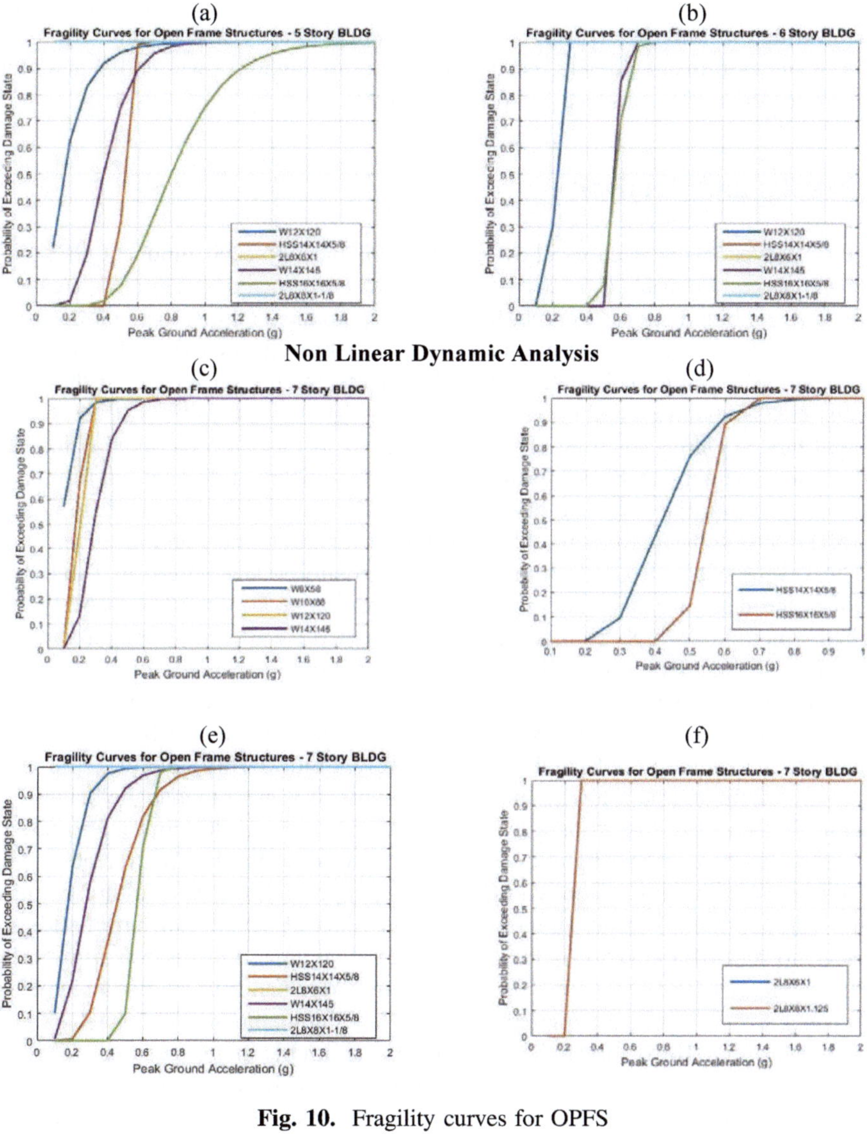

Fig. 10. Fragility curves for OPFS

Table 3. Case 2 parameters

W member section	Area (in^2)	2L member section	Area (in^2)	HSS member section	Area (in^2)
W14X145	42.7	2L8X8X1 1/8	33.6	HSS16X16x5/8	35.0
W12X120	35.2	2L8X6X1	26.2	HSS14X14X5/8	30.3
W10x88	26.0	2L6X6X1	22.0	HSS12X12X5/8	25.7
W8X58	17.1	2L6X4X7/8	16.0	HSS8X8X5/8	16.4

displacement, and acceleration for a particular damage state to occur in any of more than 36 different building types. The fragility curve theory was verified by Mander (1999) using empirical evidence. The algorithm described above was coded in a Matlab platform to produce to produce the fragility curves.

6 Results and Analysis

Fragility curves of each cases show the performance of each unique OPFS at IO. Fragility curves for each different parameter are constructed and for each case, the curves are combined into one graph. It is beneficial to have all members plotted on one graph to compare each different model. Figure 10(a) shows five different fragility curves into one graph. This data shows the curves for the OPFS where the bracing member size is increased and decreased. Figure 10(a) shows the OPFS structure where different braced element is used. The fragility curve is based on the axial deformation of the brace member at IO performance level. Significant difference performance is portrayed between the W shape brace and the HSS shape. It is important to note that W12X120 was considered as the basis of design for the models and all other bracing members were selected based on the adjusted and similar sizes listed in Table 3.

From the data, it can be interpreted that there is a 96% percent probability of the original structural, W12X120 bracing design, to reach IO after an earthquake with the Peak Ground Acceleration of 0.5 g. However, with the increase of 20% in cross sectional area and using W14X145, the probability of the structure to exceed immediate occupancy is lowered to 78%. In addition, with the use of HSS members, this probability is lowered to 30% and 8% respectively. Figures 10(b) and (c) represent the behavior of similar structure with increasing heights. It is clear that as the height of the structure is increased, the probability of the structure to reach IO limit is increased. In the 6 story structure, the probability of the original design, W12X120, to reach IO limit at 0.5 g is 100% in comparison with the 5 story graph with the 96% failure. Similarly, to Fig. 10(c) with the 7 storey structure, the probability of the HSS16, strongest bracing member, to reach IO at 0.5 g is increased to 25% compared to the 5% with 5 storey structure. Figures 10(d)–(f) shows the increase in performance based on the increase in the bracing size. It can be interpreted that other bracing members will behave similarly.

However, Fig. 10(d) shows the behavior of the double angle members. The graph only shows a single line element, since both members showed a similar failure data, and both bracing members failed at all 22 earthquake cases due to smaller cross-sectional area compared to HSS and W shapes. Additionally, only the two of the biggest sizes of 2L and HSS were considered for comparison with W shapes, because Fig. 10(e) showed the two smaller sizes of W shapes reach 100% failure for 0.5 g acceleration.

Table 4. W12X120, 7 story structure case

TABLE 4(a): Modal Parti cipating Mass Ratios									
OutputCase	StepType	StepNum	Period	UX	UY	UZ	SumUX	SumUY	SumUZ
Text	Text	Unitless	Sec	Unitless	Unitless	Unitless	Unitless	Unitless	Unitless
MODAL	Mode	1	0.35887	2.198E-17	0.7187	1.349E-07	2.198E-17	0.7187	1.349E-07
MODAL	Mode	2	0.358573	0.7198	4.55E-17	1.067E-18	0.7198	0.7187	1.349E-07
MODAL	Mode	3	0.195725	9.919E-09	6.582E-18	2.424E-17	0.7198	0.7187	1.349E-07
MODAL	Mode	4	0.189899	9.686E-18	4.777E-08	0.2901	0.7198	0.7187	0.2901
MODAL	Mode	5	0.186169	1.356E-17	1.209E-10	0.00003803	0.7198	0.7187	0.2902
MODAL	Mode	6	0.185859	6.722E-16	8.6E-09	0.01	0.7198	0.7187	0.3002
MODAL	Mode	7	0.185468	5.499E-16	1.247E-08	0.022	0.7198	0.7187	0.3222
MODAL	Mode	8	0.184826	3.866E-17	1.826E-10	0.0055	0.7198	0.7187	0.3276
MODAL	Mode	9	0.182373	9.403E-16	6.963E-11	0.0002374	0.7198	0.7187	0.3279
MODAL	Mode	10	0.182314	1.559E-15	7.197E-11	0.0004672	0.7198	0.7187	0.3283
MODAL	Mode	11	0.178175	6.264E-17	0.0024	7.183E-09	0.7198	0.7212	0.3283
MODAL	Mode	12	0.178024	6.257E-16	2.549E-07	4.15E-09	0.7198	0.7212	0.3283
MODAL	Mode	13	0.177991	1.732E-16	0.000006476	4.205E-10	0.7198	0.7212	0.3283
MODAL	Mode	14	0.177979	8.894E-16	6.469E-08	2.704E-10	0.7198	0.7212	0.3283
MODAL	Mode	15	0.17229	1.938E-16	0.0041	1.034E-08	0.7198	0.7253	0.3283
MODAL	Mode	16	0.171706	7.009E-18	0.00003757	2.995E-10	0.7198	0.7253	0.3283
MODAL	Mode	17	0.171356	2.202E-18	0.0003539	4.422E-10	0.7198	0.7257	0.3283
MODAL	Mode	18	0.116562	0.2188	1.558E-16	1.068E-15	0.9386	0.7257	0.3283
MODAL	Mode	19	0.115695	1.439E-17	0.2143	6.377E-09	0.9386	0.94	0.3283
MODAL	Mode	20	0.080484	1.257E-15	1.289E-07	0.5501	0.9386	0.94	0.8785
MODAL	Mode	21	0.065325	0.000004562	1.467E-15	3.54E-14	0.9386	0.94	0.8785
MODAL	Mode	22	0.064583	0.00009812	2.826E-14	9.365E-14	0.9387	0.94	0.8785
MODAL	Mode	23	0.064339	0.0001844	1.786E-14	1.815E-15	0.9389	0.94	0.8785
MODAL	Mode	24	0.064146	9.449E-11	4.64E-15	5.329E-13	0.9389	0.94	0.8785
MODAL	Mode	25	0.063931	0.000006847	4.377E-15	1.51E-13	0.9389	0.94	0.8785
MODAL	Mode	26	0.063323	0.0114	8.653E-15	4.683E-13	0.9503	0.94	0.8785
MODAL	Mode	27	0.061784	6.96E-08	5.918E-14	2.279E-13	0.9503	0.94	0.8785
MODAL	Mode	28	0.06174	0.0001135	2.168E-14	1.019E-13	0.9504	0.94	0.8785
MODAL	Mode	29	0.06166	4.736E-10	5.997E-16	1.621E-15	0.9504	0.94	0.8785
MODAL	Mode	30	0.061647	8.771E-07	3.783E-14	1.712E-13	0.9504	0.94	0.8785
MODAL	Mode	31	0.061638	4.052E-08	3.475E-14	1.067E-13	0.9504	0.94	0.8785
MODAL	Mode	32	0.061628	7.582E-10	1.199E-14	2.183E-14	0.9504	0.94	0.8785
MODAL	Mode	33	0.060003	1.331E-16	0.0348	1.447E-07	0.9504	0.9748	0.8785
MODAL	Mode	34	0.058786	0.0242	1.386E-14	1.458E-14	0.9746	0.9748	0.8785
MODAL	Mode	35	0.056354	1.008E-09	2.479E-14	1.105E-14	0.9746	0.9748	0.8785
MODAL	Mode	36	0.056085	1.502E-08	3.491E-15	2.703E-16	0.9746	0.9748	0.8785
MODAL	Mode	37	0.055666	3.124E-08	3.59E-14	4.653E-14	0.9746	0.9748	0.8785
MODAL	Mode	38	0.042965	4.875E-15	0.0159	1.431E-09	0.9746	0.9907	0.8785
MODAL	Mode	39	0.042524	0.0171	2.563E-15	4.815E-14	0.9917	0.9907	0.8785
MODAL	Mode	40	0.040347	3.726E-09	3.522E-19	5.567E-14	0.9917	0.9907	0.8785
MODAL	Mode	41	0.038986	1.264E-14	0.002	5.434E-09	0.9917	0.9927	0.8785

MODAL	Mode	42	0.038916	9.934E-09	3.616E-15	1.312E-13	0.9917	0.9927	0.8785
MODAL	Mode	43	0.037778	0.0008607	1.145E-14	2.478E-14	0.9926	0.9927	0.8785
MODAL	Mode	44	0.036655	6.133E-10	9.57E-14	1.802E-14	0.9926	0.9927	0.8785
MODAL	Mode	45	0.034221	5.615E-14	0.000002123	0.0398	0.9926	0.9927	0.9183
MODAL	Mode	46	0.03416	3.563E-14	0.005	0.00001636	0.9926	0.9977	0.9183
MODAL	Mode	47	0.034004	0.0051	1.619E-14	1.114E-16	0.9976	0.9977	0.9183
MODAL	Mode	48	0.03323	9.881E-10	3.493E-14	1.591E-14	0.9976	0.9977	0.9183
MODAL	Mode	49	0.032382	2.74E-14	0.000002471	1.057E-09	0.9976	0.9977	0.9183
MODAL	Mode	50	0.032274	0.00003112	3.697E-15	8.315E-14	0.9977	0.9977	0.9183
MODAL	Mode	51	0.032154	1.189E-09	5.837E-15	3.62E-14	0.9977	0.9977	0.9183
MODAL	Mode	52	0.032123	0.000003845	2.293E-17	3.835E-14	0.9977	0.9977	0.9183
MODAL	Mode	53	0.032122	2.716E-10	1.45E-14	5.766E-14	0.9977	0.9977	0.9183
MODAL	Mode	54	0.032108	4.472E-09	4.212E-15	1.73E-14	0.9977	0.9977	0.9183
MODAL	Mode	55	0.0321	0.000001054	3.123E-15	2.16E-14	0.9977	0.9977	0.9183
MODAL	Mode	56	0.032096	2.67E-08	2.343E-17	2.538E-13	0.9977	0.9977	0.9183

TABLE 4(b): Modal Load Participation Ratios				
OutputCase	ItemType	Item	Static	Dynamic
Text	Text	Text	Percent	Percent
MODAL	Acceleration	UX	99.9985	99.7672
MODAL	Acceleration	UY	99.9985	99.7711
MODAL	Acceleration	UZ	99.7712	91.8333

7 Conclusions

Fragility curves for thirty-six cases are developed for the OPFS structure with varying brace sizes and storey heights by performing Nonlinear Dynamic Analysis (Time History Analysis). The axial deformation for the brace member is considered for the performance level of Immediate Occupancy after an earthquake using FEMA 356 acceptance criteria. The following conclusions are observed:

Changing the number of stories and the bracing sizes plays an important factor in fragility curves. The number of stories and the increase in probability of failure are directly proportional. As the number of stories is increased, the probability of the structure exceeding Immediate Occupancy limit is increased. This is opposite to relation of bracing members. As the bracing sizes are increased within each structure, the probability of failure is decreased.

In the LS performance level, it is likely for the fragility curve to not reach 100% chance due to the higher value of the acceptance criteria for LS, compared to IO in accordance to the limitations set by FEMA 356. HSS braces provide the lowest probability of failure in comparison to W-shapes and 2L members. However, it is clear that the W14 and W12 will not reach 100% failure if considered within the Life Safety category. Per FEMA 356 specifications, the limit for the Immediate Occupancy failure is $0.25\Delta c$ and the limit for Life Safety is $5\Delta c$; therefore, there is a 20 times increase in the allowance limit. As a result, W14 and W12 shapes will not reach 100% failure. Considering the probability of failures based on the fragility curves provided and the extensive labor and cost associated with creating a bracing connection with HSS members due to notching and welding required in the field, W shapes are a much more feasible option to be used as bracing members due to ease of pre fabricated hole patterns and bolting in field.

References

Federal Emergency Management Agency (FEMA): Hazus–MH 2.1: Technical Manual, p. 718 (2012)

Salem, Y.S., Nasr, M.A.M.: Evaluating response modification factors of open frames steel platforms. In: Tenth US National Conference on Earthquake Engineering (2014)

Mander, J.B.: Fragility curve development for assessing the seismic vulnerability of highway bridges. Res. Prog. 1–10 (1999)

SAP2000: Integrated Structural Analysis & Design Software. Vers. 20. Computer and Structures, Inc.: Computer Software, Berkeley, CA (1997)

Erberik, M.A., Elnashai, A.S.: Seismic vulnerability of flat-slab structures (2003). https://www.ideals.illinois.edu/handle/2142/9182

Karim, K.R., Yamazaki, F.: A simplified method of constructing fragility curves for highway bridges. Earthquake Eng. Struct. Dynam. **32**(10), 1603–1626 (2003)

Analytical Fragility Curves for Pipe Rack Structure

Yasser S. Salem(✉), P. E. Tiffany Yoo, Ghad M. Gad, and Jin Sung Cho

Civil Engineering Department, Cal Poly Pomona, Pomona, CA, USA
{ysalem, thyoo, gmgad, jinsungcho}@cpp.edu

Abstract. The seismic performance of Pipe Rack structures which is a complex structure commonly found in large oil refineries is evaluated in this study. Analytical Fragility curves are developed to evaluate the performance of each pipe rack models at Life Safety damage state after an earthquake. The methodology that were used to develop the curves, are based on calculating axial deformation of the brace member and the plastic rotation of the moment frame using nonlinear time history analysis for different ground motions with different intensities and frequency contents. The fragility curves are developed based on the acceptance criteria for the brace member and the partially connection moment frame given from the FEMA 356. It was found that as the number of bay increased, the probability of the structure exceeding/reaching Life Safety performance level for the brace member significantly increased and decreased as the number of bay decreased. Similarly, as the number of stories increase, the chance of reaching LS increased, as decreasing the number of stories lowered the change of achieving LS for both moment frame and brace member. It was also found that changing the member size does not have a remarkable effect to their fragility values in Life Safety level of performance.

1 Introduction

Structural steel pipe rack structures are commonly utilized in electrical, chemical, and nuclear power plants to carry pipes, cables, and trays. Pipe are play essential part to the overall operation of the plant. Typically, the structure is a multilevel consisting moment frames in the transverse direction and brace frames in longitudinal direction called, bents and struts respectively.

These structures are known as “Nonbuilding Structures” because they carry no concrete slab or horizontal bracing to act as a diaphragm (ASCE 7-10). Therefore, this unique structure does not have same seismic performance as that of a moment frame or a braced frame.

The design procedure for pipe rack structures is governed by the following codes:

- International Building Code (IBC) Chapters 16 through 23
- Process Industry Practices (PIP) STC01015
- Minimum Design Loads for Buildings and Other Structures (ASCE/SEI 7-10) Chapter 15: Seismic Design Requirements for Nonbuilding Structures

H. Rodrigues and A. Elnashai (Eds.): GeoMEast 2018, SUCI, pp. 292–306, 2019.
https://doi.org/10.1007/978-3-030-01932-7_23

In this study, different cases with different parameters of the pipe rack structure are investigated to study the behavior of the structure under the seismic demand. Analytical fragility curves will be developed to assess the pipe rack structure vulnerability risk.

2 Sample Buildings

Structural pipe racks vary in terms of capacity, shape, and member sizes, but their main function is to carry pipe, cables, or other vessels. Generally, the design approach will depend on the job criteria. The steel pipe rack structure used in this study is constructed of Ordinary Concentrically Braced Frames (OCBF) and Ordinary Moment Frames (OMF). The moment frames are used in the transverse direction to provide open space for maintenance access and for placing pipes.

This study will focus on strutted pipe support structures made of steel to carry an average distributed load of 40psf on each level as defined in PIP STC01015. The pipe racks structure have no diaphragms or concrete slabs to allow load path between transverse bents. No other structural steels besides moment beams are placed between the bents in transverse direction. The moment connections are assumed to be constructed of bolted end plate as they are commonly used in industrial plants. A typical pipe rack structure plan, bent elevation, and longitudinal elevations can be seen in Figs. 1, 2, 3, 4, and 5. Seven different cases of pipe rack structure are considered in this study. Each case has four different parameters for the pipe rack structure. Tables 2, 3, 4, 5, 6, and 7 shows the parameters of each different cases used for the research.

Fig. 1. Pipe rack structure

Fig. 2. Pipe rack structure overview

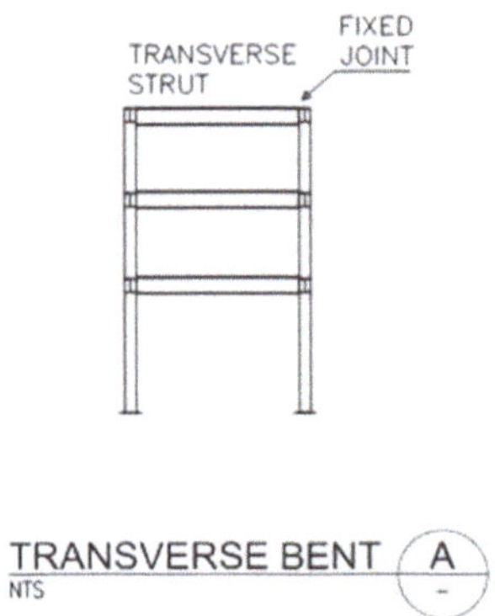

Fig. 3. Transverse view of strutted pipe rack structure

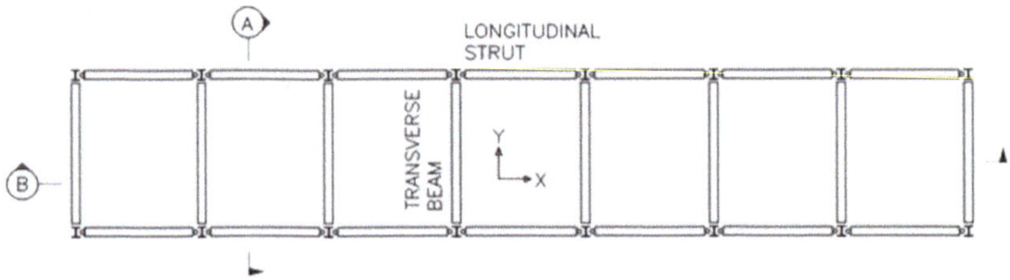

Fig. 4. Plan view of strutted pipe rack structure

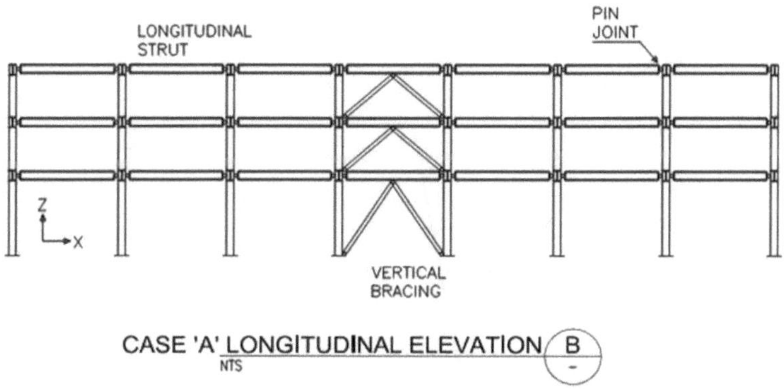

Fig. 5. Longitudinal view of strutted pipe rack structure

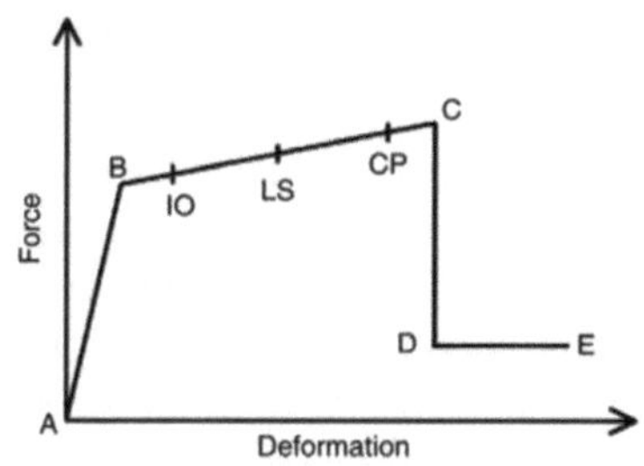

Fig. 6. Inelastic model used for beams and columns

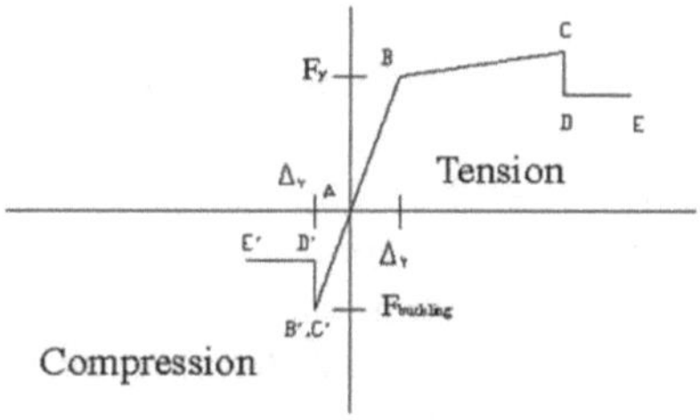

Fig. 7. Inelastic model used for bracing elements

3 Methodology

To simulate the response of these structures under seismic loads, a finite element model was developed for a typical sample model using the SAP2000 © program. The ends of the columns is pinned to model the brace frame not to take moment at the connections. The numerical model will consider a fixed bent in its analysis for the moment frame. The column's strong axis is parallel to the longitudinal direction for the simulate moment frame behaviors. In the longitudinal direction, struts connect the moment

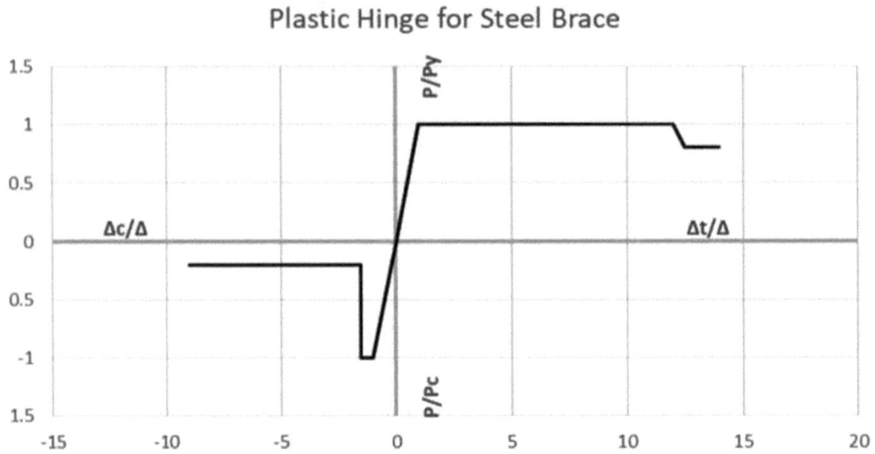

Fig. 8. Inelastic model used for 2L4x4x1/4

frame bents. In the middle bay, inverted V-bracing is placed at every level. The struts and brace members rely on axial strength and together resist applied lateral loads.

The sample structures were modeled based on the nonlinear behavior of materials and its characteristics. Pushover curves and inter-story drift ratio of the sample structures at each step of pushover analysis were produced for each case. Time history analyses were performed using In this thesis a time history was carried out to simulate the response of structure based off 22 far field and as well as near field ground motion which is selected from (PEER) ground motion database. All selected ground motions are given in Table 1. Figure 9 shows the response spectra for the 22 ground motions that were used in this study. As it can see from this graph, the selected ground motions represent a wide varieties of earthquake records that have different peak ground accelerations (PGA) and frequency contents.

Acceptance Criteria from the FEMA 356 categories different target performance after the earthquake for each of the structural members. Plastic rotation that occurs in the plastic hinge of the beam and column determines that target performance. For the axial compression member, braces, the axial deformation identifies the target performance. The FEMA 356 separates three target performances are Immediate Occupancy (IO), Life Safety (LS), and Collapse Prevention (CP). The IO level of performance targets for the structure to remain "safe to occupy" with some minor repairs. The LS targeted structure needs to remain stable and has "significant reserve capacity". Lastly IO describes the structure to remain standing, but "only barely" and any other damage or loss is acceptable. For the purpose of the research, LS is chosen to depict the behavior of the structure.

Table 1. Selected ground motions (PEER Strong Motion Database)

Earthquake	Mag. (Ms)	Station name	NEHRP Class	Distance (km)	PGA max (g)	Fault type
1994 Northridge	6.7	Beverly Hills – 14145 Mulhol	D	17.2	0.4158	Blind thrust
1994 Northridge	6.7	Canyon Country – W Lost Cany	D	12.4	0.4100	Blind thrust
1999 Duzce, Turkey	7.1	Bolu	D	12.4	0.7276	Strike-slip
1999 Hector Mine	7.1	Hector	C	12	0.2656	Strike-slip
1979 Imperial Valley	6.5	Delta	D	22.5	0.2186	Strike-slip
1979 Imperial Valley	6.5	El Centro Array #11	D	13.5	0.3640	Strike-slip
1995 Kobe, Japan	6.9	Nishi-Akashi	C	25.2	0.5030	Strike-slip
1995 Kobe, Japan	6.9	Shin-Osaka	D	28.5	0.2432	Strike-slip
1999 Kocaeli, Turkey	7.5	Duzce	D	15.4	0.2159	Strike-slip
1999 Kocaeli, Turkey	7.5	Arcelik	C	13.5	0.1502	Strike-slip
1992 Landers	7.3	Yermo Fire Station	D	23.8	0.1687	Strike-slip
1992 Landers	7.3	Coolwater	D	20	0.2754	Strike-slip
1989 Loma Prieta	6.9	Capitola	D	35.5	0.4508	Strike-slip
1989 Loma Prieta	6.9	Gilroy Array #3	D	12.8	0.5365	Strike-slip
1990 Manjil, Iran	7.4	Abbar	C	13	0.5146	Strike-slip
1987 Superstition Hills	6.5	El Centro Imp. Co.	D	18.5	0.3579	Strike-slip
1987 Superstition Hills	6.5	Poe Road (temp)	D	11.7	0.4463	Strike-slip

(*continued*)

Table 1. (*continued*)

Earthquake	Mag. (Ms)	Station name	NEHRP Class	Distance (km)	PGA max (g)	Fault type
1992 Cape Mendocino	7.0	Rio Dell Overpass	D	14.3	0.2439	Thrust
1999 Chi-Chi, Taiwan	7.6	CHY101	D	15.5	0.3100	Thrust
1999 Chi-Chi, Taiwan	7.6	TCU045	C	26.8	0.4014	Thrust
1971 San Fernando	6.6	LA - Hollywood Stor FF	D	25.9	0.2079	Thrust
1976 Friuli, Italy	6.5	Tolmezzo	C	15.8	0.3513	Thrust (part blind)

Table 2. Case 1 parameters

CASE 1							
Change in cross section of frame							
			Longitudinal		Transverse		
Case #	Story #	Height (ft)	Column	Beam	Brace	Beam	Description
Original	1	15	W10X33	W12X40	2L4X4X1/4	W10x33	
	2 & 3	5	W10X33	W12X40	2L4X4X1/4	W10x33	
1A	1	15	W12X40	W14X48	2L4X4X1/4	W10x33	Increase area & weight by 20%
	2 & 3	5	W12X40	W14X48	2L4X4X1/4	W10x33	
1B	1	15	W8x24	W10x33	2L4X4X1/4	W10x33	Decrease area & weight by 20%
	2 & 3	5	W8x24	W10x33	2L4X4X1/4	W10x33	
1C	1	15	W6X20	W10X26	2L4X4X1/4	W10x33	Decrease area & weight by 40%
	2 & 3	5	W6X20	W10X26	2L4X4X1/4	W10x33	
1D	1	15	W14X48	W16X57	2L4X4X1/4	W10x33	Increase area & weight by 40%
	2 & 3	5	W14X48	W16X57	2L4X4X1/4	W10x33	

Table 3. Case 2 parameters

CASE 2							
Change in cross section of brace							
			Longitudinal		Transverse		
Case #	Story #	Height (ft)	Column	Beam	Brace	Beam	Brace weight (plf)
Original	1	15	W10X33	W12X40	2L4X4X1/4	W10x33	6.59
	2 & 3	5	W10X33	W12X40	2L4X4X1/4	W10x33	
2A	1	15	W10x33	W12x40	HSS 4X4X.250	W10x33	6.85
	2 & 3	5	w10x33	W12X40	HSS 4X4X.250	W10x33	
2B	1	15	W10x33	W12x40	WT 4x6.5	W10x33	6.48
	2 & 3	5	w10x33	W12X40	WT 4x6.6	W10x33	
2C	1	15	W10x33	W12x40	W 6x8.5	W10x33	8.74
	2 & 3	5	w10x33	W12X40	W 6x8.5	W10x33	
2D	1	15	W10x33	W12x40	HSS 5X0.125	W10x33	6.51
	2 & 3	5	w10x33	W12X40	HSS 5X0.125	W10x33	

Table 4. Case 3 parameters

CASE 3			
Change in number of bay			
Case #	Story #	Height (ft)	# of Bays
Original	1	15	7
	2 & 3	5	
3A	1	15	5
	2 & 3	5	
3B	1	15	3
	2 & 3	5	
3C	1	15	9
	2 & 3	5	
3D	1	15	11
	2 & 3	5	

4 Analytical Model of Sample Buildings

To perform the nonlinear static analysis and nonlinear dynamic analysis plastic hinge is introduced to the structure members. Plastic hinge is assigned to each structure elements, such as beams, columns, and braces, to model the material nonlinear properties

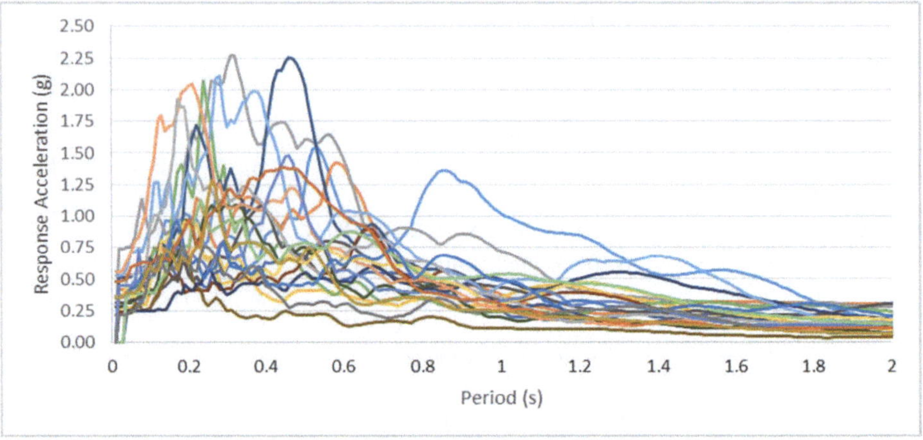

Fig. 9. Unscaled far-field spectra for ground motion records used in the non linear dynamic analysis

for SAP2000 to simulate the yielding and post-yielding behaviors. The plastic hinge properties are based on FEMA 356 standards. Figures 6 and 7 show the steel moment hinge model for beams and columns. The parameters a, b, and c are defined in FEMA 356. Strain-hardening slope in between 3 and 6% can be used.

For steel brace elements, steel axial hinge model is used. The Fig. 6 represents the hinge for the steel brace member. Unlike the steel moment hinge, the brace hinge model is not symmetrical. The initial compression slope is same as the initial tension slope (FEMA 273). The values of C, D, E and C′, D′, E′ the points called out is defined in Tables 5, 6 and 7 of FEMA 356.

Table 5. Case 4 parameters

CASE 4		
Change in number of stories		
Case #	# of Stories	Height (ft)
Original	3	30
4A	1	15
4B	4	35
4C	2	20
4D	5	40

Table 6. Case 5 parameters

CASE 5				
Change in beam span (longitudinal)				
Case #	Story #	Height (ft)	Span (ft)	Description
Original	1	15	20	
	2 & 3	5		
5A	1	15	24	Increase by 20%
	2 & 3	5		
5B	1	15	28	Increase by 40%
	2 & 3	5		
5C	1	15	16	Decrease by 20%
	2 & 3	5		
5D	1	15	12	Decrease by 40%
	2 & 3	5		

Table 7. Case 6 parameters

CASE 6				
Change in beam span (transverse)				
Case #	Story #	Height (ft)	Span (ft)	Description
Original	1	15	25	
	2 & 3	5		
6A	1	15	30	Increase by 20%
	2 & 3	5		
6B	1	15	35	Increase by 40%
	2 & 3	5		
6C	1	15	20	Decrease by 20%
	2 & 3	5		
6D	1	15	15	Decrease by 40%
	2 & 3	5		

The parameters for this plastic hinge model can be determine by using the following equations

$$P_c = 1.7F_aA \tag{1}$$

where Fa is the allowable axial stress obtained from AISC Table, 4–11 is the area of the brace cross section, and Pc represents the compressive axial force of the brace member. With Pc, deformation of the brace member can be used by using,

$$\Delta_c = \frac{P_cL}{AE} \tag{2}$$

where, L is the length of the brace member.

For the tension side of the force-deformation relation model, the axial force and the deformation is defined as

$$P_y = F_y A \tag{3}$$

$$\Delta_T = \frac{P_y L}{AE} \tag{4}$$

respectively, where the Fy is the yield stress of the steel. The Fig. 8 depicts the plastic hinge model of the brace frame used in the original structure. Double angle (2L4x4x1/4) section is used for the brace member and the length of the brace member is 18 ft. The steel materials of Fy = 50 ksi is used.

5 Fragility Modeling

Fragility curves show the probability of reaching or exceeding a specific limit state at a given level of ground shaking intensity. A limit state usually represents a damage condition, or a limitation of usage. Limit states can be defined in terms of inter-story drift ratio, strain or plastic rotation (Maniyar et al. 2009). Erberik and Elnashai (2003) explained different methodologies which may be utilized for derivation of fragility curves in their study. The first one is empirical fragility curves which are based on available damage data from previous earthquakes. The second one is judgmental fragility curves. This approach is based on expert opinions when there is not enough data to develop fragility curves. The third one is analytical fragility curves. The fragility curves which are constructed using this approach are based on engineering analysis. Probability is obtained using damage distributions simulated from analyses of structural models under increasing earthquake intensity. Hybrid methods attempt to compensate for the scarcity of observational data, subjectivity of judgmental data and modeling deficiencies of analytical procedures by combining data from different sources. In this study, analytical method is used to construct the fragility curves.

$$\mathrm{P}[\mathrm{ds}|\mathrm{Sa}] = \phi\left(\frac{1}{\beta_{\mathrm{ds}}} \cdot \ln\left(\frac{\mathrm{S_a}}{\mathrm{S_{a,ds}}}\right)\right) \tag{5}$$

Fragility curves express the probability of reaching or exceeding a specific limit state at a given level of ground shaking intensity. Lognormal distribution for damage probability is a common assumption in the literature (Karim and Yamazaki 2003). The probability of reaching or exceeding a limit state (LS) at a given earthquake intensity can be expressed as follows:

A log-normal distribution is used because all values are expected to be positive. The cumulative probability function is given by Eq. 5, where Φ is the standard log-normal cumulative distribution function; Sa is the spectral acceleration amplitude (PGA in this case); Sa, ds is the median value of spectral acceleration (PGA) at which the building reaches the threshold of damage state, ds; and βds is the normalized composite log-normal standard deviation respectively (FEMA 2012). Demand spectra and capacity curves are described probabilistically by median properties and variability parameter, βD and βC, respectively (FEMA 2012).

$$S_{a,ds} = \delta_{R,S_{ds}} \alpha_2 \cdot h \tag{6}$$

It is noted that "Structural fragility is characterized in terms of spectral displacement and by equivalent-PGA fragility curves for buildings that are components of lifelines". FEMA also explains that damage state medians are developed using functions to describe not only the drift at the damage state, but also variability in the capacity curve, demand spectrum, and the threshold of damage itself. First, damage state drift ratios are converted to spectral acceleration using Eq. 6, where δR, Sds is the drift ratio at the threshold of structural damage state, ds; α2 is the fraction of the building height at the location of pushover mode displacement; and h is the typical height, in inches, of the model building type of interest (FEMA 2012).

To account for variability, FEMA used Eq. 7 where βSds is the log-normal standard deviation that describes the total variability for structural damage state, ds; βC is the log-normal standard deviation parameter that describes the variability of the capacity spectrum; βD is the log-normal standard deviation parameter that describes the variability of the demand spectrum; βM(Sds) is the log-normal standard deviation parameter that describes the uncertainty in the estimate of the median value of the threshold of structural damage state, ds (FEMA 2012).

$$\beta_{S_{ds}} = \sqrt{\left(\mathrm{CONV}\left[\beta_C, \beta_D, S_{d,S_{ds}}\right]\right)^2 + \left(\beta_{M(S_{ds})}\right)^2} \tag{7}$$

"The variability of building response depends jointly on demand and capacity since capacity curves are nonlinear. The function "CONV" implies a complex process of convolving probability distributions of the demand spectrum and the capacity curve, respectively. Demand spectra and capacity curves are described probabilistically by median properties and variability parameter, βD and βC, respectively" (FEMA 2012). FEMA uses these median values and variability to develop median values of drift, displacement, and acceleration for a particular damage state to occur in any of more than 36 different building types. The fragility curve theory was verified by Mander (1999) using empirical evidence. The algorithm described above was coded in a Matlab platform to produce to produce the fragility curves. The curves are plotted in then the PGA corresponding to 50% chance of a certain damage state occurring is used to compare different retrofit methods and overall feasibility of each method. This will quantitatively show a relative improvement from one retrofit method to another.

6 Results and Analysis

Fragility curves of each cases show the performance of each unique pipe rack structure at LS. Fragility curves for each different parameter are constructed and for each case, the curves are combine into one graph. It is beneficial to have all four different parameters plotted on one graph to compare each different model. Figure 10(a) shows

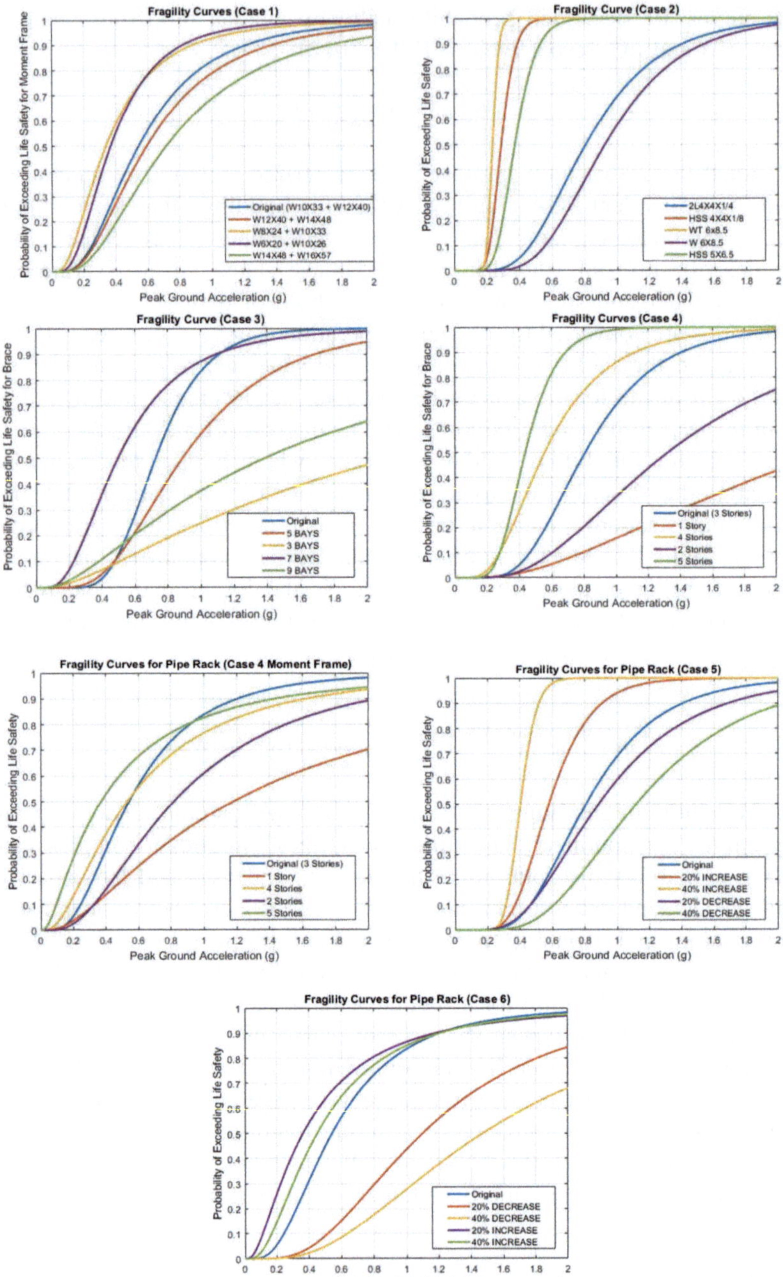

Fig. 10. Fragility curves for pipe rack structure

five different fragility curves into one graph. This data shows the curves for the pipe rack structure where the beam and column member size is increased and decreased. From the data, it can be interpreted that there is an 83% percent of the original pipe rack design will reach LS after an earthquake with the Peak Ground Acceleration of 1.0 g. However, the pipe rack design with 40% increase in beam and column size has a probability of 70% of reaching LS at the Peak Ground Acceleration of 1.0 g. This performance is based on the behavior of the moment frames in the transverse direction. Figure 10(b) shows the pipe rack structure where different braced element is used. The fragility curve is based on the axial deformation of the brace member at LS performance level. Significant difference performance is portrayed between the W shape brace and the WT shape. At a PGA of 0.6 g, wide flange brace results in a 13% chance of reaching LS, where else, the WT brace already reach 100% chance of reaching LS. For Fig. 10(c), fragility curve of the brace element is shown as the number of bay alters. As the number of bays increase, the failure mode began to change where, the plastic hinge on the axial brace member does not form (Fig. 10(d)).

Reviewing each of these fragility curves easily shows the performance of each different designs of pipe rack structure. Figure 10(e) and (f), shows the performance of the moment frame and the brace member for a same case. That is because change in number of stories affects the both moment frame and the brace member. Figure 10(g) depicts the behavior of the brace member for different strut span length, where Fig. 10(h) portrays the behavior of the moment frame for different beam length of the pipe rack structure.

7 Conclusions

Fragility curves for seven different cases are developed for the steel pipe rack structure with varying member sizes, span lengths, and story heights by performing Nonlinear Dynamic Analysis (Time History Analysis). The axial deformation for the brace member and plastic rotation of the moment frame is considered for the performance level of Life Safety after an earthquake using FEMA 356 acceptance criteria. The following conclusions are observed:

Changing the number of stories, the span length of a beam length, and the number of bays plays an important factor in fragility curves. As the number of bay increased, the probability of the structure exceeding/reaching Life Safety performance level for the brace member significantly increased and decreased as the number of bay decreased. Similarly, as the number of stories increase, the chance of reaching LS increased, as decreasing the number of stories lowered the change of achieving LS for both moment frame and brace member. The span length of the beam on both longitudinal and transverse direction have significant impact on both moment and brace frame. There is a direct relation to where increase in the beam span increases the probability of the structure reaching LS and vice versa.

Changing the member size does not have a remarkable effect to their fragility values in Life Safety level of performance.

In the LS performance level, it is likely for the fragility curve to not reach 100% chance due to the higher value of the acceptance criteria for LS, compared to IO.

References

ATC 40: Seismic Evaluation and Retrofit of Concrete Buildings ATC. Applied Technology Council 1996 1 and 2

Erberik, M.A., Elnashai, S.A.: Seismic vulnerability of flat-slab structures. Technical Report DS-9 Project (Risk Assessment Modeling) Mid-America Earthquake Center, University of Illinois at Urbana-Champaign (2003)

Federal Emergency Management Agency (FEMA): Hazus–MH 2.1: Technical Manual (2012)

Karim, K.R., Yamazaki, F.: A simplified method of constructing fragility curves for highway bridges. Earthq. Eng. Struct. D. **32**(10), 1603–1626 (2003)

Mander, J.B.: Fragility curve development for assessing the seismic vulnerability of highway bridges. Research Progress (1999)

Maniyar, M.M., Khare, R.K., Dhabal, R.P.: Probabilistic seismic performance evaluation of nonseismic, RC frame buildings. Struct. Eng. Mech. **33**(6), 725–745 (2009)

PEER Strong Motion Database. http://peer.berkeley.edu/smcat/

SAP2000: Integrated Structural Analysis & Design Software. Vers. 17. Computer and Structures, Berkeley, CA (1997). Computer Software

Ductility-Based Seismic Vulnerability Assessment of RC Bridge Piers

Abderrahmane Kibboua(✉), Fouad Kehila, Nadjib Hemaidi-Zourgui, and Mustapha Remki

National Earthquake Engineering Research Center (CGS), Algiers, Algeria
akibboua@cgs-dz.org

Abstract. In this paper, a qualitative approach, selected to estimate the seismic vulnerability index of reinforced concrete bridge piers in terms of ductility is presented. Samples of three bridges that are representative of RC bridges in Algeria were identified to carry out this study. For each structure, the sectional ductility was first computed using a multi-layered sectional analysis program (Extract). Structural ductility values were then obtained with a simplified pushover analysis carried out with a non-linear finite-element program (Seismostract). Then, the ductility demand was obtained from time-history analyses, using six ground motion records for Algiers and its surroundings (Nonlin program). Based on the Priestley theory, the ductility values were then used to elaborate a ductility index to be used as a preliminary evaluation criterion for seismic vulnerability assessment of bridge piers. The calculated ductility demand (required ductility) when compared to the ductility capacity of the selected bridge bents could be used in a seismic vulnerability assessment to identify the seismic behavior of bridge piers and to make a decision as to its safety or strengthening.

1 Introduction

The previous earthquakes such as the Loma Prieta, USA (1989), Kobe, Japan (1995), Izmit, Turkey (1999) and Chi-chi, Taiwan (1999) confirmed that bridges could be very vulnerable under dynamic (cyclic) loadings (Kibboua 2006). Seismic behaviour problems of long structures, such as bridges, are not new and have already being addressed in the past at the time of great earthquakes: Niigata (Japan) in 1964, San Fernando (California) in 1971, El Asnam (Algeria) in 1980 and Boumerdes (Algeria) in 2003. The earthquake of Kobe (January 17th, 1995) confirmed the need to understand clearly the behaviour of these structures (Kibboua et al. 2011). Many failed structures such as bridges, had originally been designed in accordance with older standard practices (Kibboua et al. 2017). Failure of a bridge structure can cause an extensive disruption to the traffic system even though each failure is limited to a particular point in the road system (Choi 2002). When bridges suffer heavy damages or collapse due to major earthquakes, emergency rescue efforts will not reach sites where assistance is needed (Hwang et al. 2000) and various business and social activities will be disrupted for a long time (Kibboua et al. 2014). In addition, road reconstruction will be useless

H. Rodrigues and A. Elnashai (Eds.): GeoMEast 2018, SUCI, pp. 307–323, 2019.
https://doi.org/10.1007/978-3-030-01932-7_24

because bridges repair in term of time is extremely longer than road repair. Thus, collapse of bridges should be prevented as much as possible (Moschonas et al. 2009).

In many regions, there was a great expansion of freeway and highway bridge structures designed by codes that are now known to provide inadequate safety, which may yet be subjected to intense shaking within their design life. As a consequence, the stock of existing bridges designed based on substandard seismic codes will exceed the number of new, well designed bridges for many years. It can thus be argued that it is important to develop methods for assessing and retrofitting existing bridges, to make a prioritization of existing bridges for retrofit and to determine which bridges are at greatest risk, and hence must be upgraded immediately, and which bridges may with little risk be left until later (Priestley et al. 1996). Thus, a significant number of existing reinforced concrete bridges requires maintenance and a permanent survey (Kibboua et al. 2008). It is important to note that in all the methods which deal with the seismic vulnerability of bridges, there is a very little consideration for bridge piers. These elements often have a great influence on the behavior of the bridge when an earthquake occurs. One of the objectives of this paper is to propose an index value that would be accounted for the ductile behavior of bridge piers and included in a preliminary rapid screening method.

Design codes have improved significantly over the years and, as a consequence, reparation and retrofitting have become necessary in order to extend the service life of many existing bridges constructed before seismic actions were adequately understood (Kibboua 2012). Since it is proven that it is too expensive to retrofit each and every bridge, efficient rapid screening methods are needed. These methods should take into account various factors such as site seismicity, seismic risks, structural vulnerability as well as social and economical impacts of a bridge failure. Most methods are designed to take into account different structural deficiencies, but the risk of a bridge pier failure is not well defined. This study presents an evaluation method, simple and efficient, based on ductility demands to establish a priority list within a group of bridges that are vulnerable in terms of displacement ductility factors of piers (Roy et al. 2004).

2 Bridges Selected for this Study

In order to evaluate the ductility demands and to propose a ductility-based evaluation index, some typical structural bridge piers in Algeria were selected for this study, considering three typical RC bridge bents. As it deals with piers, it is assumed that only the section size and the reinforcement can be changed with other conditions such as pier height, length and weight of the superstructure. The three sample bridges used to perform the analysis are listed in Tables 2 and 3.

3 Methodology

In order to characterize the seismic vulnerability of the selected bridges, a ductility study of the bridge piers was carried out. Each structure was analyzed using the following approaches:

- Sectional ductility analysis using static analysis via Extract program (Chadwell et al. 2002);
- Structural ductility analysis using pushover analysis via Seismostruct program (Seismostruct 2016);
- Ductility demand analysis using simplified non linear time history analysis via Nonlin program (Charney 1998).

3.1 Sectional Analysis

The sectional analysis of the bridge piers is carried out to get the moment curvature relationship necessary for the non linear analysis. In this respect, the cross sectional dimension of the bridge piers, the yield strength of steel σ_{sy}, the compressive strength of concrete σ'_c, the diameter of the longitudinal reinforcement bars as well as the tie reinforcement bars are taken as input parameters. The moment-curvature responses were computed with an axial load, including the superstructure gravity loads and pier weights. The sectional ductility μ_{φ}, is computed based on analysis carried out with Extract and is defined as:

$$\mu_{\varphi} = \frac{\varphi_u}{\varphi_y} \tag{1}$$

where φ_u is the ultimate curvature and φ_y is the curvature at yielding.

3.2 Structural Ductility

The structural ductility for each bridge bent was determined with an inelastic static analysis. The bridge piers were subjected to lateral loads causing successive yielding of the base and the top of the columns and subsequently their ultimate rotation. The lateral load and the resulting displacement are computed incrementally, as a function of the progressive yielding of the piers. The ultimate displacement is a function of the ultimate plastic rotation. Each bridge bent was modeled with the Seismostruct program. The size of the cap-beam and the span between the columns (piers) have an effect on the results of the analyses of the columns (piers). Actually, the probability of having damage on bridges depends not only on the structural properties (geometry, materials, etc.) but on the soil conditions as well. Therefore, in our study, the cap beam is considered as a rigid element. The structural ductility is defined as:

$$\mu_u = \frac{\Delta_u}{\Delta_y} \tag{2}$$

where Δu is the ultimate displacement and Δy is the displacement at yielding.

The yielding level of the piers is considered when yielding of the most tensioned longitudinal reinforcement bar is reached. As to the ultimate level, it is considered when the strain of the most compressed concrete fiber of the cross section of the pier reaches a strain of 0.35%.

3.3 Ductility Demand

The ductility demand was computed with a time-history analysis, using the Nonlin program. It is used to produce a time-history response of a non-linear two-dimensional framed structure subjected to ground acceleration. A bi-linear base shear-displacement relationship was used for the analysis. In the case of a flexural analysis, there is no contribution of the shear component to the displacement, however, in the case of a shear analysis, there is contribution of both the shear and the flexural components to the displacement. The ductility demand is defined as:

$$\mu_d = \frac{\Delta_d}{\Delta_y} \tag{3}$$

4 Dynamic Analysis

To perform the dynamic response analysis, the piers are modeled as single-degree-of freedom (SDOF) systems using a bilinear model (Priestley et al. 1996). Concerning strong motions, several records were obtained by the national accelerographs network of the National Earthquake Engineering Research Center (CGS). The stations at the nearest free field that recorded the main shock are located at about 20 km from the epicenter. The maximal accelerations recorded by six (06) stations are given in Table 1. The distance between the two stations at Keddara is about 100 m. The difference between the values of the maximal accelerations of the two stations was related to site effects (Belazougui 2008a, b). The analysis of the earthquake records showed that the time-length of the strong motion part is about 10 s and the frequency content is from 4 to 10 Hz (Laouami and Slimani 2013). The Fig. 1 shows the acceleration time history of the main shock recorded at the station Keddara-2. Six earthquake records were used for the study and are described in Table 1. They are spectrum compatible time histories with the response spectrum used in the Seismic Design code for Bridges in Algeria (RPOA-2008). These simulated time histories were used to match the target spectra for the bridge locations (Seismostruct 2016), and to provide a realistic representation of ground motion for the earthquake magnitudes and distances that contribute to the seismic hazard at the selected region.

Table 1. Recorded accelerations of the main shock

Station	Component (g)			Distance between the station and the epicenter (Km)
	E-W	N-S	V	
Keddara-1	0.34	0.24	0.26	20
Keddara-2	0.59	0.22	0.35	20
Hussein-Dey	0.28	0.23	0.09	36
Dar El Beida	0.55	0.46	0.16	29
Blida	0.05	0.038	0.028	72
El Affroun	0.17	0.09	0.03	86

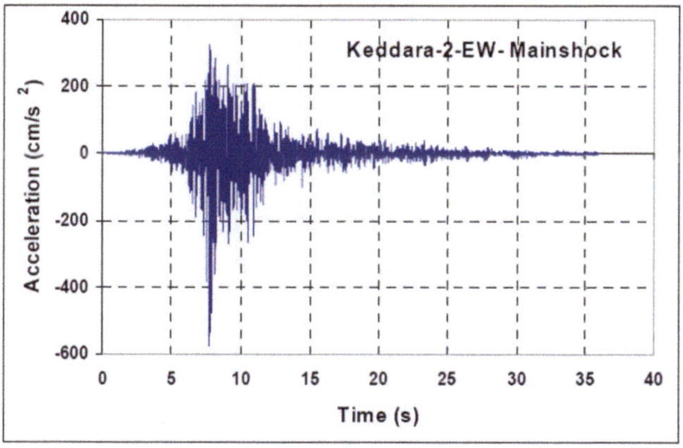

Fig. 1. Recorded acceleration at Keddara-2

5 Numerical Examples

Among typical structural bridge piers constructed in Algeria, three typical RC bridge bents were selected for the ductility analysis. As dealing with piers that were designed with the new Algerian seismic design code for bridges (RPOA 2008), it is assumed that only the size and the reinforcement of the piers can be changed with other conditions such as the height, the length and the weight of the superstructure (piers). The three sample bridge bents used to perform the analysis are listed in Tables 2 and 3.

Table 2. Selected bridges

Bridges	Number of bridge bents	Length (m)	Wide (m)	Skew (degrees)
1	1	50.50	7.50	75.50
2	1	50.52	10.00	93.00
3	2	65.90	12.50	111.00

Table 3. Bridge bents description

Bridges	Number of columns	Height (m)	Diameter (m)	Longitudinal reinforcement	Transverse reinforcement/spacing
1	2	6.40	1.40	30 HA 32	HA16/s = 15 cm
2	3	6.00	1.40	16 HA 25	HA16/s = 15 cm
3	4	6.50	1.40	16 HA 25	HA16/s = 15 cm

5.1 Bridge 1

It is a simply supported bridge with prestressed girders. It has an overall length of 50.50 m with two spans of 25.25 m each. The deck of 7.50 m width is supported by a pier and by an abutment at each end. The pier has 2 columns with a circular cross section of 1.40 m diameter each. Figure 2 gives the elevation and the lateral view of the bridge 1.

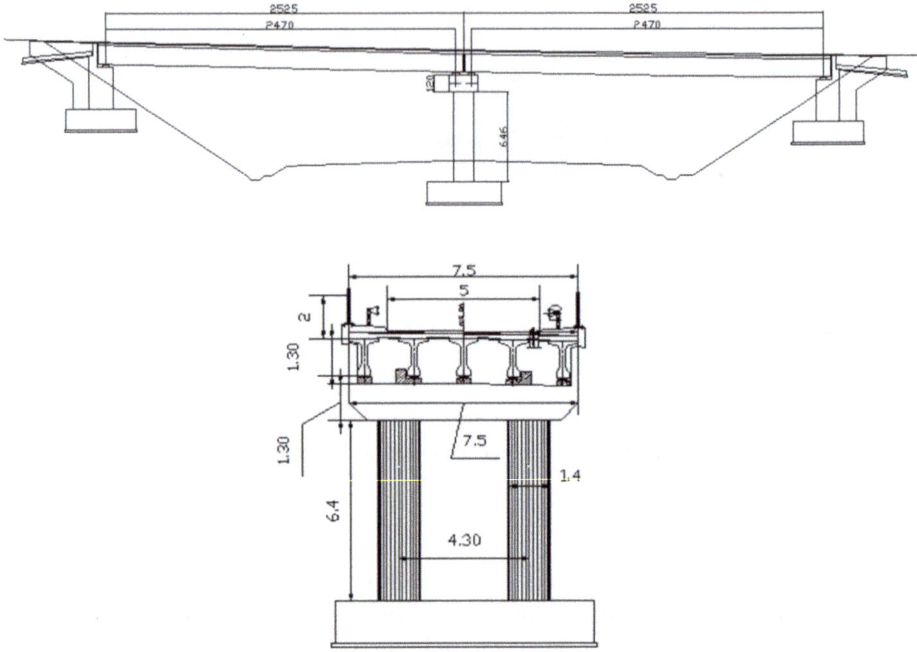

Fig. 2. Elevation and lateral view of the bridge 1

5.2 Sectional Ductility

The moment-curvature relationship for the cross section of the pier was obtained under an axial loading of 2110 KN using the Extract software. Figure 3 shows the cross section and the resulting moment curvature curve of the bridge pier.

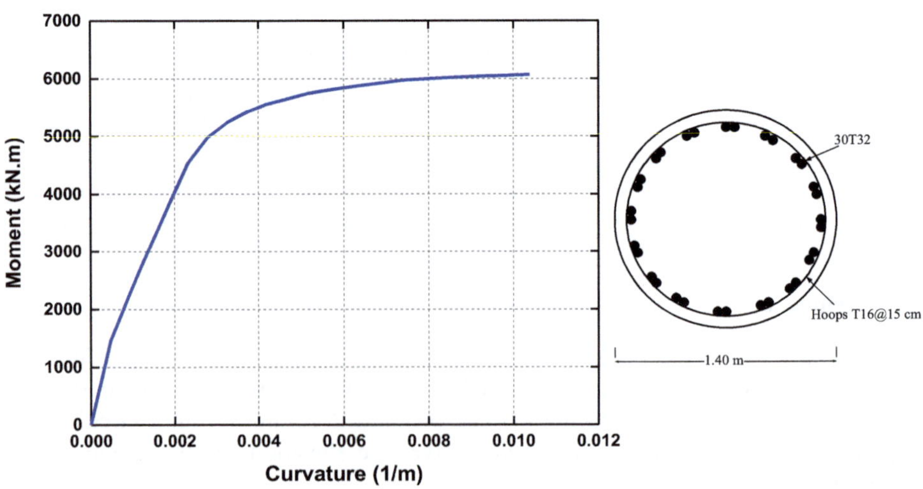

Fig. 3 Cross section and its moment-curvature relationship for the bridge pier - sample 1

5.3 Structural Ductility

In order to obtain the pier capacity of the bridge in the transverse direction, the non linear static analysis commonly called pushover analysis is used by applying a monotonically increasing horizontal load. Nonlinear concrete model of Mander et al. (1988) was used for the core and cover concrete and the bilinear Manegotto-Pinto steel model (Menegotto and Pinto 1973) was selected for the steel reinforcement bars. Plastic hinge length (Lp) was calculated according to the formula proposed by Priestley et al. (1996). Table 4 lists the mechanical characteristics of concrete and steel used in this study.

Table 4. Material properties

Material parameters	Values
Concrete	
Compressive strength	27 MPa
Tensile strength	2.22 MPa
Modulus of elasticity	33000 MPa
Strain at peak stress	0.002
Specific weight	25 KN/m^3
Steel reinforcement	
Modulus of elasticity	200000 MPa
Yield strength	400 MPa
Strain hardening parameter	0.005
Specific weight	78 KN/m^3

The 3D view of the modeled bridge bent obtained with seismostract program is shown in Fig. 4.

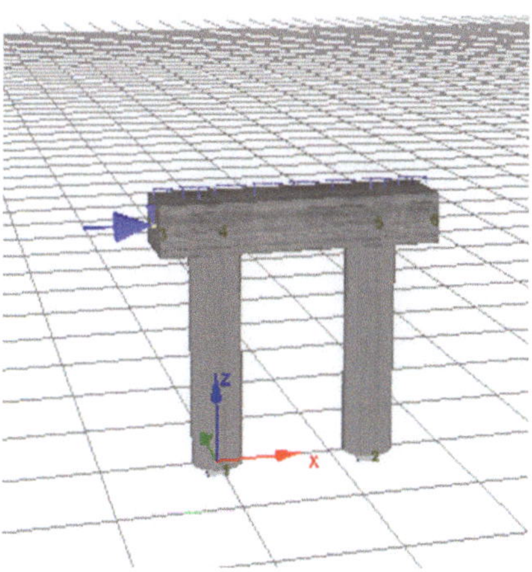

Fig. 4. 3D view of the modeled bridge bent - sample 1

The force-displacement curve (capacity curve) of the bridge pier obtained with the pushover analysis is presented in Fig. 5.

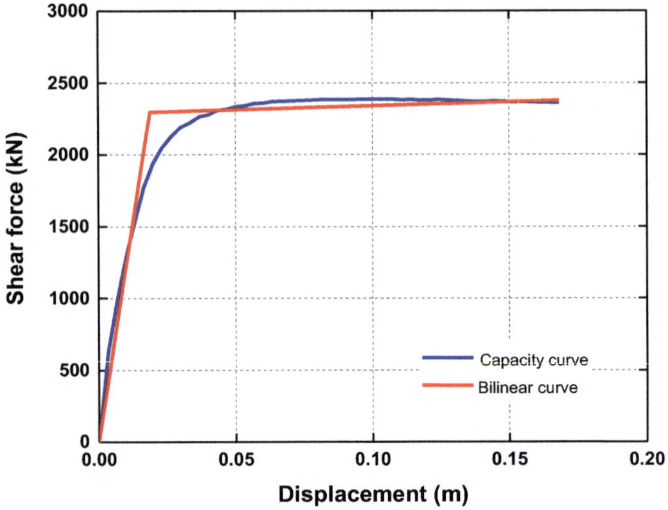

Fig. 5 Capacity curve for the bridge pier - sample 1

5.4 Bridge 2

It is a simply supported bridge with prestressed girders; it has an overall length of 50.52 m with two spans of 25.26 m each. The deck of 10.00 m width is supported by a pier and by an abutment at each end. The pier has 3 columns with a circular cross section of 1.40 m diameter each. Figure 6 gives the elevation and the lateral view of the bridge 2.

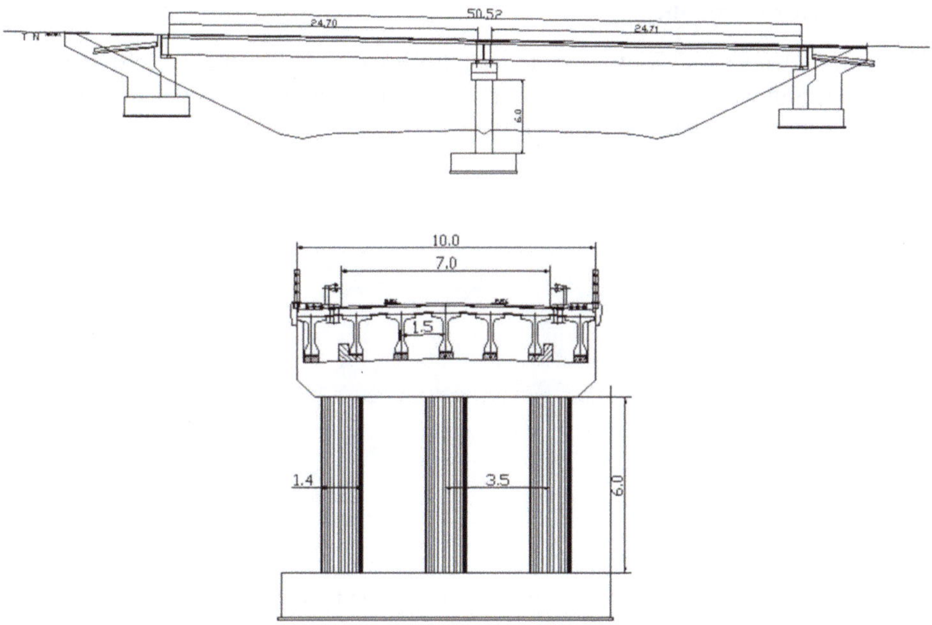

Fig. 6. Elevation and lateral view of the bridge 2

5.5 Sectional Ductility

The moment curvature curve for the cross section of the pier was obtained using the Extract software under an axial loading of 1936 KN. The Fig. 7 shows the cross section and the resulting moment curvature curve of the bridge pier.

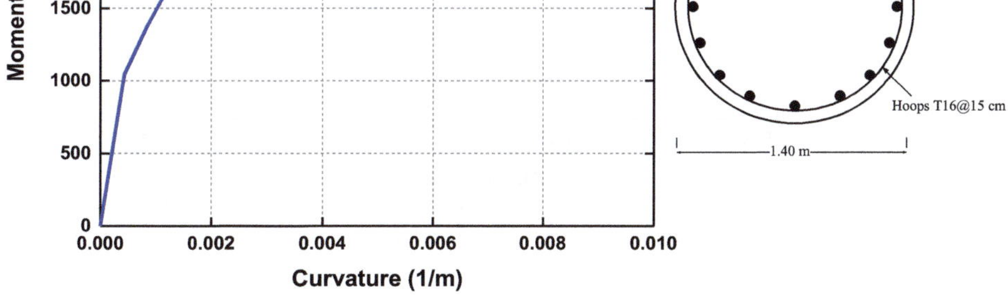

Fig. 7 Cross section and its moment-curvature relationship for the bridge pier - sample 2

5.6 Structural Ductility

Mechanical characteristics of concrete and steel used to obtain the capacity curve for bridge bent sample 2 are given in Table 4. The 3D view of the modeled bridge bent obtained with Seismostract program is shown in Fig. 8.

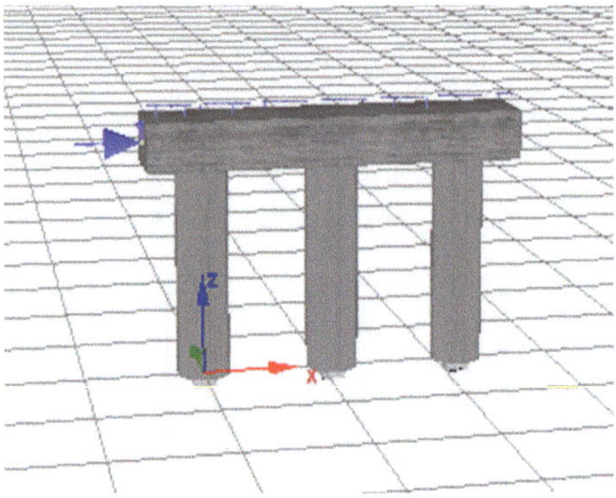

Fig. 8. 3D view of the modeled bridge bent - sample 2

The force-displacement curve (capacity curve) of the bridge pier obtained with the pushover analysis is presented in Fig. 9.

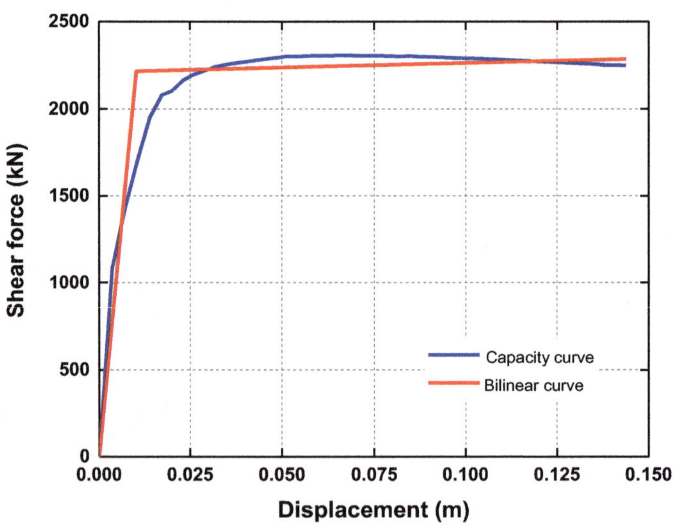

Fig. 9. Capacity curve for the bridge pier - sample 2

5.7 Bridge 3

It is a simply supported bridge with prestressed girders, it has an overall length of 65.90 m with two spans of 32.95 m each. The deck of 12.50 m width is supported by a pier and by an abutment at each end. The pier has 4 columns with a circular cross section of 1.40 m diameter each. Figure 10 gives the elevation and the lateral view of the bridge 3.

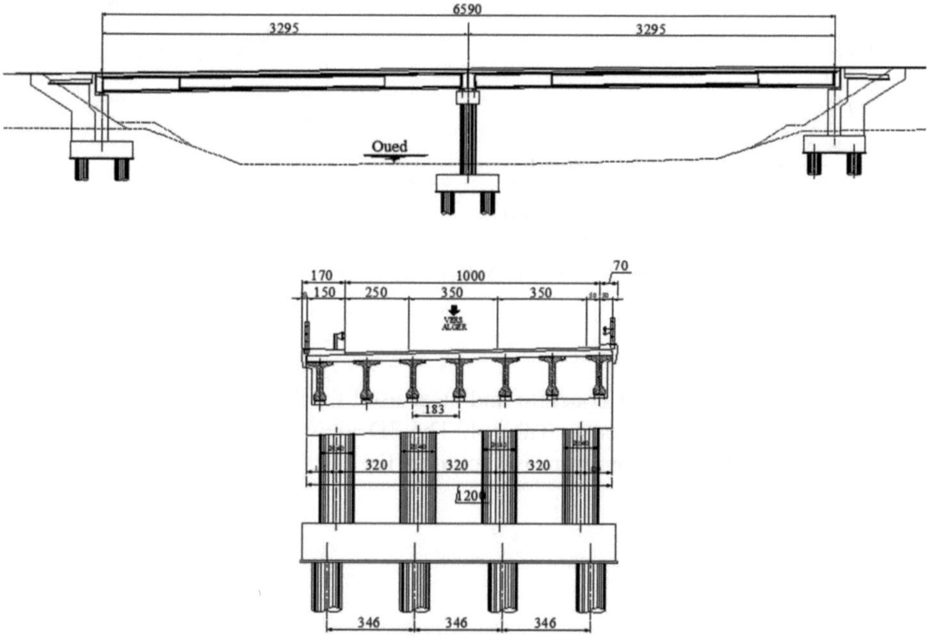

Fig. 10. Elevation and lateral view of the bridge 3

5.8 Sectional Ductility

The moment curvature curve for the cross section of the pier was obtained using the Extract software under an axial loading of 2310 KN. The Fig. 11 shows the cross section and the resulting moment curvature curve of the bridge pier.

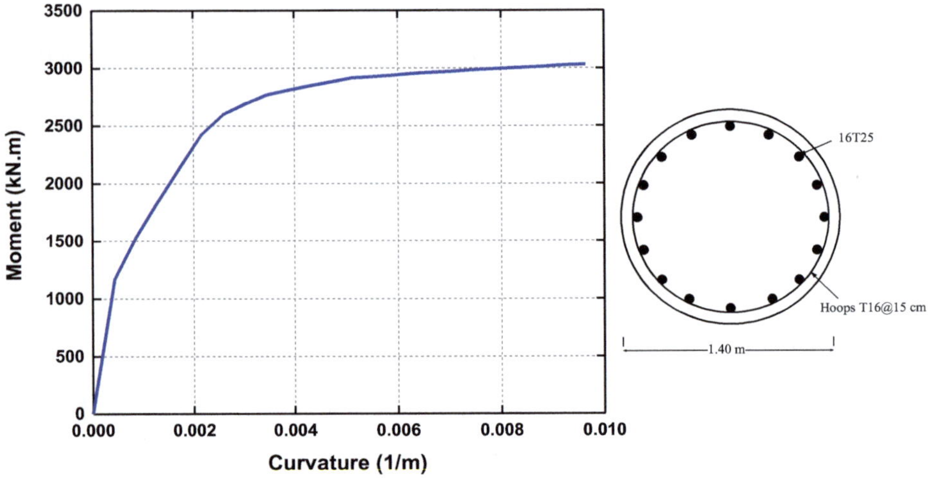

Fig. 11. Cross section and its moment-curvature relationship for the bridge pier - sample 3

5.9 Structural Ductility

Mechanical characteristics of concrete and steel used to obtain the capacity curve for bridge bent sample 3 are given in Table 4. The 3D view of the modeled bridge bent obtained with Seismostract program is shown in Fig. 12.

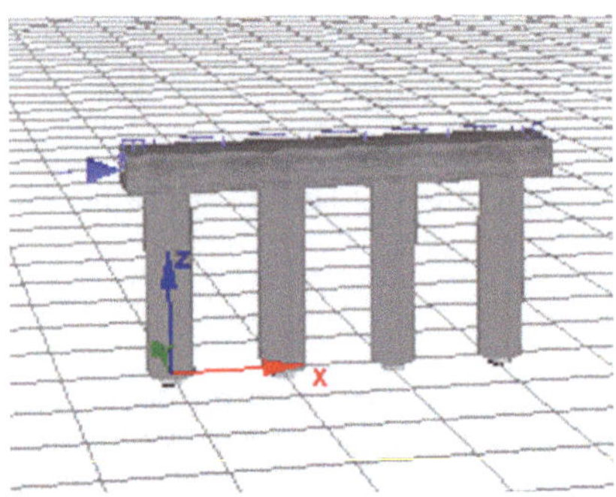

Fig. 12. 3D view of the modeled bridge bent - sample 3

The force-displacement curve (capacity curve) of the bridge pier obtained with the pushover analysis is presented in Fig. 13.

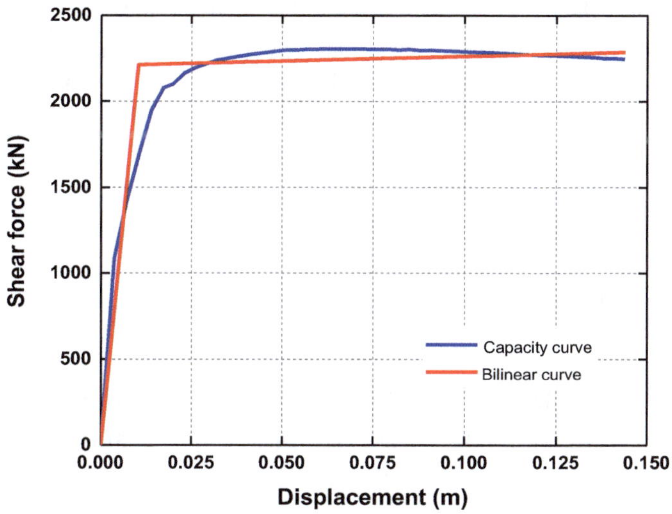

Fig. 13. Capacity curve for the bridge pier - sample 3

6 Analysis Results of Typical Bridges

For each bridge bent, the structural ductility results obtained from the pushover analysis using Seismostruct program and the ductility demand obtained from the time history analysis using Nonlin program are given in Tables 5, 6 and 7.

Table 5. Results of the ductility study for bridge 1

Seismic records	Component E-W (g)	Ductility demand $\mu_d = \frac{\Delta_d}{\Delta_y}$	Structural ductility $\mu_u = \frac{\Delta_u}{\Delta_y}$
Keddara-1	0.34	0.346	8.81
Keddara-2	0.59	0.345	
Hussein-Dey	0.28	0.419	
Dar El Beida	0.55	1.265	
Blida	0.05	0.215	
El Affroun	0.17	0.365	

The bridges exhibit significant non linear responses when subjected to each of the six selected earthquake records. As it can be seen from Tables 5, 6 and 7, non linear excursions are significant for all bridges.

On the basis of previous analyses carried out for each bridge type, the structural response to a given dynamic effect expressed through the ductility ratio is given by the following relationships.

Structural ductility (Ductility capacity): $\mu_u = \frac{\Delta_u}{\Delta_y}$

Table 6. Results of the ductility study for bridge 2

Seismic records	Component E-W (g)	Ductility demand $\mu_d = \frac{\Delta_d}{\Delta_y}$	Structural ductility $\mu_u = \frac{\Delta_u}{\Delta_y}$
Keddara-1	0.34	0.564	13.88
Keddara-2	0.59	0.560	
Hussein-Dey	0.28	0.841	
Dar El Beida	0.55	2.991	
Blida	0.05	0.266	
El Affroun	0.17	0.950	

Table 7. Results of the ductility study for bridge 3

Seismic records	Component E-W (g)	Ductility demand $\mu_d = \frac{\Delta_d}{\Delta_y}$	Structural ductility $\mu_u = \frac{\Delta_u}{\Delta_y}$
Keddara-1	0.34	0.596	13.17
Keddara-2	0.59	0.590	
Hussein-Dey	0.28	0.722	
Dar El Beida	0.55	1.691	
Blida	0.05	0.372	
El Affroun	0.17	0.611	

Ductility demand (Required ductility): $\mu_d = \frac{\Delta_d}{\Delta_y}$
where Δ_y, Δ_u and Δ_d are the yield, ultimate and demand displacements respectively.

- If $\mu_d \leq 1$: The system's response is in the elastic range.
- If $1 \leq \mu_d \leq \mu_u$: The system's response is in the nonlinear range.
- If $\mu_u \leq \mu_d$: The system's response is in the state of failure.

The maximum ductility demand (required ductility): $\mu_d \leq \mu_u$ and $\mu_d \leq 3$ to 4.
The choice of the maximum values $3 \leq \mu_u \leq 4$ for this study is based on the fact that it is applicable to the bridge networks in Algeria because the typologies of Algerian bridges are among the European typologies and especially for this type of bridges studied in this paper (prestressed concrete multi beam bridges).

7 Damage Control Limit States

Under a design level earthquake of reduced probability of occurrence in comparison with the probability of occurrence relative to a serviceability limit state, a certain amount of repairable damage may be permissible. The damage may include spalling of cover concrete requiring cover replacement, and the formation of wide flexural cracks requiring grouting to avoid later corrosion problems. However, the essential aspect of response to this limit state is that the required repair should essentially be superficial.

Fracture of transverse reinforcement or buckling of longitudinal reinforcement should not occur, and the core concrete in plastic hinge regions should not need replacement. Ground shaking of intensity likely to induce response corresponding to the damage control limit state should have a low probability of occurrence during the expected life of the structure. With well designed bridges, this limit state generally corresponds to displacement ductility factors in the range of $\mathbf{3 \leq \mu_u \leq 6}$, although an upper limit of $\mathbf{\mu_u = 4}$ would be felt to be appropriate in Europe (Priestley et al. 1996). The value of the maximum ductility demand (required ductility) $\mathbf{3 \leq \mu_u \leq 4}$ is also appropriate for Algeria but only for bridges of the same typology as the European prestressed concrete multi beam bridges. Moreover, if we want to confirm these values, in this case we have to make experimental tests using the reaction wall of the National Earthquake Engineering Research Center (CGS) laboratory.

8 Discussion

On the basis of the different obtained results from Tables 5, 6 and 7, the required ductility is compared with the ductility capacity for the six (06) different selected ground motions in order to have an idea about the seismic behavior of the structure.

The ductility demand (required ductility) of the three bridges is smaller than the structural ductility (ductility capacity). Consequently, the bridges don't need strengthening to comply with the relationships given in the previous section.

From Tables 5, 6 and 7, the responses of the bridge bent 3 in terms of the ductility demand (required ductility) are higher than those of the bridge bents 1 and 2 when subjected to the earthquakes Keddara-1, Keddara-2 and Blida, as the natural frequency of the bridge bent 3 lies, in contrary to the other bridge bents, within the frequency range corresponding to high acceleration amplitudes of the 3 earthquake motions. On the other hand, and for similar reasons, the bridge pier 2 when subjected to the earthquake motions Hussein-Dey, Dar El Beida and El Affroun, it results in larger ductility demands than it does for the bridge bents 1 and 3.

It is important to note that this is a qualitative comparison of the ductility of bridge bents for a selection of typical bridges in Algeria. This method has some limitations, like for shorter piers that might experience shear failure and never reach the computed structural ductility. For example, if a bridge pier has a low slenderness ratio (l/d) and would probably not fail in flexure, or is in the limit of failing by flexure, a shear behavior study need to be carried out to assess properly this ratio.

In a seismic evaluation context for a retrofit study, the proposed index is an interesting correlation between the ductility ratios and the seismic behaviour of bridge bents. It can be used to identify, within a given sample of bridges, the structures that are more vulnerable and that should be prioritized in a rehabilitation program. It could also be easily included in a classification procedure that would account for the structural behavior of bridge piers.

9 Conclusions

A sample of reinforced concrete bridges, representative of typical bridges constructed in Algeria was selected to carry out a ductility study. Sectional and structural ductilities were evaluated for each bridge pier. Simplified pushover and time-history analyses were carried out and the resulting ultimate displacement (the maximum expected displacement) and ductility demands were used to compute a ductility index. The calculated ductility demand (required ductility) when compared with the structural capacity (ductility capacity), could be used in a seismic vulnerability assessment to identify the seismic behavior of a bridge structure and to make a decision as to its safety or strengthening. The ductility ratios do not account for low slenderness ratio columns (short columns), and more detailed analyses of bridge bents' shear behaviour need to be completed for an appropriate application of the proposed ductility ratio. This a qualitative approach, as a simple method was selected to carry out this study and the criteria used in this procedure are intended to be only used for preliminary evaluations of the seismic vulnerability of existing bridge structures.

References

Kibboua, A.: Analyse dynamique sous vibrations ambiantes d'un pont à haubans sur l'Oued Dib à Mila. Magister Thesis, Ecole Nationale des Travaux Publics, Algiers, Algeria (2006)

Kibboua, A., Naili, M., Benouar, D., Kehila, F.: Analytical fragility curves for typical Algerian reinforced concrete bridge piers. Struct. Eng. Mech. **39**(3), 411–425 (2011)

Kibboua, A., Kehila, F., Hemaidi-Zourgui, N., Remki, M.: Comparison between fragility curves of RC bridge piers designed by old and recent Algerian codes. Eurasian J. Eng. Sci. Technol. **1**(2), 56–67 (2017)

Choi, E.: Seismic analysis and retrofit of mid-America bridges. Doctoral dissertation, School of Civil and Environmental Engineering, Georgia Institute of Technology (2002)

Hwang, H., Jernigan, J.B., Lin, Y.W.: Evaluation of seismic damage to Memphis bridges and highway systems. J. Bridg. Eng. **5**(4), 322–330 (2000)

Kibboua, A., Bechtoula, H., Mehani, Y., Naili, M.: Vulnerability assessment of reinforced concrete bridge structures in Algiers using scenario earthquakes. Bull. Earthq. Eng. **12**(2), 807–827 (2014)

Moschonas, I.F., Kappos, A.J., Panetsos, P., Papadopoulos, V., Makarios, T., Thanopoulos, P.: Seismic fragility curves for Greek bridges: methodology and case studies. Bull. Earthq. Eng. **7**(2), 439–468 (2009)

Priestley, M.N., Seible, F., Calvi, G.M.: Seismic Design and Retrofit of Bridges. Wiley, Hoboken (1996)

Kibboua, A., Farsi, M.N., Chatelain, J.L., Guillier, B., Bechtoula, H., Mehani, Y.: Modal analysis and ambient vibration measurements on Mila-Algeria cable stayed bridge. Struct. Eng. Mech. **29**(2), 171–186 (2008)

Kibboua, A.: Développement d'une méthodologie d'analyse pour la vulnérabilité sismique des piles de ponts Algériens. Doctoral dissertation. Université des Sciences et de la Technologie Houari Boumediene, Algiers, Algeria (2012)

Roy, N., Paultre, P., Proulx, J.: A bridge ductility study for seismic assessment and rehabilitation. In: 13th World Conference on Earthquake Engineering, Vancouver, Canada (2004)

Chadwell, C.B., Imbsen & Associates.: XTRACT—cross section analysis software for structural and earthquake engineering (2002). http://www.imbsen.com/xtract.htm

SeismoStruct: A computer program for processing strong motion data, Seismosoft Ltd (2016). www.seismosoft.com

Charney, F.A.: NONLIN—nonlinear dynamic time history analysis of single degree of freedom systems. Federal Emergency Management Agency Training Center, Emmitsburg, Maryland, Advanced Structural Concepts, Golden, CO and Schnabel Engineernig, Denver, Co (1998)

Belazougui, M.: Overview of last 30 years Algerian experience in the field of seismic risk reduction, history of earthquake engineering lessons learned and outcomes in today practice —perspectives from different countries. Association Française du Génie Parasismique (AFPS), Nice, France (2008a)

Belazougui, M.: Boumerdes Algeria earthquake of May 21, 2003: damage analysis and behavior of beam-column reinforced concrete structures. In: 14th World Conference on Earthquake Engineering, Beijing, China (2008b)

Laouami, N., Slimani, A.: Earthquake induced site effect in the algiers–boumerdes region: relation between spectral ratios higher peaks and observed damage during the may 21st mw 6.8 boumerdes earthquake (algeria). Pure Appl. Geophys. **170**(11), 1785–1801 (2013)

RPOA: Règles Parasismiques Applicables au Domaine des Ouvrages d'Art. Document Technique Règlementaire, Ministère des Travaux Publics, Alger, Algeria (2008)

Mander, J., Priestley, M.J.N., Park, R.: Theorical stress-strain model for confined concrete. J. Struct. Eng. **114**(8), 1804–1826 (1988)

Menegotto, M., Pinto, P.E.: Method of analysis for cyclically loaded R.C. plane frames including changes in geometry and non-elastic behaviour of elements under combined normal force and bending. In: Symposium on the Resistance and Ultimate Deformability of Structures Acted on by Well Defined Repeated Loads, International Association for Bridge and Structural Engineering, Zurich, Switzerland, pp. 15–22 (1973)

Behavior of Boxed Cold-Formed Steel as Composite Beam with Rebar as Shear Connector

Musab N. A. Salih(✉), Mahmood Md. Tahir, Shahrin Mohammad, and Yusof Ahmad

Institute for Smart Infrastructure and Innovative Construction, Faculty of Civil Engineering, Universiti Teknologi Malaysia, Johor Bahru, Malaysia
nasmusab@yahoo.com

Abstract. The use of composite systems comprising of concrete and conventional steel is commonly used in a multi-story steel frame with hot-rolled steel section. However, the use of cold-formed steel section designed as composite beam is yet to be established. Cold-formed steel (CFS) is usually categorized as a slender section which tends to buckle and deform. The strength of this section is usually reduced significantly as the section is very thin. However, all these problems can be significantly improved by designing the section as composite construction by integrating concrete and steel with the use of rebar as a shear connector. Therefore, this paper presents experimental works on the structural performance of cold-formed steel (CFS) section with self-compacting concrete (SCC) as a composite beam by means of the shear connection mechanism of the proposed using reinforcement bar. A specimen was carried out where the behavior of encased double cold-formed steel composite beam with composite slab was tested until failure. The proposed shear connector of size 12 mm was embedded in CFS and encased in SCC with a compressive strength of 40 N/mm^2. The specimen comprised of two parallel CFS attached together to form encased beam with a concrete slab. Based on the experimental results it was found that the proposed composite system using bent-up rebar as shear connector showed slightly higher results than expected with ultimate moment capacity of 358.3 kNm for experimental and 341 kNm for predicted results. The failure mode was recorded as ductile which make it suitable to be used in the seismic zone.

Keywords: Cold-formed steel · Composite beam · Encased sections Self-compacting concrete · Shear connector

1 Introduction

Over the past 2 decades, the expanding utilization of cold-formed steel in the construction industry has brought substantial progress in research to ensure strength and reliability of constructed steel structures. However, the main restrictive aspect of CFS is

H. Rodrigues and A. Elnashai (Eds.): GeoMEast 2018, SUCI, pp. 324–332, 2019.
https://doi.org/10.1007/978-3-030-01932-7_25

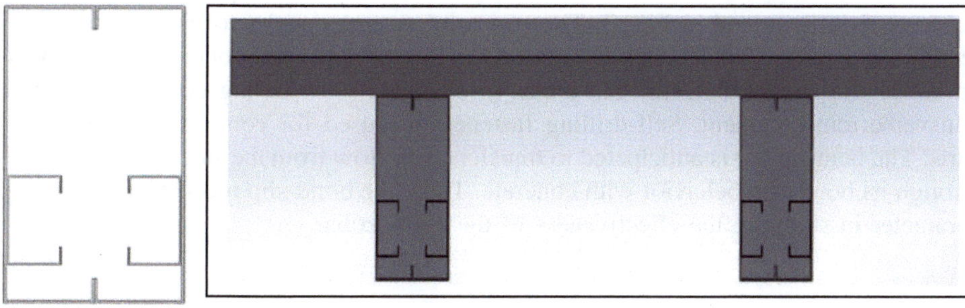

Fig. 1. CFS box section

the thinness of the section that makes it vulnerable to lateral-torsional, lateral-distortional torsional, distortional, and local buckling problems (Irwan et al. 2011). Therefore, orienting two CFS lipped channel sections toe-to-toe to form a boxed section restores symmetricity and minimizes lateral-torsional buckling effect as in Fig. 1. These two advantages retain by the system, encourage the use of CFS sections in a wider range of structural applications (Irwan et al. 2011). The research includes the study of buckling properties of cold-formed steel members, the failure mode of the members, and strength and stiffness of the cold-formed steel member.

A composite floor structure regularly consists of a reinforced concrete slab that is supported on a set of steel beams. Steel and concrete parts should be fully connected such that shear flow can be transferred between them. The composite action reliability is provided by shear connectors. The fundamental concept in composite construction is to use the advantages of both concrete and steel materials while evading their characteristic disadvantages. For this idea to work, steel and concrete parts should be fully connected so that no delamination and/or slip can arise between the two parts.

Therefore, to take advantage of both concrete and steel, they can conjoin to form a composite structure. The composite structure that combines steel and concrete is composite slab which is more robust, stiffer and stronger than using an individual material. The composite structure can benefit from both the advantages of concrete and steel together. The capability of the composite slab to bear loads governed by the degree of connection between the concrete and the steel. Therefore, whenever the connection between these two materials is improved, the stiffness and capacity of composite slab increased.

Ammar et al. (2012) performed a research on ductility and strength of concrete encased composite beams. Their study showed that the ductility of the encased beam was very high due to the high percentage of steel area and this was one of the promising features for use in seismic zones.

This paper investigates a structural performance of CFS with Self-Compacting Concrete (SCC) as a composite beam system. A full-scale testing has been performed to study the performance of a new type of shear connector for the proposed composite beam system illustrated in Fig. 2. The structural system consists of two toe-to-toe C-sections, a corrugated steel deck (which acts as formwork for the concrete slab), a bent-up bar as the shear connector and a cast-in-place self-compacting concrete slab with transverse reinforcement. Self-drilling fasteners are used for connecting all the steel parts. The bent-up bar is anticipated to transfer shear flow from the steel joist to the slab through its bond-slip behavior with concrete. Thus, the bond-slip mechanism is the key parameter in studying the effectiveness of the bent-up bar.

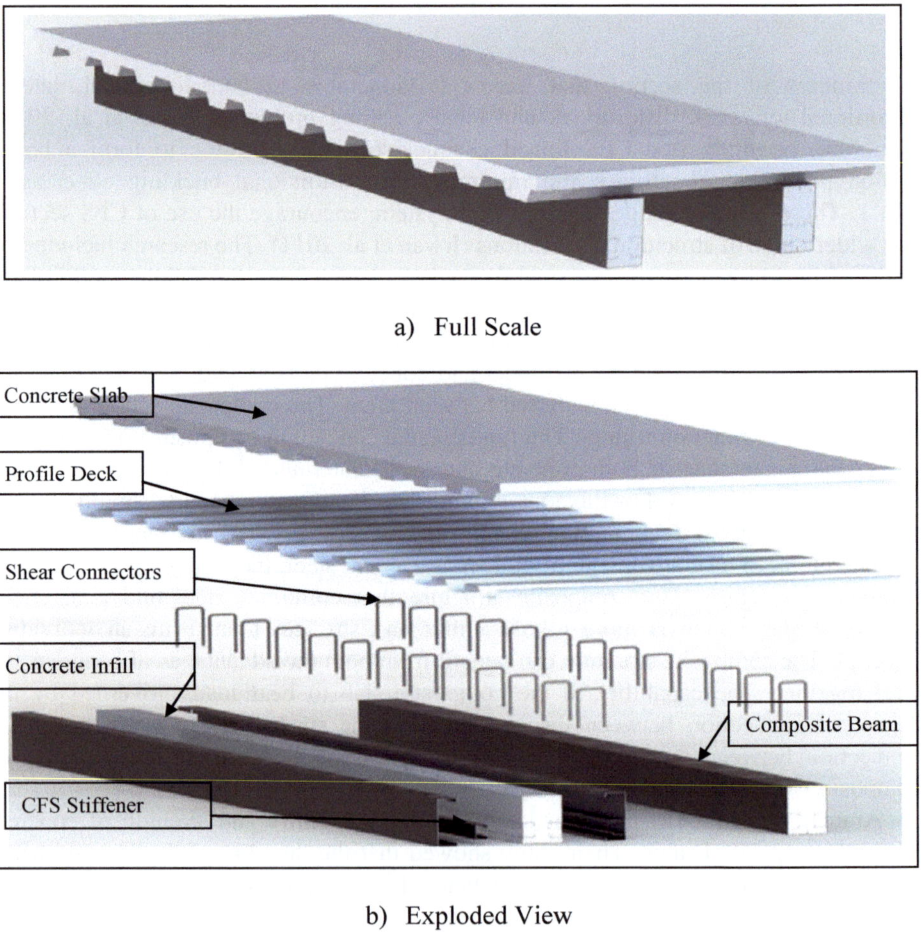

Fig. 2. The proposed composite system

2 Materials and Methods

2.1 Material

CFS of lipped channel section toe-to-toe with the depth of 250 mm, the width of 150 mm and lipped depth of 20 mm and a thickness of 2.4 mm with a yield strength of 450 N/mm^2 is used for the proposed encased composite beam with a modulus of elasticity taken as 210 kN/mm^2. The proposed shear connector is a bent-up rebar of 12 mm diameter with a yield strength of 275 N/mm^2. Welded wire mesh A142 with a diameter of 6 mm with a spacing of 200 mm in both directions was used as reinforcement mesh and SCC of grade 40 N/mm^2 at 28 days respectively are the material used in this study. The profile steel deck yield strength of 350 MPa, the nominal thickness is 1.0 mm and profiled sheeting rib height of 50 mm.

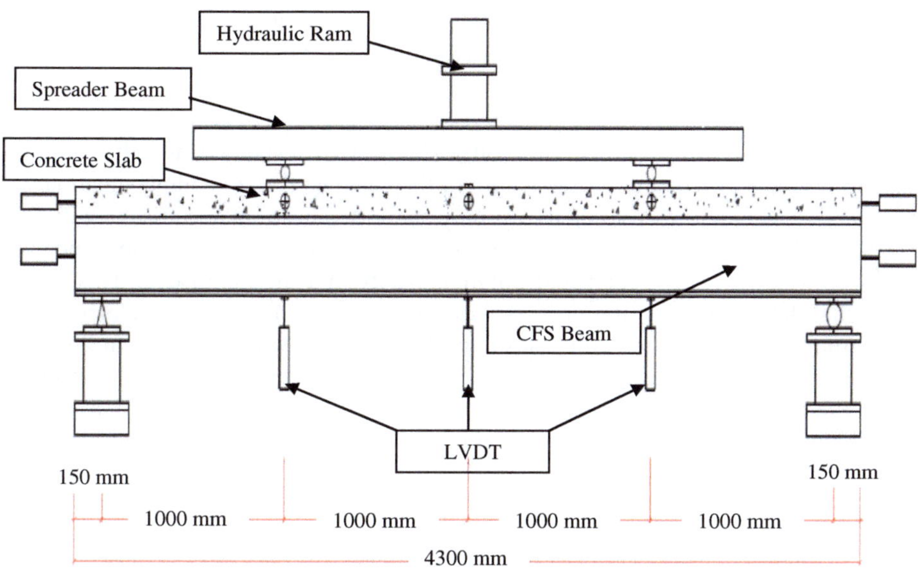

Fig. 3. Schematic diagram of a test arrangement

2.2 Test Setup and Instrumentation

The sample was supported with a roller at one end and a pinned support at the other end to represent a simply supported beam. Deflections of the specimen were monitored using six linear variable displacement transducers (LVDT's) under the bottom flanges of the CFS section at mid-span and quarter-spans of the test specimen as in Fig. 3. Slips were also monitored using two LVDT's attached to the centerline of the concrete slab faces at the supports sides i.e. roller and pin.

2.3 Testing Procedure

A hydraulic jack machine with capacity of 2000 kN and 100 kN load cell as used in the setting up of a full-scale testing. The test specimen was subjected to four points bending test where the loads were applied on the 4 m beam using a hydraulic jack at the mid-span and were converted into a two-point load using a spreader beam of 2 m spacing which was supported by two roller plates. This distance was still within the standard distance of a quarter of the length of the beam so that a bending situation was assured. The specimen was set up as a simply supported beam in accordance with EC4 (EN1994-1-1 2004). The load was applied at an increment of 10 kN. The specimen was loaded up to 20% of its anticipated failure capacity and then unloaded. This was to ensure that the instrumentation process was satisfactory and that the specimen was in equilibrium state prior to the proper testing. The specimen was loaded again this time, not up to 20% of its predicted capacity. The load was increased until the specimen failed. The testing of the specimen was stopped when there was a significant drop in the applied load or when a large deformation of the test specimen was spotted (Table 1).

Table 1. Composite specimen dimensions

Specimen ID	TT-250-12-12
Length (mm)	4000
Slab width (mm)	1500
Slab thickness (mm)	100
Shear connector size (mm)	12
Shear connector longitudinal spacing (mm)	332
The depth of CFS (mm)	250
Transverse spacing between beams (mm)	750

Table 2. Experimental and theoretical capacities of composite specimen

Specimen ID	TT-250-12-12
f_{ck} at test day (N/mm^2)	60.55
Ultimate load P_u, exp (kN)	713.6
Mid-span deflection at P_u, exp (mm)	45.5
Ultimate moment M_u, exp (kNm)	358.3
Interpolation method M_u, theory (kNm)	337.26
M_u, exp/M_u, theory	0.94
Stress block method M_u, theory (kNm)	365.58
M_u, exp/M_u, theory	1.02

3 Experimental Results and Validation

The failure modes observed in the tested full-scale composite beam specimen could be due to concrete cracking or crushing or both and in some cases combined with CFS web buckling underneath the point load position. But, at the point loads positions, a combination of maximum moment and maximum shear could be experienced. Table 2 shows the maximum bending moment the specimens could undergo underneath the point loads positions.

The load carrying capacity of the tested specimen or Load against mid-span deflections curves for specimens is shown in Fig. 4. The ultimate loads (Pu, exp.) of the specimen were 713.6 kN, and with a 45.5 mm recorded mid-span deflections at an ultimate loads level. The specimens in this category exhibited a similar mode of failure. The failure mode initiated was the formation of longitudinal cracks along the line of shear connectors on the surface of the concrete slab and transverse cracks at loading and support zones. The load was monotonically increased until failure occurred.

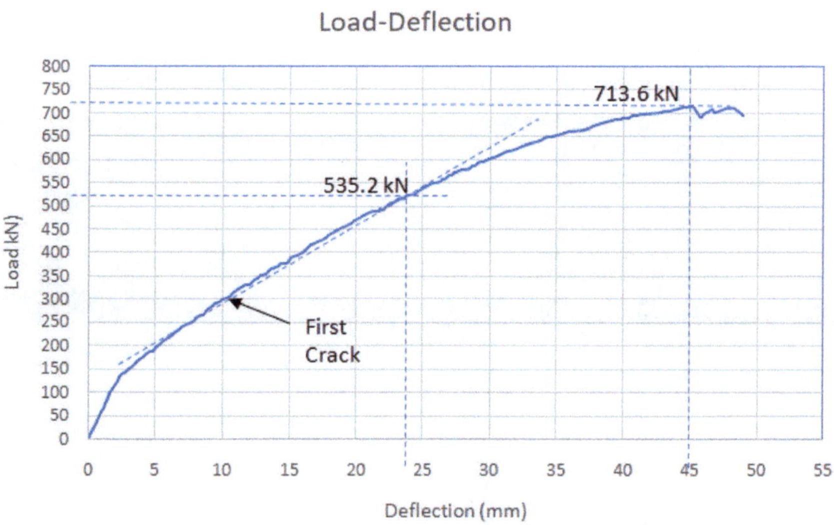

Fig. 4. Load-deflection curve

The first crack occurred at 330 kN on the concrete slab surface on top of the support zone. The specimen failed due to local buckling of the CFS section when the ultimate was reached as shown in Fig. 5.

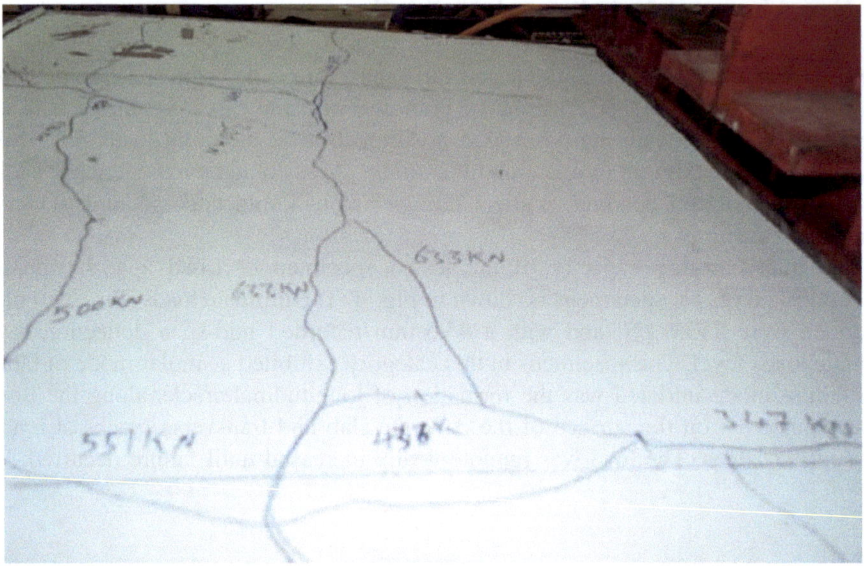

a) Transverse and Longitudinal Cracks in Concrete Slab Surface

b) Local Buckling on CFS Web

Fig. 5. Failure modes of composite specimens

3.1 Predicted Section Capacity

The theoretical analysis and validation of the experimental results consisted of determination of the composite beams flexural capacity. This includes the evaluation of the predicted plastic moment capacity (Mu, theory) of the test specimens based on the recommendations of EC4. For a composite section with the partial shear connection, the full plastic capacity of the composite section has not been reached. In the approach adopted from EC4, the moment capacity of the composite beam section designed with the partial shear connection is determined by using two methods namely; Linear-interaction method (or interpolation method) and stress block method (or exact method).

According to Eurocode 3 (EC3), the effective area should be determined to assume that the cross-section is only subjected to stresses due to uniform axial compression. Meanwhile, the effective section modulus should be determined to assume the cross-section is only subjected to bending stresses. The predicted strength capacity of the section is depending on the section classification and effective area due to the thinness of the CFS section used it is classified as Class 4.

4 Conclusions

The present experimental study is conducted to gain knowledge about the resistance of the composite system by means of using a partially encased CFS beam and an innovative type of shear connector. The main conclusions of this investigation are as follows:

1. The results show that it is possible to predict the ultimate capacity of the proposed composite system by using the procedure adopted from EC4.
2. Using bent-up rebar as shear connector gives a high degree of composite action ratio which means attaining higher yield strength of the composite section.
3. The partially encased steel beam is an efficient technique for the enhancing its flexural capacity.

Acknowledgments. The work reported in this study was graciously supported by Universiti Teknologi Malaysia – Construction Research Center (UTM-CRC) and Institute for Smart Infrastructure and Innovative Construction (*ISIIC*) with Grant Number 4B235 under the supervision of Prof. Mahmood Md. Tahir and the help of Dr. Shek Poi. The author remains indebted for the support and help and collaboration to all members of the team.

References

Irwan, J.M., Hanizah, A.H., Azmi, I., Koh, H.B.: A large-scale test of symmetric cold-formed steel (CFS)–concrete composite beams with BTTST enhancement. J. Constr. Steel Res. **67**(4), 720–726 (2011)

Kvočáka, V., Drab, L.: Partially-encased composite thin-walled steel beams. Procedia Eng. **40**, 91–95 (2012)

EN1994: Design of Composite Steel and Concrete Structures-Part 1-1: General Rules and Rules for Buildings. European Committee for Standardization, Brussels (2004)

Nonlinear Finite Element Analysis of CFRP Shear Walls Subject to Blast Loading

J. L. Ajimituhuo and O. S. Abejide(✉)

Department of Civil Engineering, Ahmadu Bello University, Zaria, Nigeria
abejideos@yahoo.com

Abstract. Designers are drawn to a fresh challenge of improving the capacity of military and civil infrastructures in order to protect lives and properties during blast attacks which is becoming frequent in occurrence throughout the world. However, there seem to remain an ambiguity as to the accurate assessment of threat levels associated with any blast case which may be due to shortage in experimental data of blast loading and its effects on structures as a result of cost. This study therefore evaluates the nonlinear dynamic response of a shear wall structure member subject to blast loading. The extent of damage or response of the wall as per stresses, strains, pressures and displacements were effectively measured with the aid of finite element (FE) which is simulated in ABAQUS software. The laminated wall performed better with an average of over 90% improvement over the control wall in all the response categories considered. Also, while the walls may be damaged, the CFRP laminates contains it in order to save lives.

Keywords: Blast load · Trinitrolouene (TNT) · Shear wall · CFRP

1 Introduction

Blast explosions have also become common due to terrorist attacks all over the world in recent times. These blast explosions have posed dangerous threats on existing structures as they exert heavy impulses on the structures within a very short time. Because of the catastrophic nature of blast explosion attacks, designers have been drawn to a fresh challenge of improving the capacity of military and civil infrastructures in order to protect human lives and properties during the attacks. Moreover, Manmohan and Vasant (2014) also reported that blast-induced threats cannot be assessed accurately, as it is not possible to account for every threat level to which structures may be subjected. Only agreeable threat levels are arrived at after a detailed load assessment. Though other explosive cases such as gas explosion; whether domestic or in laboratories exist, drastic measures need be taken to ensure safety of structure. This is because of rate of loss of human lives and property as a result of collapse associated with blast loading is alarming.

Meanwhile, conventional structures are not usually designed to withstand and resist blast loads and this has led to a destructive damage whenever blast explosions occur. Draganić and Sigmund (2012) commented that Eurocodes (EN 1991–1–7) have not dealt with this type of load and as such, further elaborations are needed as engineers

H. Rodrigues and A. Elnashai (Eds.): GeoMEast 2018, SUCI, pp. 333–345, 2019.
https://doi.org/10.1007/978-3-030-01932-7_26

have no guidelines on how to design or evaluate structures for the blast phenomenon. This may be due to shortage of sufficient experimental data. Therefore, experimental studies with intentions to showcase the dynamic response of structural elements under blast loadings as well as a detailed understanding of the blast phenomenon itself need to be highly promoted. Though engineering in the military circles seem to have been acquainted with blast resistant designs, there is also need to study and incorporate this idea while building structures with large occupancy like hospitals, shopping malls, trade centres, etc. This is because of the unprecedented occurrence of terrorist attacks involving blasts in many parts of the world in recent times. Therefore, proper implementation schemes ought to be developed to ensure safety of lives and properties.

Moreover, researches and studies so far prove that the bulk of studies have been carried out on reinforced concrete shear walls in buildings. These studies include interests on optimum positioning of the walls to obtain higher resisting effect (Sejal 2014; Varsha 2014; Anil and Yajdani 2015; Suresh and Yadav 2015), and on the strengthening of the shear walls. There is now therefore, need to conduct researches on the non-linear analyses of shear walls and its corresponding structural safety when improved with Fibre Reinforced Plastic (FRP) composite laminates after been subjected to blast loading. The nonlinear analysis helps us to measure probabilistically its failure mode and in turn, tells us how efficient our successful idea of incorporating an external reinforcement (CFRP) improves shear walls resistance in buildings will be. This is the core focus of this study.

1.1 Background Study on FRP Laminated Shear Walls

The incorporation of FRP composites in the shear wall structure is an innovative and recent practice in civil engineering. The application of composite materials in shear walls vary from strengthening, retrofitting and repair of existing shear walls, to innovative design ideas of shear walls for special purposes.

Li et al. (2005) conducted a finite element modelling of cyclic behaviour of shear wall structures retrofitted using GFRP. In their research, the authors adopted the SPRING element to simulate the constraint on deformation provided by FRP strips and concluded that the model predicts the failure mode and the overall hysteretic behaviour of GFRP retrofitted shear wall structures reasonably well. Fariborz and Masoud (2013) conducted an analytical study on the strengthened steel plate shear walls by FRP laminates. They investigated the nonlinear behaviour of composited steel plate shear walls by means of glass fibre reinforced polymer (GFRP) laminates under quasi-static loading. In that regard, they selected and simulated using finite element method, numbers of the tested half scaled one-story un-stiffened steel plate shear walls (SPWs) based on the available experimental data in the literature. Their results indicate that the GFRP laminate increases the ultimate shear capacities of SPWs, and the secant shear stiffness of the system, especially if the principal orientation of GFRP laminates lie in the direction of tension field line. This issue of orientation has also been investigated by Norris et al. (1997), and they concluded that strength enhancement and failure mode were related to the direction of the reinforcing fibre. They buttressed that brittle failure modes associated with the use of CFRP might be avoided by using particular combinations of fibres and orientations.

Other experimental studies particularly with the use of Carbon Fibre Reinforced Polymer (CFRP) include seismic strengthening and repair of reinforced concrete shear walls by Lombard et al. (2000). They presented the results obtained in a feasibility study on the effectiveness of using externally bonded carbon fibre tow sheets for the seismic strengthening and repair of reinforced concrete shear wall. They arrayed four 2.0 × 1.5 × 0.1 m reinforced concrete shear wall test specimens (which included a control wall, a repaired wall and two strengthened walls). The test specimens were loaded to failure in the in-plane direction according to a predetermined quasi-static cyclic loading sequence. Based on the test results and the good correlation between the measured data and the analytical predictions (they developed), it was concluded that the new retrofit technique is an effective seismic strengthening and repair procedure for reinforced concrete shear walls. The carbon fibre system can be used to recover the initial elastic stiffness and to increase the yield load and ultimate flexural capacity of seismically damaged walls, as well as to enhance the stiffness and strength capacity in the strengthening application of undamaged walls. Cruz-Noguez et al. (2014) also conducted a similar study using CFRP tow sheets (though, different wall specimen set-up) for seismic retrofitting and concluded that the carbon fibre system can be used to recover (in repair applications) or increase (in strengthening applications) the initial elastic stiffness and the maximum flexural capacity of flexurally dominant walls. They also proposed an experimental program that shows promising results in terms of increasing the ultimate strength and ductility of the system that is currently underway which will result in changing the mechanism of failure from brittle diagonal shear to a ductile one.

However, loss of composite action (de-bonding) is always associated with FRP-shear walls and thereby leading to premature failure. De-bonding mechanism plays a major role in the response of FRP-reinforced shear walls and hence, anchoring systems are a critical part of any retrofitting scheme, although they are usually not considered when using conventional section analysis. David and Cruz-Noguez (2013), discussed a simple design procedure for the design of an innovative tube anchor, suitable in practise. Other failures associated with laminated shear wall include yielding of the reinforcement steel (followed by rupture of the FRP laminate), and failure of the concrete (Esfahani et al. 2007).

Cyclic tests on CFRP-Retrofitted Reinforced concrete shear wall panels were also conducted by Galal and El-Sokkary (2012). They concluded that the FRP-rehabilitated wall panel performed efficiently showing an improved flexural behaviour compared to the control wall. The control wall was able to sustain a displacement ductility of 10.0 measured at an average lateral load of 59 kN. The retrofitted wall showed an increase in the flexural capacity of 80% compared to the control wall accompanied by a decrease of the wall's displacement ductility. Kheyroddin and Naderpour (2008), also performed nonlinear finite element analysis of composite reinforced concrete (RC) shear walls. From their study, they postulated that steel jacketing could be an alternative for the boundary elements of RC shear walls, but the application of CFRP sheets is an effective seismic strengthening procedure. Using CFRP sheets, the flexural and shear strength would be increased by applying the CFRP sheets with the fibres oriented in the vertical or horizontal direction. The sheets also increase the pre-cracked stiffness, the cracking load (up to 35%) and the ultimate flexural capacity (up to 18%) of the RC walls.

More particularly on blast mitigation and resistance, many research studies have been conducted and more are still underway. This is because of the need for experimental and accurate computational results, which would be of consideration for the preparation of blast criteria in building codes. Lantz et al. (2016) made a State-of-the-art review on the blast protection of unreinforced masonry walls. In their study, they used several retrofitting materials and found out that FRP and polyurea are the two most widely studied retrofitting techniques because of their effectiveness, lightweight, practicality of application, and cost. Myers et al. (2004) confirmed this by demonstrating that FRP composites offer great benefits for the strengthening of masonry walls to resist blast loads. FRP systems have been proven to increase the out-of-plane flexure capacity of un-reinforced masonry walls (URM) elements to resist a higher level of blast threat levels.

Other researches on the strengthening of masonry wall systems with FRPs include that of Bazan and Charles (2010), Buchan and Chen (2007), Verma et al. (2016), Carney and Myers (2005), and Ehsani and Pena (2009). In general, their findings were that FRP strengthening of URMs is an effective blast explosion method. The FRP retrofits increases the endurance capacity of the wall structures and also prevent fragmentation. Loss of life and property damage due to high speed projectiles of the URM as a result of the blast pressure intensity is also prevented when strengthened with CFRP.

Razaqpur et al. (2006), also conducted an experimental and analytic test to check the response of reinforced concrete panels reinforced with externally bonded GFRP laminates. In their test, control panels against GFRP retrofitted ones were set up and subjected to blast. They concluded that in an average sense, the retrofitted panels had higher residual strength than the companion control panels, but there was no consistent trend. They further stressed that from their result, it is difficult to draw definite conclusions about the blast mitigation effectiveness of the GFRP bonding laminates. Verma et al. (2016) also observed the limitation of GFRP system and reported its deterioration in performance when exposed to saline or alkaline conditions.

1.2 Nonlinear Analysis Modelling of Composite Shear Wall in FE Using ABAQUS Software

Kheyroddin and Naderpour (2008) captured that for composite wall systems, four kinds of analysis models are usually used. They are (i) equivalent frame models, (ii) multi-spring models, (iii) fibre section models, (iv) continuum element model.

This continuum finite element model exists in many advanced finite element analysis software (including ABAQUS adopted for this study). It is very suitable for three dimensional situations and its input parameters are easy to specify. Other advantages of the continuum finite element method in connection to this study include the fact that it provides a more physical description of the nonlinearities that occur in RC shear walls. The possibility of modelling the distribution of diagonal cracking and local crushing makes such models more realistic (Kheyroddin and Naderpour 2008). Continuum elements also capture important behavioural responses such as axial-flexure interaction, inelastic shear deformation, and the steel confining effect of concrete behaviour, concrete compression softening, and concrete tension stiffening (Spacone and El-Tawil 2004).

1.3 Mechanism of Blast Load

The mechanism of a blast is such that after detonation, the ambient pressure increases almost instantaneously and promptly begins to decay, forming a nearly triangular overpressure pulse. The peak pressure is called the peak positive over pressure. It represents the pressure seen at a point in space when the shock wave is unimpeded in its motion. The duration of the positive overpressure is called the positive phase. The peak over pressure and positive phase duration determine the specific impulse of the blast wave. All three blast wave parameters influence the property damage and injury that the blast wave can cause (Bajic et al. 2009). The Fig. 1, below best explains the mechanism.

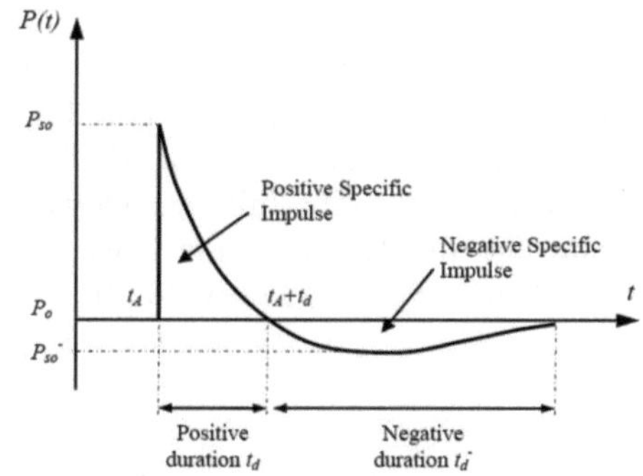

Fig. 1. Relationship of wave pressure to time history (Mohammed et al. 2014). Detonation takes place at time t = 0. After time t_A, the blast wave arrives at the point and pressure instantaneously increases from ambient pressure, P_o to peak overpressure, P_{so} caused by the detonation. At time $t_A + t_d$, the pressure returns to ambient pressure, P_o which is positive phase, is over and followed by negative phase, P_{so}

The Fig. 1 above expresses the simplest behavioural form of a blast wave and it is termed the Friedlander waveform. The equation for a Friedlander waveform describes the pressure of the blast wave as a function of time (Dewey 2010).

That is,

$$P(t) = P_{so}\left(e^{\left(-t/t^{x}\right)}\right)\left(1 - \left(t/t^{x}\right)\right) \tag{1}$$

There are common sets of equations which are normally used for calculation of blast wave from explosive charges, called the Sadovsky's formulas. Regardless on the nature of explosion, the equation converts the magnitude on trinitrotoluene

(TNT) equivalent. Below are the blast parameters determination for overpressure, positive phase duration and impulse (Bajic et al. 2009).

$$\Delta P_m = 1.02\frac{q^{\frac{1}{3}}}{r} + 4.36\frac{q^{\frac{2}{3}}}{r^2} + 14\frac{q}{r^3}\,(\text{bar}) \tag{2}$$

$$\tau = 1.2\sqrt[6]{q}\sqrt{r}\,(\text{ms}) \tag{3}$$

$$I = 200\frac{q^{\frac{2}{3}}}{r}\ (\text{Pa}) \tag{4}$$

Where q is the explosive mass in kg and r is the (standoff) distance in m, ΔP_m is blast overpressure, τ is positive phase duration and I is the impulse.

1.4 Characterization of the Blast Load (Pressure)

According to Sadovsky's formular, Eq. (2), blast pressure ΔP_m is a function of the mass of the bomb specimen, q, and the standoff distance, r. This formular works regardless of the nature of explosion and depends only on trinitrotoluene (TNT) equivalent of explosive charge. The explosive type for the blast load case is a mixture of ammonium nitrate and fuel oil (ANFO), which is first converted to its equivalent mass in Trinitroluene (TNT). The full characteristic of the explosive type and parameters to determine the full length of the blast duration according to the Friedlander waveform are presented below (Dewey 2010).

(a) See Table 1.

Table 1. Parameters for explosive

Explosive	ANFO
Mass	108.9 kg
TNT equivalence factor	0.88
TNT equivalent mass	87.9 kg
Explosive charge	3717 kJ/kg
Stand-off distance	9.14

(b) Calculations using Sadovsky's Formular

I. $\Delta P_m = 1.02\frac{(87.9)^{\frac{1}{3}}}{9.14} + 3.9\frac{(87.9)^{\frac{2}{3}}}{9.14^2} + 13\frac{(87.9)}{9.14^3}$

$\Delta P_m = 3.1397\,\text{bar} = 314\,\text{kN/m}^2$

II. $\tau = 1.2\sqrt[6]{87.9}\sqrt{9.14} = 0.01\,\text{s}\,(\text{approx.})$

III. $I = 200\frac{87.9^{\frac{2}{3}}}{9.14} = 432.6\,\text{N/m}^2$

The Table 2 below shows the pressure–time calculations using Friedlander's equation (Dewey 2010).

(c) See Table 2.

Table 2. Pressure-time calculation using Friedlander equation

Time (ms)	1	2	3	4	5	6	7	8	9	10	11	12
Pressure (kN/m^2)	0	−42.5	−31.3	−17.3	−17.3	−8.5	−3.9	−1.7	−0.7	−0.1	−0.05	−0.02

(d) The waveform of Friedlander is presented in Fig. 2.

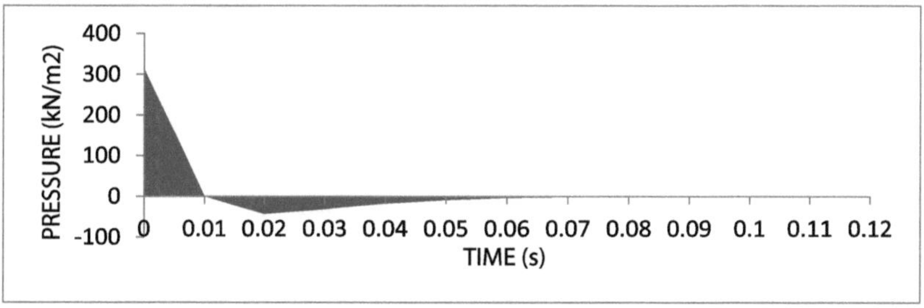

Fig. 2. Relationship of blast pressure to time (Friedlander's waveform)

1.5 Finite Element Method Coded in ABAQUS

Below is a summary of the steps involved in the run of ABAQUS software for shear wall-modelling.

(a) *Part module:* this module provides the tools for creating, editing and managing the elements of the FRP shear wall as well as the control wall. These elements are the steel reinforcement, concrete wall and CFRP composite wrap.

(b) *Property module:* this module specifies the different characters and material properties of the elements created in the part module. The table below gives the summary of parameters used in the part module session.

Table 3. Input parameters for FE assessment in ABAQUS

Materials	Compressive strength	Yield strength	Young's modulus	Density	Poisson's ratio
Concrete	30	–	31	2400	0.2
Steel	–	410	210E+09	7880	0.33
CFRP	–	2250	1.5E+09	1500	0.28

(c) *Assembly module:* the element parts of the model are assembled in this module.
(d) *Step module:* this module is used to create the analysis type and sequence. Specifying the sequence provides a convenient way to capture changes in the loading and boundary conditions of the model
(e) *Interaction module:* this module is a step-dependent module. According to the type of connection used, the contact between the concrete shear wall with steel reinforcement bar and CFRP laminates are defined.
(f) *Load module:* this module is also step dependent. The boundary conditions and load types are specified. For this study, the walls are fixed on all four sides and blast load (simulated as pressure), was used. The magnitude of the amplitude will be set with respect to the calculated pressure in the Table 3 above in a tabular form, as shown in Plate I.

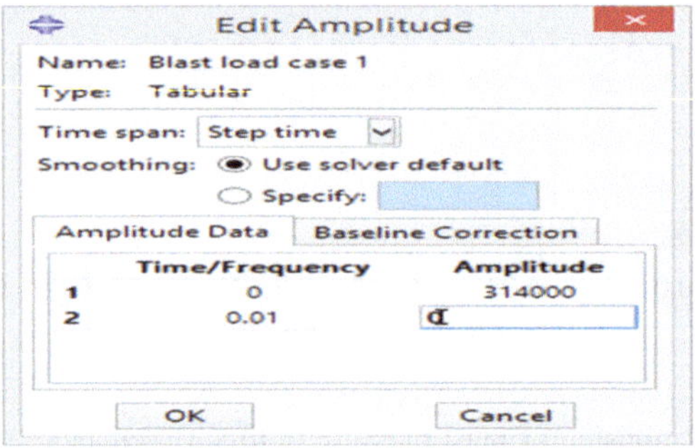

Plate I. Input of the blast load parameter

This helps to obtain the positive shock wave effect of the blast which is responsible for the structural damage as reported by Bajic et al. (2009).
(g) *Mesh module:* according to the simplicity of modelling, sizing the mesh to achieve the most accurate result is necessary. For easy analysis, the meshes are assigned to all the various elements before they are assembled (Plate II).
(h) *Job module:* this module helps to specify analysis type, create the job and submit it for analysis, while one monitors its progress.

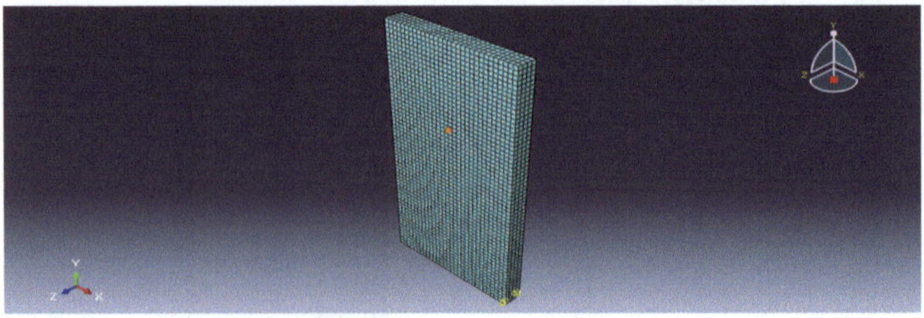

Plate II. The meshing process in ABAQUS

2 Results and Discussion

The source of the load case considered is the experimental load in the study performed by Ehsani and Pena (2009). All other necessary characterizations have been duly outlined. The ABAQUS software presents the structural mechanics in terms of stress patterns, load-deformations characteristics, the analysis of the loads and how they exert pressure on the shear wall structure.

The results presented in the Plates III–X below are all obtained from an identical node (No. 1500) for both laminated and un-laminated shear walls. They are namely (see Appendix); (a) displacement, (b) maximum principal stress, (c) pressure and (d) maximum principal strain. The boundary condition used is FIXED for all four edges of the wall and all results with regard to the experimental load case considered. From Plates III–X, the maximum and minimum values of all the categories of results considered are presented at time 0.01 s, that is, end of the positive wave duration. Their corresponding relationships (mostly sinusoidal) for all the categories further explains the exact point of maximum and minimum values for all of the categories of results considered.

The Plates III–X show the maximum values of results for the shear walls with their coloured pictorial severe locations. It is also noticed that for the laminated shear wall (Plates III–VI), the surfaces as presented by the plate was blue/greenish for most of the results which signifies minimum/mid values. This is because the structural interaction of the shear wall is modelled to transfer the loads and stresses from the surface of impact (CFRP layer) through to the sandwiched reinforced concrete wall, and then to the end CFRP layer. This thus defines the entire laminated shear wall structure as a composite.

Furthermore, values on the plates' legend are obtained as at when time of detonation is 0.01 s and so does not give the full details of the performance of the shear walls (with and without CFRP). For displacement, maximum principal stress and pressure, the shear wall without external reinforcement slightly performed better than the one with CFRP external reinforcement.

In Figs. 3, 4, 5, 6, 7, 8, 9 and 10, the CFRP laminated shear wall showed remarkable improvements in all of the results presented as compared to the shear wall without laminates. It is also noticed from the plots that the time to attain the maximum values in each category of responses of the two shear walls vary. The maximum values of the results for the shear walls are the computational tools which aid designers the actual parameters for retrofitting and strengthening of such structures in order to militate the effect of blast loadings and other severe actions. The Table 4 below shows the comparison of the maximum values of responses for the shear walls.

From Table 4, the percentage difference was found to be very large in all of the cases considered. This depicts the extent of strength and stiffness improvement on the shear wall when it is laminated with CFRP. This huge improvement explains the experimental result of Ehsani and Pena (2009). In their experiment, the laminated wall deflected only about 9 mm after the blast, while the control wall crushed into pieces leaving, no chance of survival of lives behind it. Therefore, this finite element method as modelled in ABAQUS enables us to know the detailed responses/exact point of damage within the destructive shock wave phase.

Table 4. Tables of maximum values of displacement, stress, pressure and strain

Results (max values)	With laminate	Without laminate	Percentage improvement (%)
Displacement (m)	60.0e−6	2.7e−03	97.8
Max. principal stress (Pa)	1.4e06	11.8e06	88.1
Pressure (Pa)	0.19e−06	25.0e06	99.2
Max. principal strain (m)	17.5e−06	0.475e−3	96.3

3 Conclusion

From the results obtained, the following conclusions can be drawn:

(i). Finite element method as coded in ABAQUS CAE adopted for this study can be used to effectively simulate the blast load – nonlinear response of shear wall structures.

(ii). The modelling show that the application of externally bonded CFRP laminates is effective in resisting several forms of deformations. The resistance of the laminated shear wall against displacement and strain was improved by 97.8% and 96.3% respectively, for the reinforced concrete shear wall laminated with CFRP.

(iii). The modelling indicates that the application of externally bonded CFRP laminates on shear walls reduces the effect of pressures and stresses exerted on the wall as a result of the blast load. The effect of maximum principal stresses and pressures on the wall as a result of the blast were reduced by 88.1% and 99.2% respectively when CFRP laminates were introduced.

Appendix

Reinforced Concrete Shear Walls Without Laminate (Graphical Results)

See Plates III, IV, V, VI and Figs. 3, 4, 5 and 6.

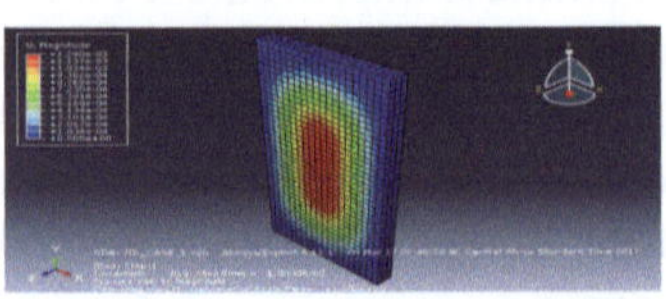

Plate III. Displacement at 0.01 s

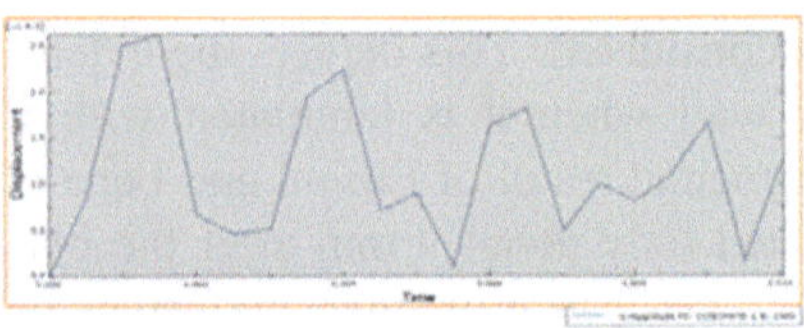

Fig. 3. Displacement plot

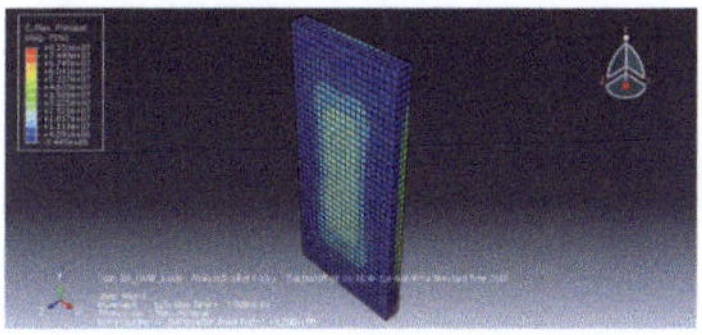

Plate IV. Principal stress at 0.01 s

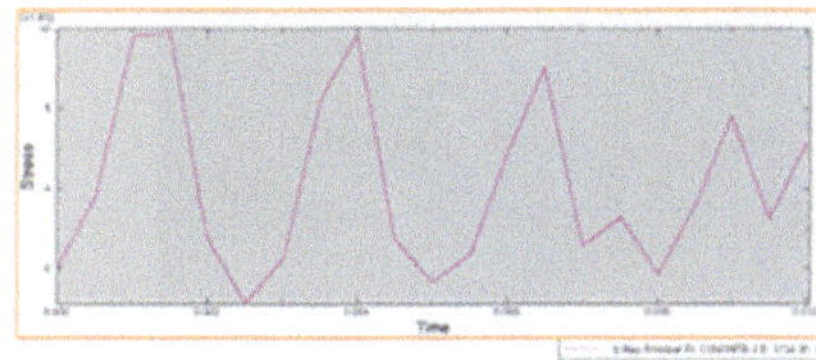

Fig. 4. Principal stress plot

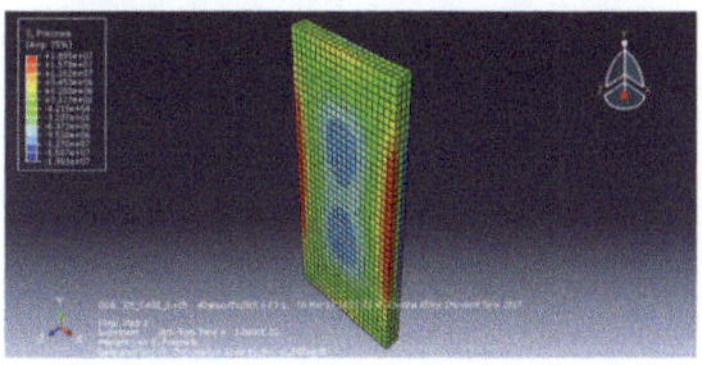

Plate V. Pressure at 0.01 s

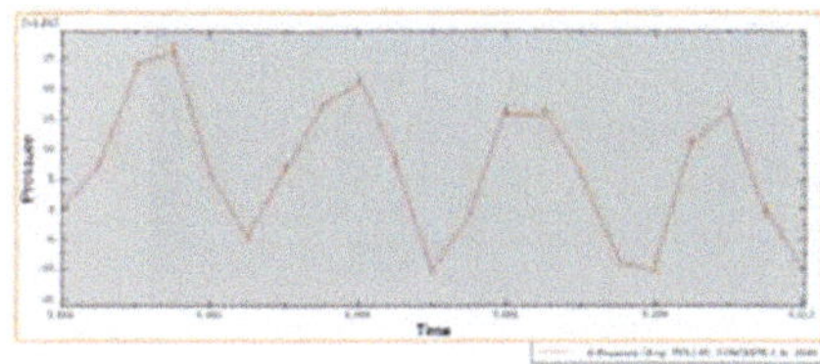

Fig. 5. Pressure plot

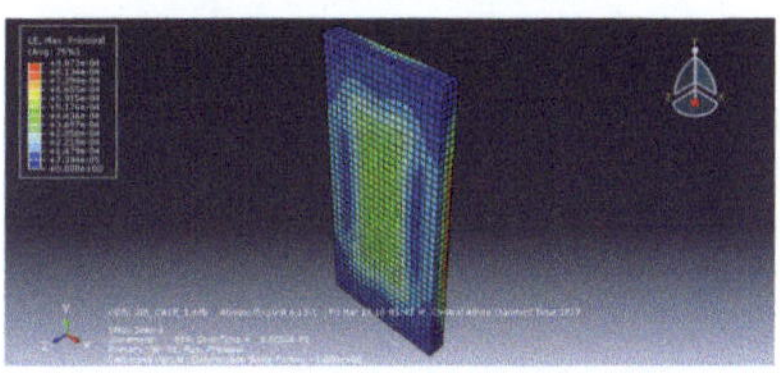

Plate VI. Principal strain at 0.01 s

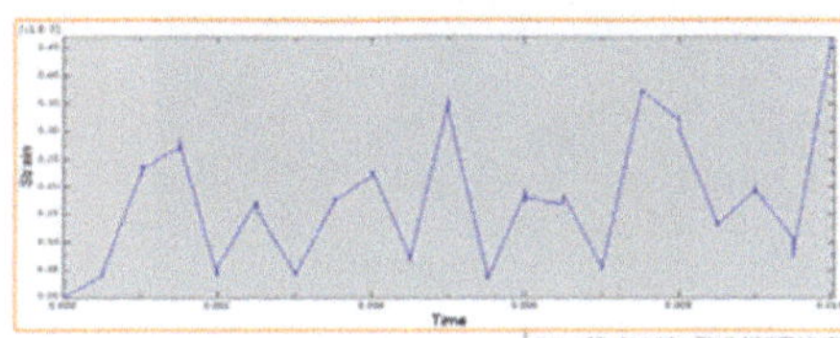

Fig. 6. Principal strain plot

Reinforced Concrete Shear Walls with External CFRP Laminate (Graphical Results)

See Plates VII, VIII, IX and X and Figs. 7, 8, 9 and 10.

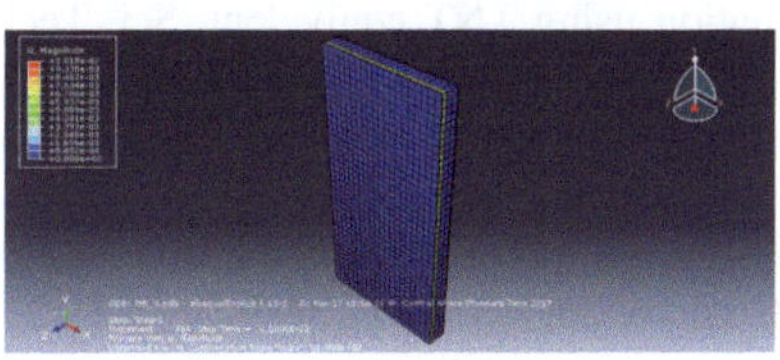

Plate VII. Displacement at 0.01 s

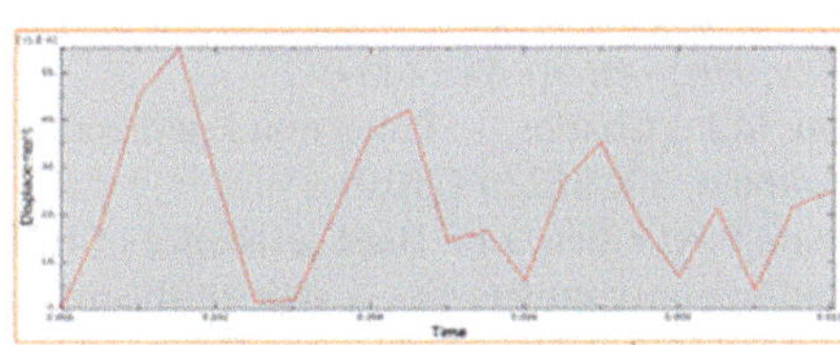

Fig. 7. Displacement plot

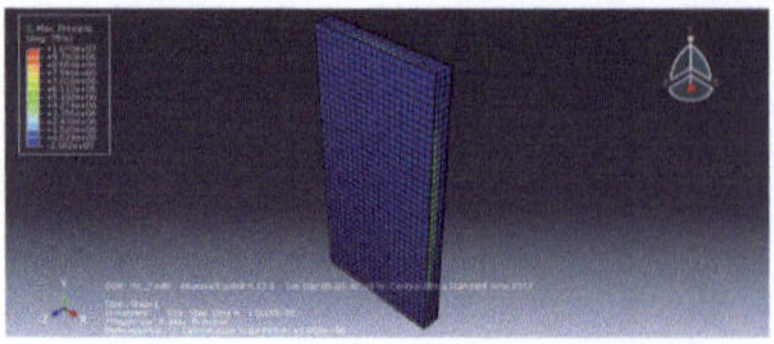

Plate VIII. Principal stress at 0.01 s

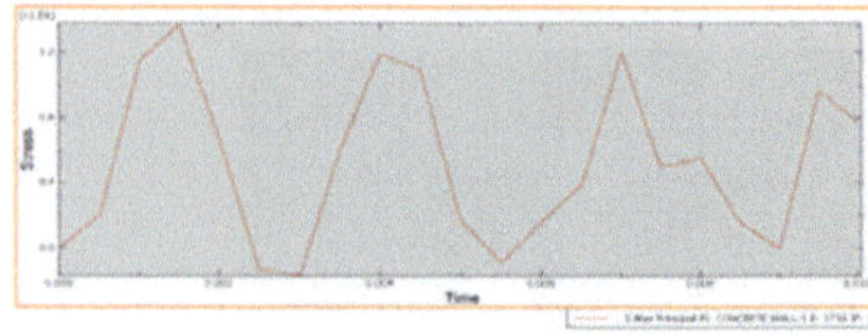

Fig. 8. Principal stress plot

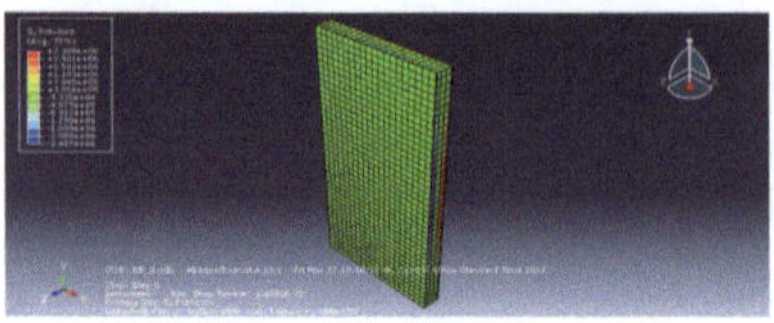

Plate IX. Pressure at 0.01 s

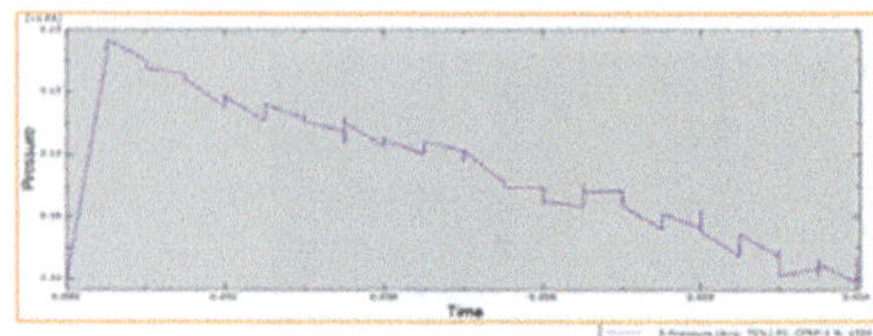

Fig. 9. Pressure plot

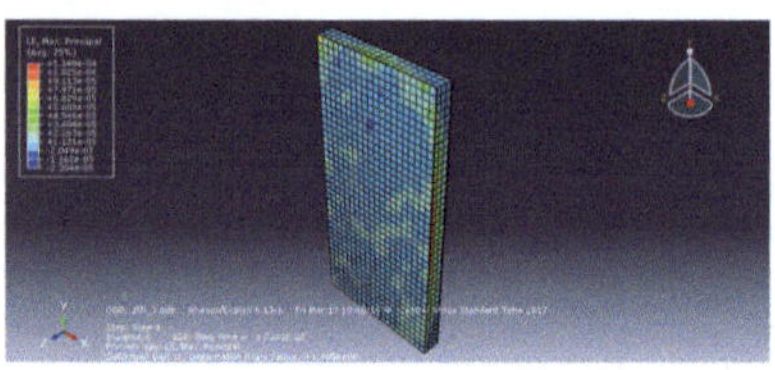

Plate X. Principal strain at 0.01 s

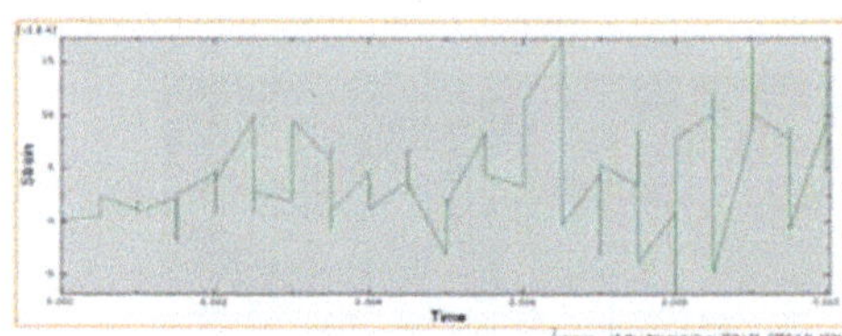

Fig. 10. Principal strain plot

References

Anil, B., Yajdani, S.K.: Seismic analysis of RC framed building for different position of shear wall. IJIRSET: Int. J. Innov. Res. Sci. Eng. Technol. **4**(5), 3346–3353 (2015). eISSN: 2319-8753, pISSN: 2347-6710

Bajic, Z., Bogdanov, J., Jeremic, R.: Blast effects evaluation using TNT equivalent. Sci. Tech. Rev. **59**(3–4), 50–54 (2009)

Bazan, M.L., Charles, J.: Protection Engineering Consultants, 4203 Gardendale, Suite C112, San Antonio, TX 78229 (2010)

Buchan, P.A., Chen, J.F.: Blast resistance of FRP composites and polymer strengthened concrete and masonry structures—a state-of-the-art review. Compos. Part B: Eng. **38**(5–6), 509–522 (2007). http://www.elsevier.com/locate/compositesb

Carney, P., Myers, J.J.: Out-of-plane static and blast resistance of unreinforced masonry wall connections strengthened with FRP. Am. Concrete Inst. Spec. Publ. **230**, 229–248 (2005)

Cruz-Noguez, C.A., Lau, D.T., Sherwood, E.G., Lombard, J., Hiotakis, S.: Seismic behaviour of RC shear walls strengthened for in-plane bending using externally. J. Compos. Constr. (2014)

David, T.L., Cruz-Noguez, A.C.: Developments on seismic retrofit of shear walls with FRP. In: 5th International Conference on Advances in Experimental Structural Engineering, 8–9 November 2013, Taipei, Taiwan (2013)

Dewey, J.M.: The shape of the blast wave: studies of the Friedlander equation. Presented at the 21st International Symposium on Military Aspects of Blast and Shock, Israel (2010)

Draganić, H., Sigmund, V.: Blast loading on structures. Tech. Gaz. **19**(3), 643–652 (2012). ISSN 1330-3651

Ehsani, M., Pena, C.: Blast loading retrofit of unreinforced masonry walls with carbon fibre reinforced polymer (CFRP) fabrics. STRUCTURE Magazine (2009). http://www.structuremag.org

Esfahani, M., Kianoush, M., Tajari, A.: Flexural behaviour of reinforced concrete beams strengthened by CFRP sheets. Eng. Struct. **2007**(29), 2428–2444 (2007)

Fariborz, N., Masoud, K.: Analytical study on the strengthened steel plate shear walls by FRP laminate. Proc. Eng. **54**, 377–386 (2013). http://www.sciencedirect.com

Galal, K., El-Sokkary, H.: Cyclic tests on FRP-retrofitted RC shear wall panels. In: The 15th World Conference on Earthquake Engineering (WCEE), Lisboa (2012)

Kheyroddin, A., Naderpour, H.: Nonlinear finite element analysis of composite RC shear walls. Iran. J. Sci. Technol. Trans. B Eng. **32**(B2), 79–89 (2008)

Lantz, L., Joshua, M., Cook, W., Claudia, M.D.W.: Blast protection of unreinforced masonry walls: a state-of-the-art review. Adv. Civ. Eng. **2016**, Article ID 8958429 (2016)

Lombard, J., Lau, D.T., Humar, J., Foo, S., Cheung, M.: Seismic strengthening and repair of reinforced concrete shear walls. In: Proceedings of 12th World Conference on Earthquake Engineering (2000)

Li, Z.J.T., Balendra, K.H.T., Kong, K.H.: Finite element modelling of cyclic behaviour of shear wall structure retrofitted using GFRP. Am. Concr. Inst. **230**, 1305–1324 (2005)

Mohammed, A.L., Rafika, N.R., Norazman, M.N., Ariffin, I., Muhammad, A.Y., Ng, C.P.: Simulation of reinforced concrete blast wall subjected to air blast loading. J. Asian Sci. Res. **4** (9), 522–533 (2014)

Myers, J.J., Abdeldjelil, B., Khaled, A., El-Domiaty, K.A.: Blast resistance of FRP retrofitted un-reinforced masonry (URM) walls with and without arching action. Masonry Soc. J. **22**, 9–26 (2004)

Norris, T., Saadatamanesh, H., Ehsani, M.: Shear and flexural strengthening of RC beams with carbon fiber sheets. J. Struct. Eng. ASCE **123**(7), 903–911 (1997)

Suresh, M.R., Yadav, S.: The optimum location of shear wall in high rise R.C. building under lateral building. Int. J. Res. Eng. Technol. (2015). eISSN: 2319-1163, pISSN: 2321-7308

Manmohan, D.G., Vasant, M.: Blast-resistant design of structures. Res.: Pract. Period. Struct. Des. Constr. May 2014. www.researchgate.net/publication/262374833

Razaqpur, G.A., Tolba, A., Contestabile, E.: blast loading response of reinforced concrete panels reinforced with externally bonded GFRP laminates. Compos. Part B **38**(2007), 535–546 (2006)

Sejal, B.: Optimization of a multi-storey-building by optimum positioning of shear wall. IJRET Int. J. Res. Eng. Technol. (2014)

Spacone, E., El-Tawil, S.: Nonlinear analysis of steel-concrete composite structures: state of-the-art. ASCE J. Struct. Eng. **130**(2), 159–168 (2004)

Varsha, H.R.: Comparative study of strength of RC shear wall at different location on multi-storied residential building. Int. J. Civ. Eng. Res. **5**(4), 391–400 (2014). ISSN: 2278-3652

Verma, S., Choudhury, M., Saha, P.: Blast resistant design of structure. IJRET: Int. J. Res. Eng. Technol. (2016). eISSN: 2319-1163, pISSN: 2321-7308

New Technology of Concrete Surface Improvement and Soil Properties

Larisa Svatovskaya[1], Maxim Sychov[2], and Ivan Drobychev[1(✉)]

[1] Emperor Alexander I St. Petersburg State Transport University (PGUPS), Moskovsky pr. 9, St. Petersburg 190031, Russia
drobyshev-i@inbox.ru

[2] Petersburg State Technology Institute (Technological University), Moskovsky pr. 26, St. Petersburg 190031, Russia

Abstract. The paper deals with technologies of concrete surface improvement. Concrete surface is important because of its hardness, durability and other problems. Besides, it is necessary to protect it from aggressive environment. The purpose of the article is to introduce a few techniques of silica sol use in order to change concrete surface properties. The absorption, irrigation and watering processes were applied as the techniques. Both silica sol and its mixes with solutions contained Ca(II) were used as the substances. The experimental methods were realized during the study. The experiment results have shown that hardness of heavy concrete and foam concrete surface was increased up to 40–50% and the level of soil strengthening was 0.2 MPa. It means that if silica sol and solutions containing Ca(II) are used the soil strength is increased more than 40–50%.

1 Introduction

The main idea of the paper is to improve properties of a mineral stone surface using silica sol as a substance for absorption, irrigation, watering and other techniques to obtain layers on a concrete surface through penetration silica sol into pores and capillaries.

Nowadays it is possible to distinguish at least two important problems concerned with a mineral stone surface. They are technical properties during operation and geoecoprotective properties for the lithosphere preservation. Hardness and other properties of a mineral stone surface can be observed as technical ones. Among geoecoprotective properties penetrability is of great importance [1–5]. In any case, for improvement both of technical properties and of geoecoprotective ones a mineral stone surface must be strengthened by means of some substances that can impart strength to a surface even without penetration.

Silica sol was used for new technologies of a concrete surface improvement. In the Table 1 silica sol properties as a base for new technologies are shown [6–10].

Taking into consideration thermodynamic calculations presented in the Table 2, one can see that the reactions between surface active agents and silica sol are allowed, the data from the Table 3 having been used for the calculations of thermodynamic parameters.

H. Rodrigues and A. Elnashai (Eds.): GeoMEast 2018, SUCI, pp. 346–353, 2019.
https://doi.org/10.1007/978-3-030-01932-7_27

Table 1. Properties of silica sol.

Property	Property improvement
Sol → gel, sol conversion into gel	Glue ability and binding capability of dispersion
Reactions with Ca(II) of material	Hardness of concrete
Complex of conversion and reactions	Strengthening of a geosystem

Table 2. Reactions of silica sol on a mineral stone surface.

Reactions on a mineral stone surface	pH	ΔG°_{298}, KJ/mol
$CaSO_4 \cdot 2H_2O + 2(SiO_2 \cdot H_2O) + 2OH + Ca^{2+} = CaO \cdot 2SiO_2 \cdot 2H_2O + CaSO_4 + 3H_2O$	>7	−248
$Ca(OH)_2 + 2(SiO_2 \cdot H_2O) = CaO \cdot 2SiO_2 \cdot 2H_2O + H_2O$	>7	−168,4
$2Ca(OH)_2 + SiO_2 \cdot H_2O = 2CaO \cdot SiO_2 \cdot 1{,}17H_2O + 1{,}83H_2O$	>7	−95,32
$2Ca(OH)_2 \cdot SiO_2 \cdot 1{,}17H_2O + 2(SiO_2 \cdot H_2O) = 2CaO \cdot 3SiO_2 \cdot 2{,}5H_2O + 0{,}67H_2O$	>7	−740,8
$Ca^{2+} + 2(SiO_2 \cdot H_2O) + 2OH^{-} = CaO \cdot 2SiO_2 \cdot 2H_2O + H_2O$	>7	−201,6
$6Ca^{2+} + 3(2SiO_2 \cdot 3H_2O) + 12OH^{-} = 6CaO \cdot 6SiO_2 \cdot H_2O + 14H_2O$	>7	−287,9

Table 3. Parameters of ΔG°_{298} substances.

Substance	ΔG°_{298}, KJ/mol
$CaO \cdot 2SiO_2 \cdot 2H_2O$	−2869,1
$CaSO_4 \cdot 2H_2O$	−1794,01
$CaSO_4$	−1319
$Ca(OH)_2$	−895,89
$2CaO \cdot SiO_2 \cdot 1{,}17H_2O$	−2478,3
$SiO_2 \cdot H_2O$	−1073,6
OH^{-}	−157,1
$CaSO_4$	−1319
Ca(II)	−552,5
H_2O	−237
$2SiO_2 \cdot 3H_2O$	−2426,6
$6CaO \cdot 6SiO_2 \cdot H_2O$	−9444,29

It should be noted that other substances like ones given in the Table 2 can be used

The Table 4 demonstrates the techniques of silica sol action on a construction geosystem. It is worth mentioning that any surface strengthening is named preservation technology in this article.

Table 4. Preservation technologies of a mineral geosystem strengthening by means of silica sol.

Mineral geosystem as a base for a surface preservation	Construction geosystem	The aim of silica sol use	Techniques of silica sol use	Details of use
1. Calcium sulphate, $CaSO_4 \cdot 2H_2O$	• Products • Structures	• Strengthening of a surface • The lithosphere protection from pollutants absorbed by an artificial stone Protection from the environment impact	• Absorption • Irrigation	• Silica sol can be used several times (a few impregnations)
2. Calcium silicate hydrates, $4CaO \cdot nSiO_2 \cdot xH_2O$, calcium aluminate hydrates and other binding systems	• Concrete products • Structures	• strengthening of a surface • The lithosphere protection from pollutants absorbed by an artificial stone • Protection from the environment impact	• Absorption • Irrigation	• Silica sol can be used several times (a few impregnations)
3. Soils and structures saturated with mineral dispersions	• Mass of dispersed materials	• Strengthening • Antidusting	• Irrigation • Watering • Spraying	• Silica sol can be used several times

2 Methods

1.5–30% silica sol was used as a substance for absorption, irrigation and watering, particle sizes having been nearly 10 nm, pH – 10. The first solution has been of 30% concentration and then it has been diluted up to 1…1.5% Three techniques were applied in the research, namely: absorption – for products and structures, irrigation and watering – for structures and soils.

Under absorption technology after 28 days of hardening the products and structures were sunk into 1.5–30% silica sol for 1 day in order to saturate them to a full level. Heavy concrete products had grade up B30 and foam concrete ones had the same concrete grade B30 with an average density 400–600 kg/m^3. The material consumption for foam concrete is illustrated in the Table 5, mechanical properties of the concrete having been tested according to the Russian standards, 10180-78; 7076-99; 21520-89.

Table 5. The material consumption for foam concrete having a density 400–600 kg/m^3.

Average density, kg/m^3	Consumption of materials per 1 m^3 of foam concrete, kg						W/C	Concrete flowability, cm
	cement	Sand	Water in solution	Foam admixture*	Foam volume	Water in foam		
400	160	160	67	2.64	820	68	0.84	32
500	210	210	86	2.49	763	62	0.7	28
600	260	259	104	2.33	712	58	0.62	24

*Protein based foam admixture was used

As absorption was the method of article sinking into the 1…3% silica sol solution for full saturation. The sandy soils with different particles were irrigated by 30% silica sol several times up impregnations to a full saturation level and then in 1 week soils were tested for strength. For soils complex of silica sol and solutions contained Ca(II) was used as well. For that two solutions have been used. One solution was silica sol and the second solution was one with Ca(II) ions.

3 Results and Discussing Points

In the Tables 6 and 7 the results of silica sol effects on the products are indicated and one can establish that strengthening of a system takes place due to silica sol action.

Table 6. Results of silica sol absorption.

Concrete grade	Hardness, +Δ%	Improvement of properties, Δ%*				
		Compressive strength, +Δ	Bending strength, +Δ	Water absorption, −Δ	Abrasion, −Δ	Freeze resistance cycle, +Δ
B15, B20, B25, B30	40–50	15–30	20–35	50–65	20–30	75–90

*1.5…3% solution of silica soil

Table 7. The improvement of foam concrete properties when applying silica sol absorption technology.

Density of foam concrete, kg/m^3	Change of foam concrete properties, %; after 28 days of hardening			
	Compressive strength, +Δ%	Hardness, +Δ%	Freeze resistance, cycle, +Δ%	Water absorption, −Δ%
400	10–15	30	10–25	65–35
500	10–15	40	10–25	60–30
600	15–20	50	10–25	60–30

The experiment results have shown that hardness of heavy concrete and foam concrete surface was increased to 40–50%. It is very important to pay attention to water absorption reduction which occurred during geoecochemical processes with silica sol use. In the experiments foam concrete was applied as a "stone sponge" for oil products. Therefore when a surface was treated by silica sol there was not any extraction of oil products from foam concrete. On the other hand, silica sol use preserved concrete from the environment impact that is why these techniques are named "preservation technologies" [11–16].

It is known that in construction earth mass must be strengthened for some aims. In the experiments soils were irrigated by silica sol several times up to a full saturation level a few treatments. After a time not less than 1 week the strength was tested and as a result soil strengthening was nearly 0.2 MPa (Table 8).

Table 8. Soil strengthening with silica sol irrigation technology.

Soil, nature	Strengthening*
Sandy	0.2 MPa, one time treatment

*30% solution of silica soil

So, the effects of silica sol are an alternative to Na_2SiO_3 action but in comparison with the latter silica sol is the purest substance for the lithosphere. It means that silica sol use technique is the best one for an application.

Thus, it is evident that silica sol properties influence on a surface hardness both in concrete and in soils. However, they can be improved by means of the solutions use, for example solutions containing Ca(II), in order to form calcium silicate hydrates in a stone. Such a reaction is more efficient for strength increase because when solutions containing Ca(II) are mixed with silica sol their interaction causes more than 50% soil strengthening and it is the technique that a mineral stone surface properties improve. In the proceedings [17–26] the details of these effects are described.

The Table 9 is presented the surface teat of concrete after the absorption of silica sol. According to the Table 9, a layer of nearly 3.5…5 mm is provided the improvement of concrete, and CSH-phase is the same as in the theoretical Table 2.

Table 9. Surface teat of concrete after the absorption of silica sol.

Max depth of new layer	3.5–5 mm	New phase	Formula CSH

4 Conclusions

To sum up

1. The technology of a surface properties improvement when applying silica sol was demonstrated, the level of such an enhancement having been nearly 40–50%.
2. The technology of soil strengthening by means of 30% silica sol solution is shown which was used several times was proposed. The level of soil strengthening was nearly 0.2 MPa.
3. The level of soil strengthening could be increased to 50% when used the solutions contained Ca(II) and silica sol together.

References

1. Svatovskaya, L.B., Kabanov, A.A., Sychov, M.M.: The improvement of foam concrete geoecoprotective properties in transport construction. In: IOP Conference Series: Earth and Environmental Science, vol. 90 (2017) 1755-1315 90 012010. (Larisa Svatovskaya et al. 2017 IOP Conf. Ser.: Earth Environ. Sci. 90 012010). https://doi.org/10.1088/1755-1315/90/1/012010
2. Svatovskaya, L.B., Kabanov, A.A., Sychov, M.M.: Lithosynthesis of the properties in the transport construction on the cement base. In: IOP Conference Series: Earth and Environmental Science, vol. 90 (2017) 1755-1315 90 012009. (Larisa Svatovskaya et al. 2017 IOP Conf. Ser.: Earth Environ. Sci. 90 012009) https://doi.org/10.1088/1755-1315/90/1/012009
3. Svatovskaya, L.B., Kabanov, A.A., Sychov, M.M.: Soling, aerating and phosphating for soil strengthening and detoxication. Proc. Eng. **189**, 398–403 (2017). https://doi.org/10.1016/j.proeng.2017.05.063
4. Svatovskaya, L.B., Urov, O.V., Kabanov, A.A.: Geoecoprotective technology of transport construction using silica sol absorption method. Proc. Eng. **189**, 454–458 (2017). https://doi.org/10.1016/j.proeng.2017.05.073
5. Svatovskaya, L.B., Shershneva, M.V., Baidarashvili, M.M., Yakimova, N.I., Khitrov, A.V.: Foam concrete construction demolished waste. In: Proceedings of the International Conference on Sustainable Waste Management and Recycling: Construction Demolition Waste, London, pp. 199–203 (2004)
6. Nonat, A.: The structure and stoichiometry of C–S–H. Cem. Concr. Res. **34**, 1521–1528 (2004)
7. Garbev, K., Stemmermann, P., Black, L., Breen, C., Yarwood, J., Gasharova, B.: Structural features of C–S–H (I) and its carbonation in air—A Raman spectroscopic study. Part I: Fresh phases. J. Am. Ceram. Soc. **90**(3), 900–907 (2007)
8. Nonat, A.: Interactions between chemical evolution (hydration) and physical evolution (setting) in the case of tricalcium silicate. Mater. Struct. **27**(4), 187–195 (1994)
9. Neville, A.: Water and concrete: a love-hate relationship. Concr. Int. **22**(12), 34–38 (2000)

10. Mehta, P.K., Monteiro, P.J.M.: Concrete; Microstructure, Properties and Materials, 3rd edn. McGraw-Hill, New York (2006)
11. Svatovskaya, L.B., Sakharova, A.S, Baidarashvilly, M.M., Petriaev, A.V.: Building wastes and cement clinker using in the geoecoprotective technologies in transport construction. In: Proceedings of the 14th International Conference of International Association for Computer Methods and Recent Advances in Geomechanics, Kyoto, p. 152 (2014)
12. Sychova, A.M., Svatovskaya, L.B., Mjakin, S.V., Vasiljeva, I.V.: Modification of fillers for cements. In: Electron Beam Modification of Solids: Mechanisms, Common Features and Promising Applications, pp. 35–37 (2009)
13. Sychova, A.M., Svatovskaya, L.B., Mjakin, S.V., Vasiljeva, I.V.: Activation of aqueous phase at cement and concrete solidification. In: Electron Beam Modification of Solids: Mechanisms, Common Features and Promising Applications, pp. 39–47 (2009)
14. Maslennikova, L.L., Svatovskaya, L.B., Mjakin, S.V., Vasiljeva, I.V.: Activation of reactions at solid–solid interfaces. Improvement of ceramics materials. In: Electron Beam Modification of Solids: Mechanisms, Common Features and Promising Applications, pp. 57–61 (2009)
15. Svatovskaya, L.B., Yakimova, N.I., Trunskaya, O.Y., Rusanova, E.V., Krylova, N.B.: New complex ecotechnology for oil demolished waste. In: Proceedings of the International Conference on Sustainable Waste Management and Recycling: Construction Demolition Waste, London (2004)
16. Sychova, A.M., Solomahin, A., Kotovich, V., Svatovskaya, L.B., Kamenev, Y.: Improving of the monolithic foamconcrete quality for used in the high-rise constructions. E3S Web Conf. **33**, 02058 (2018). https://doi.org/10.1051/e3sconf/20183302058
17. Gusev, N., Svatovskaya, L.B., Kucherenko, A.: Effect of changing of the parameters of the cable network of monitoring systems of high-rise buildings on the basis of string converters on their operability. E3S Web Conf. **33**, 02069 (2018). https://doi.org/10.1051/e3sconf/20183302069
18. Sychova, A., Sychov, M., Rusanova, E.: A method of obtaining geonoiseprotective foam concrete for use on railway transport. Proc. Eng. **189**, 681–687 (2017). https://doi.org/10.1016/j.proeng.2017.05.108
19. Sychova, A., Solomahin, A., Hitrov, A.: The increase of the durability and geoprotective properties of the railway subgrade. Proc. Eng. **189**, 688–694 (2017). https://doi.org/10.1016/j.proeng.2017.05.109
20. Kavetsky, A., Yakubova, G., Sychov, M., Lin, Q., Walter, G., Chan, D., Yousaf, S., Socarras, H., Abrefah, J., Bower, K.: Tritium charged capacitor. Nucl. Sci. Eng. **159**, 321–329 (2008)
21. Kavetsky, A.G., Meleshkov, S.P., Sychov, M.M.: Conversion of radioactive decay energy to electricity. In: Polymers, Phosphors and Voltaics for Radioisotope Microbatteries. CRC Press, Chapter. 1, pp. 1-38 (2002)
22. Sychov, M.M.: Modification of zinc sulfide phosphors by irradiation with gamma-ray photons and electrons. Semiconductors **40**(9), 1016–1020 (2006)
23. Bakhmetyev, V.V., Mjakin, S.V., Korsakov, V.G., Abyzov, A.M., Sychov, M.M.: A study of the surface and luminescence properties of ZnS: Mn^{2+} nanosized phosphors. Glass Phys. Chem. **37**(5), 549–554 (2011)
24. Sychov, M.M., Ogurtsov, K.A., Lebedev, V.T., Kulvelis, Y.V., Török, G.Y., Sokolov, A.E., Trunov, V.A., Bakhmetyev, V.V., Kotomin, A.A., Dushenok, S.A., Kozlov, A.S.: Effect of the Cu content and ZnS treatment on the characteristics of synthesized ZnS: (Cu, Cl) electroluminescent phosphors. Semiconductors **46**(5), 696–700 (2012)

25. Gravit, M.V., Gumenjuk, V.I., Sychov, M.M., Nedyshkin, O.: Estimation of the pores dimensions of intumescent coatings for improvement of the fire resistance of building constructions. Proc. Eng. **117**, 119–125 (2015)
26. Mjakin, S.V., Minakova, T.S., Bakhmetyev, V.V., Sychov, M.M.: Effect of the surfaces of $Zn_3(PO_4)_2$: Mn^{2+} phosphors on their luminescent properties. Russ. J. Phys. Chem. A. **90**(1), 240–245 (2016)

Probabilistic Shear Strength of Bolted Joints in Offshore

J. L. Ajimituhuo and O. S. Abejide(✉)

Department of Civil Engineering, Ahmadu Bello University, Zaria, Nigeria
abejideos@yahoo.com

Abstract. This study evaluates the formulations for prediction of ultimate shear capacity of steel joints in offshore structures in order to ascertain the safety of the structures in service using First Order Reliability Method (FORM) under the European code, EC3:1,8 (2003), design format. Two modes of failure were considered which depend on the embedding strength of the plates and bolts that influences the joints capacities and the failure due to bearing. It is generally observed that the increase in bolt diameter, bolt strength with corresponding steel grades, fasteners thickness improves the safety of the joints, while undue increases in load on the offshore structure may result in safety reduction, but not collapse. The provision for joints design in structures in EC3 (2003) is very robust and efficient. However, economy commensurate with structural safety can be achieved in the current formulations.

Keywords: Offshore · Structural safety · Shear strength · Steel bolts
Plate thickness

1 Introduction

Offshore probabilistic design is preceded by a sequence of activities that result in the selection of a field development system that best fits the field characteristics and economics. Before feasible alternatives for producing oil and gas from an offshore field are identified and the most desirable production scheme is selected, exploratory work defining the reservoir characteristics have to be completed. First, a decision has to be made whether an offshore location has the potential for hydrocarbon reserves. This assessment is achieved through a study of geological formations. The basic strength of members in ships and offshore structures include support members (for example, stiffness and plate girder), plates and stiffened panels. During their lifetime, the structures that are constructed using these members are likely subjected to various types of loadings or deformations that are for the most part operational, but may in some cases, be extreme or even accidental (Moan 2005). The sources of such loadings and deformations include fabrication-related initial imperfections (for example, initial distortions, welding residual stresses, softening in the affected zone of welded aluminum structures); abnormal waves, winds and currents; dynamic pressure loads arising from sloshing, slamming or green water, low temperature in Arctic operation; cryogenic conditions resulting from liquefied natural gas cargo, ultra-high pressures in ultra-deep water, elevated temperature due to fire, blasts load due to explosion, impact

H. Rodrigues and A. Elnashai (Eds.): GeoMEast 2018, SUCI, pp. 354–369, 2019.
https://doi.org/10.1007/978-3-030-01932-7_28

loads associated with collision, grounding or dropped objects, and age related degradation such as corrosion, fatigue cracking and local denting damage (Paik 2017). These phenomena may occur in offshore structures, while they are in service. It is interesting to mention that all of them commonly give rise to non-linear structural consequences that involve both geometric and material non-linearity. In the past, criteria and procedure for the structural design of offshore platforms were primarily based on allowable stresses and simplified buckling checks for structural components. However, the ability to correctly assess the true margin of safety would also inevitably lead to improvements in related regulations and design requirements (Paik 2017).

Circular hollow sections combine excellent structural properties with an architecturally attractive shape. This has resulted in many applications in buildings, halls, bridges, barriers, masts, towers, offshore and special applications, such as glass houses, radio telescopes, sign gantries, parapets, cranes, jibs, sculptures, etc. (Wardenier et al. 2002). Offshore oil production is one of the most visible of these applications and represents a significant challenge to the design engineer.

Nowadays, the existing offshore steel structures are no longer safe with regards to overload from excessive wind, wave and current, in particular that the load pattern assumed in design has been significantly changed (Ersdal and Hörnlund 2008). The safety of structures for an expected design life generally designed according to established Standard or Codes and method has become deration, particularly due to fatigue and other continuous degradation mechanisms of structural materials, joint connections and foundation systems such as soil subsidence to unknown extent during decades in harsh weather (Ersdal and Hörnlund 2008). Tools to assess the structural safety of welded columns in an offshore structure especially intended to be used beyond its original design life, has become a major task. Two assessment approaches for this offshore structural strength were suggested (Ersdal and Hörnlund 2008), namely non-linear static or quasi-static pushover method and risk-based method.

It is for this reason that the structural reliability-based method is employed herein.

This presentation evaluates the probabilistic shear strength of bolted columns in offshore and as formulated in design codes.

The objectives of this study include: (i) providing a probabilistic assessment of the offshore steel shear capacity as formulated in codes; (ii) providing an acceptable level of safety by defining minimum requirements for offshore columns structures and structural components and (iii) obtain optimum design criteria that will be widely accepted as per ISSC (2012), API (2000), ISO 19902 (2004) and EC3 (2003) recommendations.

There are various limitations that affect the assessment of steel joints in offshore structures. Only the shear strength capacity of bolted columns is considered herein, using a risk-based method as formulated in the First Order Reliability Method (FORM).

Some of these limitations are stated as follows: (i) fatigue assessment for structural members and columns are not considered; (ii) earthquake and dynamic loads are excluded; (iii) environmental loads including wave inundated load on deck are treated as static loads and (iv) secondary structures are modeled to generate weight and environmental loads.

2 Historical Development of Offshore Structures

Offshore steel structures are often called jackets, which originates from the early days of the offshore industry when a trussed structure jacket was placed over the piles to provide lateral stiffness to enable it withstand wave current and wind forces. Offshore jacket structures are installed in deep water depths. These structures are often exposed to corrosive action of salt water as well as extremely low temperatures (Armand *et al.* 2009). They are usually built on concrete or steel legs, or both, anchored directly on to the seabed, supporting a deck with space for drilling rigs, production facilities and crew quarters. Such structures by their virtue of immobility are designed for very long time use. Various types of structures are used, steel jackets, concrete caissons, floating steel and even floating concrete. Steel jackets are vertical sections made of tubular steel members, and are usually piled into the seabed. Concrete caisson structures, pioneered by the Condeep concept, often have in-built oil storage in tanks below the sea surface and these tanks were often used as a floating capability, allowing them to be built close to shore and then floated to their final position where they are sunk to the sea bed. Fixed platforms are economically feasible for installation in water depths up to about 520 m (Banerjee et al. 2007). Offshore exploration and production of minerals is advancing into deeper waters at a fast pace. Many deep water structures have already been installed worldwide. New oil and gas fields are being discovered in ultra-deep water. Many of these fields are small and their economic development is a challenge today to the offshore engineers. This has initiated the development of new structures and concepts. Many of these structures are unique in many respects and their efficient and economic design and installation are a challenge to the offshore community. There are two important features that need to be taken into consideration for a proper assessment to be done and these are durability and ultimate strength of steel.

The corrosion of steel can be considered as an electrochemical process that occurs in stages and requires the simultaneous presence of water and oxygen. In the absence of either, corrosion does not occur (Baddoo 2008). Marine environments are recognized to be very corrosive for mild and alloy steels. For reasons of economy, such steels are still the preferred materials for many engineering structures, such as, offshore structures. As the age of a structure increases, the loss of the structure's thickness caused by corrosion induces a decrease in the ultimate strength of the structure and potentially a loss of integrity through pit, whose loss is particularly critical when protective coating systems, such as galvanizing or painting coatings, are ineffective. Conventionally, a deterministic approach is used to assess the remaining life of the offshore structures, but it is now recognized that the probabilistic method is a better approach because it is more robust, reliable and justifies safety with associated economy. Several probabilistic models have been proposed with the majority of them focused on predicting the amount of pitting corrosion for a structure with different ages. Jelovica et al. (2014) evaluated the amount of reduction in the buckling strength of sandwich plates as a result of general corrosion. The main findings were that a 0.5-mm decrease in plate thickness will cause 25.5% decrease in their buckling strength, which is quite severe and worrisome. However, corrosion will definitely be an important hazard for the structure in cases where the corrosion protection is not sufficient for the extended life.

Hence, the corrosion effects on ageing structures in a life extension would need a specific investigation (Hørnlund *et al.* 2005).

The techniques in Finite Element Method (FEM) have matured for ultimate strength evaluation of plated structural components. Many researchers have applied Finite Element Method (FEM) to predict ultimate strength of unstiffened plates and stiffened plates, such as Zhao and Zhang (2007) and Paik (2008). In these applications, both geometric and material nonlinearities were considered. It may be said that it is fairly straightforward to use Finite Element Method (FEM) for ultimate strength prediction of plates and stiffened plates. The last decade sees many developments and innovations of tubular connections in the offshore industry. Such applications include the more widespread adoption of thick-walled sections in offshore and onshore structures, internally or externally reinforced tubular connections, *etc*. Recent research effort also focuses more on the failure assessment of tubular connections with initial defects, since fatigue induced cracks remain as a potential threat for offshore steel platforms in the event of extreme environmental loading. These practical concerns in the industry do not find corresponding theoretical background in the literature or design codes (API 2000; ISO 19902 2004). Zhao 2005 has noted that the chord stress effect for Circular Hollow Section (CHS) and Rectangular Hollow Section (RHS) joints still remain as an issue to be solved. Therefore, there is need for a more detailed understanding on the ultimate strength of tubular connections with due emphasis on larger wall thickness, presence of initial defects, provision of reinforcement, and the effect of chord stresses, for a safe and economical design.

In offshore structures, the service life is usually assumed to be equal to the initiation time, when corrosion is caused by chloride ingress. In terms of the consequences of localized corrosion, the period of propagation is traditionally not taken into account (Hobbacher 2009). Crude Monte Carlo simulation offers a direct method for estimating the failure probability. In essence, the technique involves sampling a set of values of the basic variables at random from the probability density function, and evaluating the failure function for the values to see if failure occurs. The drawback with crude Monte Carlo simulation is the computational effort involved. In order to produce a reasonably accurate estimate of the failure probability large data from at least $100/P_f$ trials may be required. This is quite cumbersome and thus, knowledge of the failure region (for example from first-order methods) can be exploited to significantly improve the efficiency of Monte Carlo simulation by tailoring the sampling scheme to the particular situation. However, if used intelligently, Monte Carlo methods are a readily understood and easily applied tool. They can be used to generate accurate answers to problems, and to such other ones that cannot be accurately modeled using first- or second-order methods. Because corrosion is a function of many variables, and of uncertain nature, a probabilistic model is more appropriate to describe the expected corrosion (Melchers 1999).

Many developments and innovations of tubular connections in the offshore industry were witnessed in recent times. Such applications include the more widespread adoption of thick-walled sections in offshore and onshore structures, internally or externally reinforced tubular connections, *etc*. Recent research effort also focuses more on the failure assessment of tubular connections with initial defects, since fatigue induced cracks remain as a potential threat for offshore steel platforms in the event of extreme environmental loading. These practical concerns in the industry do not find corresponding theoretical background in the literature or design codes (API 2000; ISO 19902 2004). Zhao (2005) has noted that the chord stress effect for Circular Hollow Section (CHS) and Rectangular Hollow Section (RHS) joints still remains as an issue to be solved for the upcoming version of the International Institute of Welding (IIW) design guidelines.

The European standard (prEN 1993-1-8: 2003), has suggested application rules to determine the static resistances of uni-planar and multi-planar joints in lattice structures composed of circular, square or rectangular hollow sections, and of uni-planar joints in lattice structures composed of combinations of hollow sections with open sections. In particular, the EC3 (2003) provision for rotation capacity in these sections: (i) do not establish any specific procedures for the evaluation of the rotation capacity; and (ii) optimum capacity of bolted or welded joints. However, they state the need to ensure adequate rotation capacity either by testing in accordance with EN 1990 (2004).

Alternatively, using appropriate calculation models based on the result of tests, for end-plate beam-to-column steel joints, these conditions specified in the code basically suggests that either the columns web panels in shear controls the behavior of the joint or, alternatively, the end plate or the column flanges in bending, which is reproduced as Eq. (1) (Da Silva 2008):

$$\frac{d}{tw} < 69\varepsilon \tag{1}$$

$$\text{Where } \varepsilon = \sqrt{\frac{235}{f_y}}.$$

$$t < 0.36d$$

2.1 Shear Connections in Single Plates

The static resistances of the joints are expressed in terms of maximum design axial and/or moment resistances for the brace members. The application rules are valid both for hot finished hollow sections to EN 10210 (1993) and for cold formed hollow sections to EN 10219 (1993), if the dimensions of the structural hollow sections fulfill the necessary requirements. The nominal wall thickness of hollow sections should be limited to a minimum of 2.5 mm and should not be greater than 25 mm unless special measures have been taken to ensure that the through thickness properties of the material adequate. These types of joints are covered by standard prEN 1993-1-8 (2003). The application rules given in prEN 1993-1-8 (2003) may be used only where all the given conditions are satisfied.

2.1.1 Bolted Connections

Even if bolted connections to hollow sections are used to assemble prefabricated elements or space structures, the most used method to assemble CHS members is welding, especially for trusses. According to prEN 1993-1-8 (2003), the design joint resistances of connections between hollow sections and of connections of hollow sections to open sections, should be based on standard failure modes as applicable and specified in the code.

Connecting two hollow section members or a hollow section and an open profile or a plate directly to each other by bolting can be difficult unless the joint is located close to the open end of a hollow section member. Otherwise, it is necessary to take measures, such as cutting a hand access hole in the structural hollow section member to enable the bolt to be tightened from the inside or using "through" or "blind" bolts. The reason for this special situation is evident, as the hollow section allows free access only to the outside; any access to the inside is restricted. Bolted connections remain nonetheless desirable in many cases in spite of the unique condition of non-accessibility to the inside of a hollow section. However, in these cases, the hollow sections can be joined indirectly using flange or capping plates, which makes it possible to effect such bolted connections in a simple and economical manner. The main methods of assembly by bolting are described below. Bolted connections are mostly detachable. They are selected for the on site assembly in order to avoid site welding, which may cause welding errors due to environmental difficulties. Site welding is also more costly than site bolting. This is presented in accordance with "Design guide for structural hollow sections in mechanical applications" of Wardenier et al. (2002) and "Guide on the use of bolts: single sided blind bolting systems of Yeomans et al. (2002). The main types of bolted connections for hollow section structures are: Bolted knee joints, Flange connections, Splice joints, Joints with fork ends, Screwed tensioner, Through bolting, Bolted connections with flattened ends, Hinged support, Column bases, Fish plate connections, Bolted subassemblies, and Fixing bolts through hand access holes (CIDECT 1995). These connections are realized using intermediate connecting steel devices, which are welded on the hollow section members, the bolted connections themselves being designed as normal connections according to prEN 1993-1-8 (2003). For this reason, design of hollow section connections does not imply specific requirements.

2.2 Design Formulations

The design formulations of EC3 (2003) as earlier recommended in prEN 1993-1-8 (2003) are as follows:

Shear resistance for all fasteners

$$\mathrm{F}_{\mathrm{v,Rd}} = \frac{n\alpha v f_{ub} A}{\gamma M_2} \tag{2a}$$

Bearing resistance

$$\mathrm{F}_{\mathrm{b,Rd}} = \frac{nk1\alpha b f_{ub} A}{\gamma M_2} \tag{2b}$$

Where γM_2 is partial safety factor for joints given as 1.25 for all fasteners and γM_2 is partial safety factor for all hybrid connections given as 1.25, where the shear plane passes through the threaded portion of the bolt (A is the tensile area of the bolt A_s):

$\alpha_v = 0.6$ for Classes 4.6, 5.6 and 8.8 bolts
$\alpha_v = 0.5$ for Classes 4.8, 5.8, 6.8 and 10.9 bolts

Where the shear plane passes through the unthreaded portion of the bolt (A is the gross cross section of the bolt) $\alpha_v = 0.6$

$\alpha_v = 1.0$, $K_1 = 2.5$

f_{ub} is the ultimate tensile strength for bolts.

The yield strength f_{yb} and the ultimate tensile strength f_{ub} for bolt classes 4.6, 5.6, 6.8, 8.8 and 10.9 are given in Table 1. These values should be adopted as characteristics values in design calculations. Also Table 2 indicates the categories of bolts used in connections in offshore structures.

Table 1. Nominal values of the yield strength, f_{yb}, and ultimate tensile strength, f_{ub}, for bolts (Eurocode3 1-8, 2003)

Bolt classes	4.6	5.6	6.8	8.8	10.9
f_{yb} (N/mm^2)	240	300	480	640	900
f_{ub} (N/mm^2)	400	500	600	800	1000

Table 2. Categories of shear connections in bolts

Category	Criteria	Remarks
Bearing type: A	$F_{v,Ed} \leq F_{v,Rd}$ $F_{v,Ed} \leq F_{b,Rd}$	No preloading required for bolt classes from 4.6 to 10.9 may be used
Slip resistant at serviceability: B	$F_{v,Edser} \leq F_{v,Rdser}$ $F_{v,Ed} \leq F_{v,Rd}$ $F_{v,Ed} \leq F_{b,Rd}$	Preloaded 8.8 or 10.9 bolts should be used. For slip resistance at serviceability
Slip-resistant at ultimate: C	$F_{v,Ed} \leq F_{s,Rd}$ $F_{v,Ed} \leq F_{v,Rd}$ $F_{v,Ed} \leq F_{b,Rd}$	Preloaded 8.8 or 10.9 bolts should be used. For slip resistance at ultimate

These formulations for the design and structural safety of bolts in steel structures will be evaluated in a probabilistic setting using appropriate probabilistic methods.

3 Methodology

3.1 Probabilistic Reliability Assessment Using Form

Considering optimization of the structural element here only and as usual as regards other cases, it can be said that, a structural element will fail if its resistance R is less than the applied load S acting on it (Melchers 1999). The probability of failure (P_f) of the structural element can be stated in any of the following ways:

$$P_f = P(R - S) \tag{3a}$$

R = Resistance,
S = Applied load on structure.

$$P_f = P(R - S) \leq 0) \tag{3b}$$

$$= P(R|S) \leq 1) \tag{4}$$

$$= P(\ln R - \ln S) \leq 1) \tag{5}$$

It can also be expressed in general as:

$$P_f = P\ (G(R, S)) \leq 0 \tag{6}$$

Where G(X) is the limit state function and the probability of failure (P_f) is identical with the limit state violation. For any random variable, X, the cumulative distribution function $F_x(X)$ is given by:

$$F_x(X) = \int_{-\infty}^{x} F_x(y)dy \tag{7}$$

Provided that $X \geq Y$.

It follows that for the common, but special case when R and S are independent, the expression for the probability of failure can be given as:

$$P_f = P(R - S) \leq 0) = \int_{-\infty}^{+\infty} F_R(rds) \tag{8}$$

It could also be written as:

$$P_f = P(R - S) \leq 0) = \int_{-\infty}^{+\infty} F_R(X)drds \tag{9}$$

This integral is known as convolution integral and $F_R(X)$ is the probability that $R \leq x$, or is the probability of the actual resistance, R, of the member less than some values of x.

The term $F_S(x)$ represent, the probability that the load effect, S, acting on the member has a value between x and $x + \Delta x \to 0$.

Considering the possible value of x, the total failure probability is obtained by:

$$P_f = \int_{-\infty}^{+\infty} F_S(x)(F_R(X))dx \tag{10}$$

This is the sum of all cases of resistance in which the applied load exceeds the resistance. The convolution integral may be integrated for a few distributions of R and S. This is easy when R and S are normal variables with mean μ_R, μ_s and variance $\sigma R_2, \sigma S_2$, respectively.

The safety margin will be:

$$Z = R - S \tag{11}$$

Therefore, the mean and variance will be given by:

$$\mu_z = E(z) = \mu_R - \mu_S \tag{12a}$$

$$\sigma_Z^2 = \text{var}(\mu) = \sigma_R^2 + \sigma_S^2 \tag{12b}$$

$$\sigma_z = \sqrt{\left(\sigma_R^2 + \sigma_S^2\right)}$$

Using well known rule for addition/subtraction of normal random variables Eq. (7) becomes:

$$P_f = (R - S) \leq 0) = P(z \leq 0) = ø\left[\frac{\mu z}{\sigma z}\right] \tag{13a}$$

Where ø is the standard normal distribution function, the reliability index, β, is defined as the ratio μ_z/σ_z.

$$P_f = ø\left(\mu_z/\sigma_z\right) = ø(-\beta) \tag{13b}$$

Hence, P_f increases with decreases in resistance and reliability indices are one of the commonly used probabilistic measures of safety. First Order Reliability Method (FORM) is one of most common basic techniques and is applicable to all probabilistic problems. It is usually preferred because of its independence on the simulation number (Webster 2001). In order to overcome the invariance problem with the failure function, it was found that it is necessary to transform the basic variables into independent standard normal variables. The First-Order Reliability Method is used by independent standard normal variables of whether transformation of independent variables can be undertaken from the cumulative probability of the distribution that is from the identity. The transformation is generally undertaken automatically within most reliability analysis software packages. However, for non-standard distribution functions, the transformation may need to be undertaken explicitly. First-order second-moment methods, involve estimating the failure probability by *linearizing* the failure surface at the closest point to the origin in standard normal space. It is usually necessary to iterate in order to determine the closest point to the origin, and a number of iterative and optimization techniques are available. The basic variable transformation and first-order reliability estimates are in two basic variables. The distance from the origin to the beta-point is equal to the first-order reliability index. This is sometimes referred to as the geometrical reliability index.

3.2 Limit State Functions

A limit state equation is a state beyond which structures no longer possess at least one of their characteristics that is, the failure at the state region.

3.2.1 Analysis of Structures Based on EC3 (2004)

The limit state equation is as follows

$$G(x) = R - S \tag{14}$$

R = resistance shear force and S = applied shear force.

Hence,

$$G(x) = F_{V,RD} - F_{V,ED} \tag{15}$$

Failure of the Joint due to Bolt Shearing

For the connection to be safe against shear failure the equation below must be satisfied.

$$F_{V,ED} \leq F_{V,RD} \tag{16}$$

$$F_{V,RD} = \frac{n \propto v f_{ub} \pi d^2}{4\gamma M_2}, \quad n = 4 \tag{17}$$

and

$$F_{V,ED} = \frac{Vl}{2} \tag{18}$$

The maximum design shear load, V, prescribed in relevant codes (API 2000) for offshore structures is given as:

$$V = (1.35G_k + 1.7Q_k) \tag{19}$$

by substituting for V in the above Eq. (18) we obtain

$$V = \frac{(1.35G_k + 1.7Q_k)l}{2} \tag{20}$$

Let $\alpha = \frac{G_k}{Q_k}$ be the load ratio.

The applied shear force is then

$$F_{V,ED} = \frac{Q_k(1.35 \propto + 1.5)l}{2} \tag{21}$$

Hence, failure due to bolt shear is given by

$$G(x) = 1.26 f_{us} d^2 - Q_k(0.675\alpha + 0.75)l \tag{22}$$

Failure due to bolt bearing

$$F_{b,RD} = \frac{n_{k1} \propto_b f_{ub} td}{4\gamma M2} \tag{23}$$

$$G(X) = f_{ub}td - Q_k(0.675 \propto + 0.75)l \tag{24}$$

The results obtained and discussions there from the first order reliability analysis as coded in FORM 5 are given in Sect. 4.

4 Results and discussion

4.1 Results

The probabilistic model generated in Table 3 were analyzed using the First Order Reliability Method (FORM) to give values of structural safety indices (β) and probability of failure (P_f) in shear and bearing capacity of bolted steel joints in offshore structures. An algorithm developed in FORTRAN module was used for the above failure mode with various ultimate strengths of steel, load ratios, nominal areas of bolts, diameter of bolts and thickness of fasteners.

Table 3. Probabilistic parameters of the basic variables under shear mode of failure, bearing mode of failure

S/no	Basic variables	Basic variables Ex	Probability distribution function	Coefficient of variation	Standard deviation Sx
1	Ultimate shear strength of bolt f_{us} (N/mm^2)	600	Log-normal	0.03	18.00
2	Ultimate bearing strength of bolt f_{ub} (N/mm^2)	600	Log-normal	0.03	18.00
3	Imposed load Q_k (KN/m^2)	10.00	Log-normal	0.03	0.30
4	Diameter of the bolt (mm)	25.00	Normal	0.05	1.25
5	Length (mm)	1500	Normal	0.05	75.00
6	Thickness of fasteners t (mm)	40.00	Normal	0.03	1.20
7	N	4.00	Normal	0.05	0.20

4.2 Discussion

The failure due to bolt shearing as illustrated from Figs. 1, 2, 3, 4, 5 and 6; shows that the structural safety indices increases with increases in bolt diameters and bolt strength, while the safety indices decreases with increases in load ratios. It is observed also that for the bolt diameters 20 mm, 22 mm and 24 mm there is significant increases in structural safety levels (that is, the probability that the structure will fail in shear is minimal or greater resistance to applied loads).

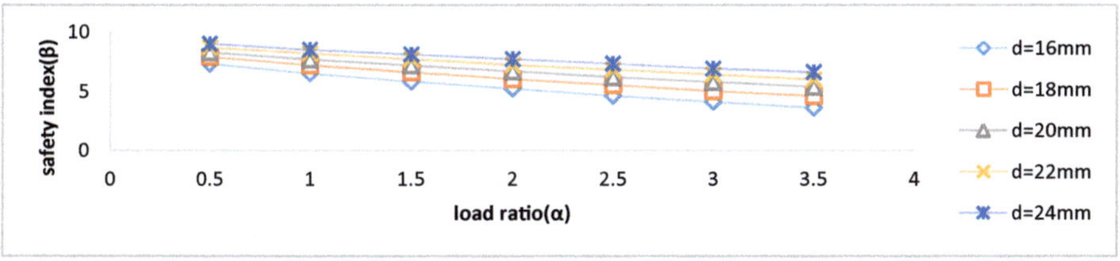

Fig. 1. Relationship of safety index (β) to load ratio (α) and bolt diameter (bolt grade 6.8, L = 2000 mm, f_{us} = 600 N/mm^2)

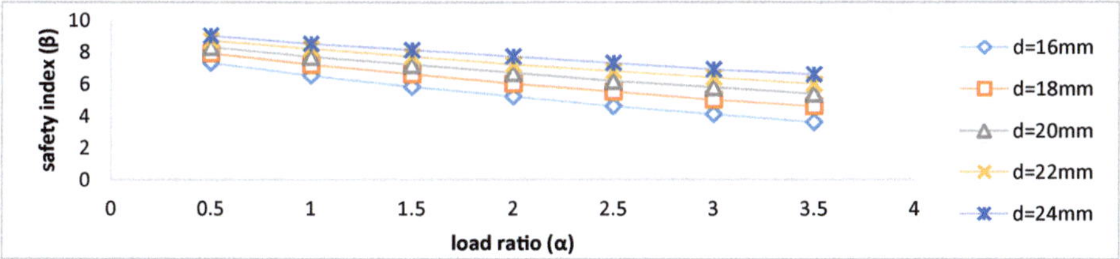

Fig. 2. Relationship of safety index (β) to load ratio (α) and bolt diameter (bolt grade 5.6, L = 1500 mm, f_{us} = 500 N/mm^2)

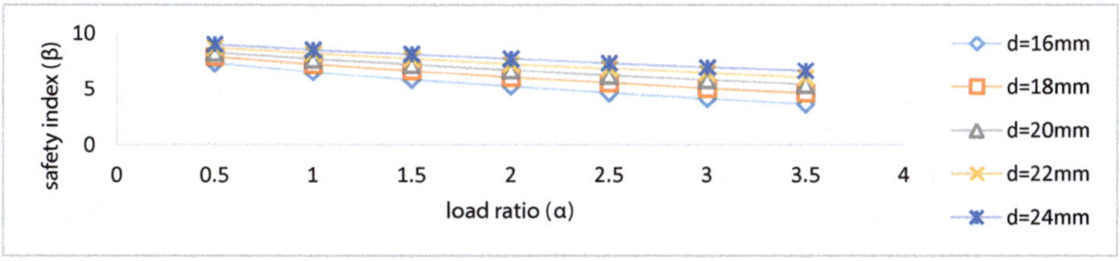

Fig. 3. Relationship of safety index (β) to load ratio (α) and bolt diameter (bolt grade 4.6, L = 1000 mm, f_{us} = 400 N/mm^2)

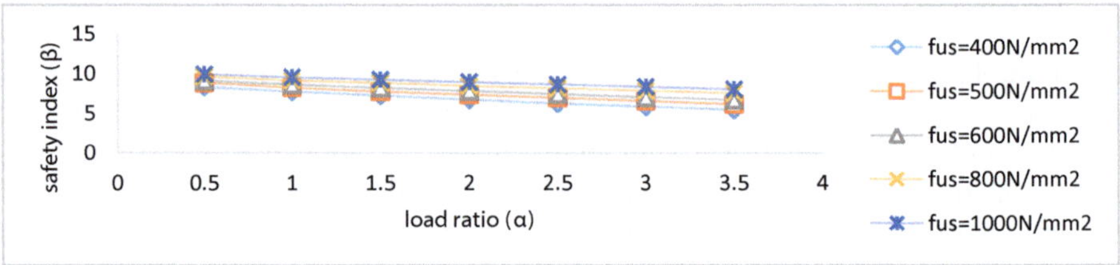

Fig. 4. Relationship of safety index (β) to load ratio (α) and 18 mm bolt diameter, L = 2000 mm and varying f_{us} in shear.

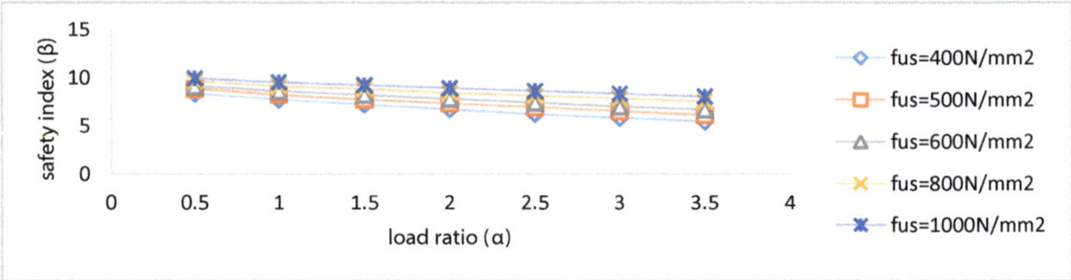

Fig. 5. Relationship of safety index (β) to load ratio (α) and 20 mm bolt diameter, L = 1500 mm and varying f_{us} in shear.

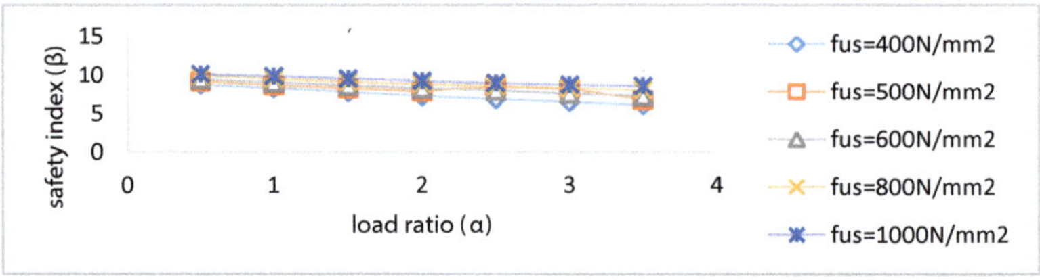

Fig. 6. Relationship of safety index (β) to load ratio (α) and 22 mm bolt diameter, L = 1000 mm and varying f_{us} in shear.

For the failure due to bearing of bolt as illustrated in Fig. 7, 8, 9, 10, 11 and 12, it is observed that the structural safety indices increases with increases in bolt diameters, bearing strengths and thickness of plates and members in joints, while the safety indices decreases with increases in load ratios. It is noted too that for the bolt diameters 20 mm, 22 mm and 24 mm, there are significant increases in structural safety levels; but at 10 mm or lower plate thickness, the safety of the structure is not guaranteed as in the case of other plate thicknesses of 20 mm, 30 mm, 40 mm and 50 mm. This suggests that bolted steel joints of members in offshore should be sufficiently thick; perhaps 16 mm thick plates should be used as minimum.

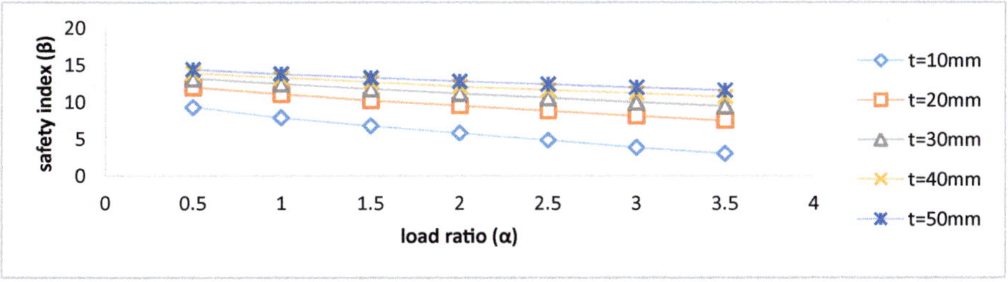

Fig. 7. Relationship of safety index (β) to load ratio (α) and bolt diameter (bolt grade 6.8, L = 2000 mm, f_{ub} = 600 N/mm^2)

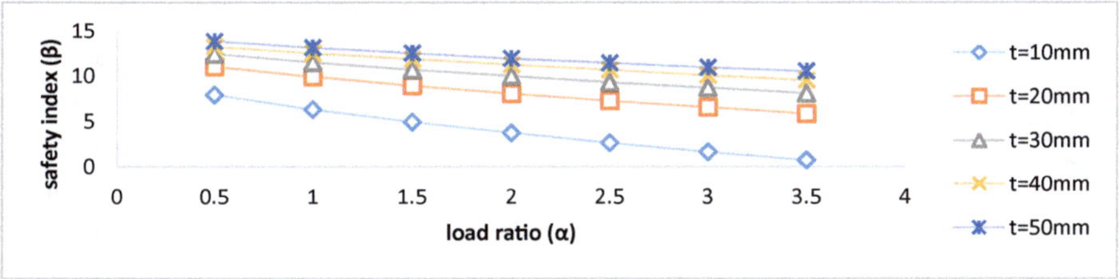

Fig. 8. Relationship of safety index (β) to load ratio (α) and bolt diameter (bolt grade 5.6, L = 1500 mm, f_{ub} = 500 N/mm^2)

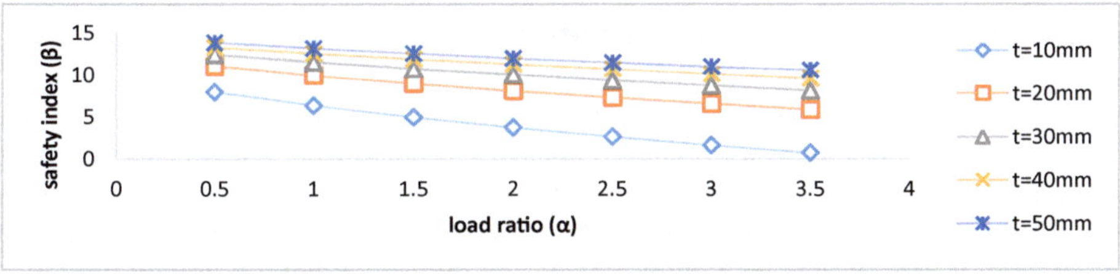

Fig. 9. Relationship of safety index (β) to load ratio (α) and bolt diameter (bolt grade 4.6, L = 1000 mm, f_{ub} = 400 N/mm^2)

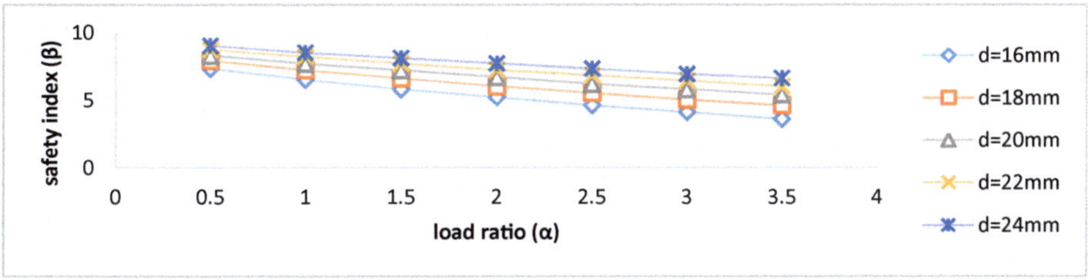

Fig. 10. Relationship of safety index (β) to load ratio (α) and bolt diameter (bolt grade 4.6, L = 1000 mm, f_{ub} = 400 N/mm^2)

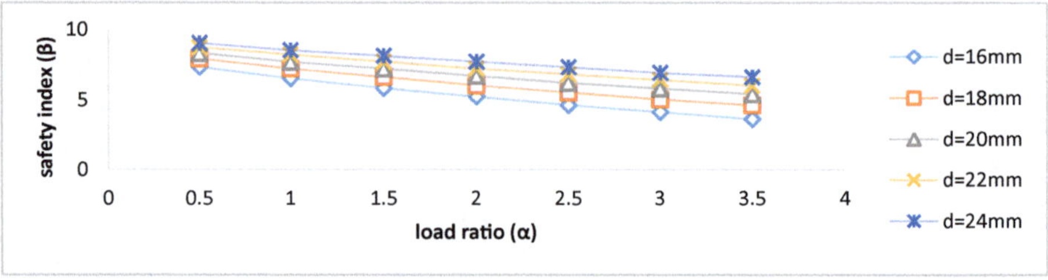

Fig. 11. Relationship of safety index (β) to load ratio (α) and 30 mm bolt thickness L = 1500 mm, f_{ub} = 500 N/mm^2 and varying diameter in bearing

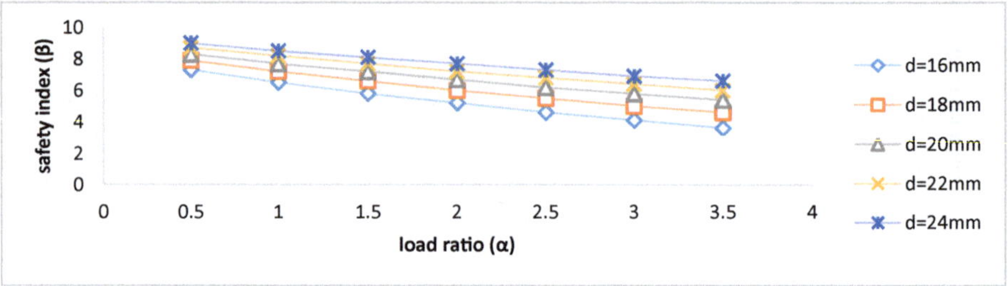

Fig. 12. Relationship of safety index (β) to load ratio (α) and 20 mm bolt thickness L = 1000 mm, f_{ub} = 400 N/mm^2 and varying diameter in bearing

5 Conclusion and Recommendation

This study presents the results of investigation of the structural safety levels of steel joints for two failure modes for bolted joints using probabilistic assessments. The level of failure which is associated with each of the failure modes were computed using the First Order Reliability Method (FORM). It was observed that the safety indices decreases with increases in load ratios which is an indication that the structural safety of steel joints is dependent on the bolt diameter, bolt strength, thickness of fasteners and the joint load. However, the formulations as evaluated is robustly specified or configured as the safety indices are well above the minimum for safety and economy as recommended by the JCSS (2005). Thus, economy commensurate with safety may still be achieved when prompt and adequate maintenance of the existing structure is carried out and for upgrade for increased structural capacity.

In order to obtain an optimum design life, as the recommendations of EC3 (2003) and JCSS (2005). It is recommended herein that the use of high-strength bolt of grade 6.8 and 24 mm diameter; and with a fastener thickness of 40 mm which not only have high resistance but also for its safety and economy is essentially structurally safe for joints in offshore steel structures.

References

American Petroleum Institute (API): Recommended practice for planning, design and constructing fixed offshore platforms, API RP2A-WSD, 21st Edn (2000)

Armand, M., Endres, F., MacFarlane, D.R., Ohno, H., Scrosati, B.: Ionic-liquid materials for the electrochemical challenges of the future. Nat. Mater. **8**(8), 621 (2009)

Baddoo, N.R.: Stainless steel in construction: a review of research, applications, challenges and opportunities. J. Constr. Steel Res. **64**(11), 1199–1206 (2008)

Banerjee, T., Sahoo, R.K., Rath, S.S., Kumar, R., Khanna, A.: Multicomponent liquid-liquid equilibria prediction for aromatic extraction systems using COSMO-RS. Ind. Eng. Chem. Res. **46**(4), 1292–1304 (2007)

Da Silva, L.S.: Towards a consistent design approach for steel joints under generalized loading. J. Constr. Steel Res. **64**(9), 1059–1075 (2008)

Ersdal, G., Hörnlund, E.: Assessment of offshore structures for life extension. In: ASME 2008 27th International Conference on Offshore Mechanics and Arctic Engineering, pp. 277–284. American Society of Mechanical Engineers (2008)

Fricke, W., Bronsart, R.: 2012 International Ship and Offshore Structures Congress (ISSC), Rostock (2012)

Hobbacher, A.: Recommendations for Fatigue Design of Welded Joints and Components. Welding Research Council, New York (2009)

Jelovica, J., Romanoff, J., Remes, H.: Influence of general corrosion on buckling strength of laser welded web-core sandwich plates. J. Constr. Steel Res. **101**, 342–350 (2014)

Melchers, R.E.: Corrosion uncertainty modelling for steel structures. J. Constr. Steel Res. **52**(1), 3–19 (1999)

Moan, T.: Safety of offshore structures. Center for Offshore Research and Engineering, National University of Singapore, Singapore, Report, 4 (2005)

Paik, J.K.: Residual ultimate strength of steel plates with longitudinal cracks under axial compression–experiments. Ocean Eng. **35**(17–18), 1775–1783 (2008)

Paik, J.K.: Recent advances and future trends on plasticity and impact mechanics of ships and offshore structures. Proc. Eng. **173**, 17–24 (2017)

Van der Vegte, G.J., Makino, Y.: Ultimate strength formulation for axially loaded CHS uniplanar T-joints. In: The Fifteenth International Offshore and Polar Engineering Conference. International Society of Offshore and Polar Engineers (2005)

Wardenier, J., Packer, J.A., Zhao, X.L., Van der Vegte, G.J.: Hollow sections in structural applications. Bouwen met staal, Rotterdam (2002)

Yeomans, N.F.: Guide on the use of bolts: single sided (blind) bolting systems. Serie CIDECT Construction with Hollow Sections. ECCS-TC, 10-01 (2002)

Zhao, X.L., Zhang, L.: State-of-the-art review on FRP strengthened steel structures. Eng. Struct. **29**(8), 1808–1823 (2007)

XRDF, SEM and Compressive Strength Properties of a New Alkali Activated Fly Ash Concrete Mortar

Hassan Al-Nageim[1(✉)], Aaraf Al-Khuzai[1], Jonethan Draker[2], Jennifer Croft[2], Linda Seton[3], and Nicola Dempster[3]

[1] Civil Engineering Department, Liverpool John Moores University (LJMU), Liverpool, UK
H.K.Alnageim@ljmu.ac.uk
[2] Bioresource Services, United Utilities, Warrington, UK
[3] Chemistry and Bimolecular Sciences, LJMU, Liverpool, UK

Abstract. This paper reports the results of laboratory studies on the mechanical and chemical properties of alkali activated fly ash concrete mortar. A new hydraulically bound cementitious material made from environmentally friendly waste fly ash and subjected to physico–chemical activation has been developed in this study. In the production of the new cement, fly ash initially has been activated by grinding the fly ash using mechanical grinding and the addition of FGD and or GGBS. The new cementatious materials containing 35% Ordinary Portland Cement (OPC) and 65% of grinded fly ash. The mix then further activated by the use of NaOH of 4 M molarity. NaOH has been added to the mix at 1%, 2% and 3% of fly ash. Alkali activated fly ash where then subjected to curing at normal room temperature for one day.

The paper report the compressive strength of the results at age of 3, 7 and 28 days.

Samples shown significant improvement in the compressive strength. The reasons for the development of strength of the new material products was studied and explained using both SEM and XRDF analysis and reported in this study.

Keywords: Cementitious materials · Fly ashes · Bottom ash
Cement mortar compressive strength

1 Introduction

Mechano-chemical activation of fly ashes using grinding and various alkalis has been studied extensively, see the work of Thomas et al. (1999a, b), Sadique et al. (2013), Sabir (1995), Hasparyk et al. (2000). Simultaneous application of mechanical and chemical activation of fly ash was reported to produce strength in excess of that of reference mortar at 20% replacement level by Hasparyk et al. The improvements were also greater when compared with same replacement level of silica fume (SF). A previous study by Thomas reported that cement clinker ground with reactive silica-based complex admixture (RSA) induced changes within the system, which led to an

H. Rodrigues and A. Elnashai (Eds.): GeoMEast 2018, SUCI, pp. 370–380, 2019.
https://doi.org/10.1007/978-3-030-01932-7_29

acceleration of the hydration of high performance cement, Thomas, et al. Extensive research has been conducted with binary and ternary blends using different additions with cement (Durand et al. 1990; Jones et al. 2004). In the case of blending it is obviously of importance to investigate how soon the incorporated pozzolanic material reacts with calcium hydroxide to form hydration products. For such a blending operation, homogeneous nucleation may occur between various particle sized fractions leading to the development of a dense microstructure of hardened product with higher durability. A higher packing density and increased hydration product in ternary blends have been reported by various researchers Rukzon et al. (2009) and Toutanji and El-Korchi (1995). Furthermore, a blend of fly ash and silica fume with cement (ternary blend) was found very effective to control alkali silica reaction (ASR) rather than using high amounts of fly ash or silica fume Yogendran et al. (1987).

The chemical activation is a particular procedure by which the powder is mixed with specific chemical activators and then the mixture is cured under a certain temperature to create solid materials. In general, the main reaction of any pozzolanic material is the breakdown of the glassy part by attack of OH^- ions on the SiO_2 and AlO_2–SiO_2 framework Rasa et al. (2009). The glassy constituent of the fly ash is changed into well-compacted cement. The main reaction product formed in alkali activation is an amorphous alumina-silicate gel. This product, also considered as a "zeolite precursor", is an X-ray amorphous material, Knofel (1971).

During mechanical activation, the external dynamic forces due to comminution induce the solid to vibrationally and electronically excited structures and destabilise the electronic structure of bonding.

The most convenient method for mechano-chemical reaction is using a mortar and pestle which stimulates the solid state reaction through grinding. Grinding leads to the generation of larger surface area, but the agglomeration phenomena (for mechano-chemical reaction during grinding) generally for long duration grinding, inversely affects the pozzolanic property).

In this paper, a new cementatious materials containing 35% Ordinary Portland Cement (OPC) and 65% of grinded fly ash[with the aid of adding 5% FGD during the grinding] where activated by the use of NaOH of 4 M molarity. NaOH has been added to the mix at 1%, 2% and 3% of fly ash and the mixes left to cure at laboratory normal room temperature for 1 day.

The paper reports the compressive strength of the studied mixes at age of 3, 7 and 28 days. The reasons for the development of strength of the new material products was studied and explained using both ESM and XRDF analysis.

2 Materials

A waste fly ash originating from industrial sources in the UK has been collected and used for formulation of new blended secondary cementitious material (SCM) at laboratory scale. The source materials of the fly ashes resulting after combustion between 700 °C and 1200 °C in their industrial operation using fluidized bed combustion system. Fluidized beds use calcium hydroxide and activated carbon to capture sulphur and nitric oxide (NOx) released during combustion. The chemical and mineralogical analysis of ground fly ashes with the aid of FGD has been illustrated in Table 1.

Table 1. Major oxides of ground materials by EDXRF analysis

Chemical compounds	OPC	FA Ground with 5% FGD
CaO	62.379	10.538
SiO_2	26.639	30.233
Al_2O_3	2.435	3.071
MgO	1.572	2.498
Fe_2O_3	1.745	15.236
SO_3	2.588	-
K_2O	0.724	1.285
TiO_2	0.385	1.189
Na_2O	1.533	1.512

Due to the core objective of this study which is based on using a very high percent replacement of OPC in normal cement sand mortar, a 65% of ground fly ash replacement of ordinary Portland cement were used. Commercially available GGBS and ordinary Portland cement (OPC) type CEM-II have been used throughout the research. Sand passing through 2 mm sieve was adopted for making mortar samples. Liquid Sodium hydroxide (NaOH) alkali with 4 Molarity in water has been used as alkali activator.

3 Experimental Method

The elemental composition of used materials (major oxides and trace elements) have been determined by Shimadzu EDX 720. Energy dispersive X-ray fluorescence (EDXRF) spectrometer and mineralogical characterisation were carried out by x-ray diffraction (XRD) using a Rigaku Miniflex diffractometer (Miniflex gonimeter) with CuK X-ray radiation, voltage 30 kV, and current 15 mA at scanning speed of 2.0 deg/min in continuous scan mode. Beckmen coulter laser diffraction particle size analyser has been used for determining the particle size distribution (PSD) of normal and activated fly ash particles in aqueous liquid module (ALM) mode. Scan electron microscope (SEM) in association with energy dispersive spectrometer (EDS) has been utilised for microstructural analysis. The physical and chemical properties of fly ashes and their changes for applying mechanical activation (grinding) will be analysed. Then four types of fly ash-additives blends composed of FA+OPC+FGD [mix1], FA+OPC +GGBS+FGD [mix 2], Mix1 with 1, 2 and 3 percent of 4 M NaOH solution [named as mixes 3, 4, 5 in the paper], and mix 2 with 1% 4 M NaOH solution [mix 6] were formulated and their physic-chemical properties were analysed and compared with control cement (OPC). The optimisation of blend perimeter such as cement binder content: sand and, water cement binder ratio were found to be 1:2.25 and 0.45 respectively. Moreover, the microstructure of hydrated products after 3 and 28 days of hydration and their associated strength development have been examined.

40 mm cube test specimens have been prepared in Hobart mixer. The samples were demoulded after 24 h and kept in sealed humid environment at 20 °C for designated period. Control AUTOMAX5 automatic compression tester has been used for determining the strength of mortar cubes at designated time with 0.2 MPa/s loading rate. The average strength value of three specimens prepared for the binder content and specified curing age were taken for analysis throughout the study.

4 Results and Discussion

4.1 Particle Size Distribution and Chemical Properties of the New SCM Products

The blend of secondary cementitious materials (SCM) satisfied the core aim of this study namely producing a low compressive strength SCM of a range of 15 MPa to 25 MPa and containing low percentage content of ordinary Portland cement are marked as mixes 1 to 6 with mix contents as follows:

Mix	Contents
1	35%OPC+65%FA+(5%FGD)
2	35%OPC+50%FA+(5%FGD)+15%GGBS
3	35%OPC+65%FA+(5%FGD)+1% NAOH
4	35%OPC+65%FA+(5%FGD)+2% NAOH
5	35%OPC+65%FA+(5%FGD)+3% NAOH
6	35%OPC+50%FA+(5%FGD)+15%GGBS+1% NAOH

All fly ash particles have been mechanically activated by grinding with a mortar and pestle grinder. Dry FGD gypsum at the rate of 5% has been used as grinding aid. The physical modification in terms of particle size for the comminution has been shown in Figs. 1 and 2.

The increase in the fineness after grinding of fly ash, see Fig. 1, where the course particles of fly ash reduced in all size range to be equal or finer that OPC powder will provide more exposed surface to participate into hydration in association with dense interlocking.

The comparative PSD, physical and chemical analysis of the blends and reference cement have been shown in Fig. 2 and Table 2 respectively. The PSD of mixes 1 to 6 are very similar to OPC. Though all mixes blended products have very analogous

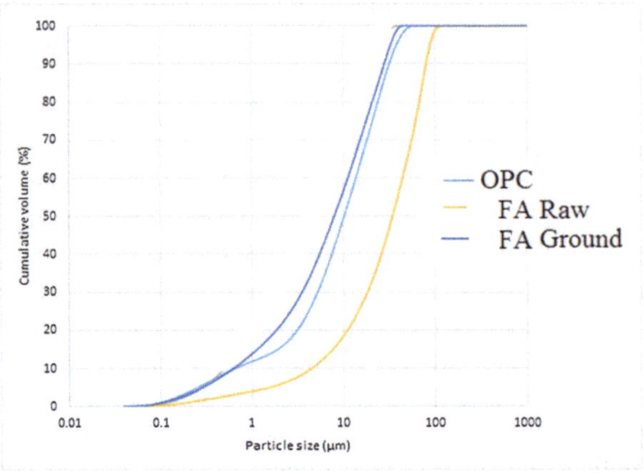

Fig. 1. Particle size distribution, PSD, for the fly compared with OPC

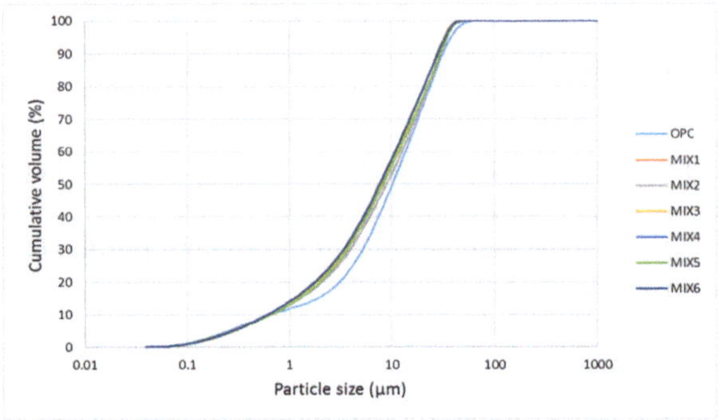

Fig. 2. Particle size distribution, PSD, for mixes compared with OPC

chemical composition in terms of major oxides but the disparity in terms of CaO, SiO_2, Fe_2O_3 contents and fineness will drive the hydration kinetics upon mixing among mixes blends. the SiO_2 is expected to consume the Na+ and K+ ion. However, there is strong disagreement exist regarding the effect of Na_2O and K_2O on strength development of concrete see the work of Sasidhar et al. (2017), Sabir (1995), Knofel (1971) and Sadique et al. (2013).

4.2 Chemical Activation and Compressive Strength of the New SCM Products

In this study it was decided to use 65% replacement of OPC with treated fly ash. The use of 1% to 43% of 4M NaOH to activate fly ash for high strength cement mortar is widely used. But there is strong disagreement regarding the Molarity and dosage of NaOH. Previous study showed the optimum NaOH of 1–3% will provide a good

Table 2. Major oxides of ground materials by EDXRF analysis

Chemical comp.	OPC	FA	MIX 1	MIX 2	MIX 3	MIX 6
CaO	62.379	10.538	32.768	27.534	35.138	34.535
SiO_2	26.639	30.233	33.736	33.234	30.234	31.285
Al_2O_3	2.435	3.071	3.411	3.370	2.808	2.937
MgO	1.572	2.498	2.559	2.492	2.215	2.252
Fe_2O_3	1.745	15.236	6.028	9.038	6.921	6.393
SO_3	2.588	-	2.959	3.193	1.982	2.778
K_2O	0.724	1.285	1.193	1.338	0.972	1.188
TiO_2	0.385	1.189	0.855	0.918	1.012	0.793
Na_2O	1.533	1.512	1.652	1.554	1.579	1.485
SiO_2 + Al_2O3+ Fe_2O_3	30.819		43.175	45.642	39.96	40.615
CaO/SiO2 **[BS EN requirement>2]**	2.34		0.91	0.83	1.16	1.1

Table 3. The new blends of cement sand mortar

Mix No	Mix details
Phase 1: Study of compressive strength of SCM containing 35% OPC and FA activated by FGD and [FGD+GGBS]- Physio-Chemical activation	
Control	100%OPC
1	35%OPC+65%FA+(5%FGD)
2	35%OPC+50%FA+(5%FGD)+15%GGBS
Phase 2: Study of compressive strength of SCM mix 1 chemically activated by 1% to 3% of NaOH with 4 M	
3	35%OPC+65%FA+(5%FGD)+1% NAOH
4	35%OPC+65%FA+(5%FGD)+2% NAOH
5	35%OPC+65%FA+(5%FGD)+3% NAOH
Phase 3: Study compressive strength of SCM mix 2 chemically activated by 1% of NaOH with 4 M	
6	35%OPC+50%FA+(5%FGD)+15%GGBS+1% NAOH

compressive strength enhancement to cement mortar, Sasidhar et al. As a minimum Molarity and percentage of NaOH solution is core objective of this current study. Therefore, initially 1% to 3% of 4 M NaOH has been used for the chemical activation. In this study, the development of the new blends of cement sand mortar run in three phases as shown in Table 3.

It can be Seen that the Compressive Strength of Both Mixes 1 and 2 are Very Close to the Target Compressive Strength Range of the Blended Products of 15–25 N/M m^2

In phase two of the research, a further activation to mix 1 has been tried using chemical activation by mixing the blended cementations materials with 1%, 2% and 3% of 4 M NaOH chemical solutions. After mixing for 2 min using normal laboratory mixer, the mix left to cure in the laboratory for one day at room temperature of approximately 20 °C, then chemically activated products are used in preparing the samples. The three mixes produced a compressive strength of 20.59, 27.58 and 25.55 at the percentage of chemical solution addition of 1%, 2% and 3% respectively. All these values are within the core objective target of the mortars compressive strength of 15–25 MPa.

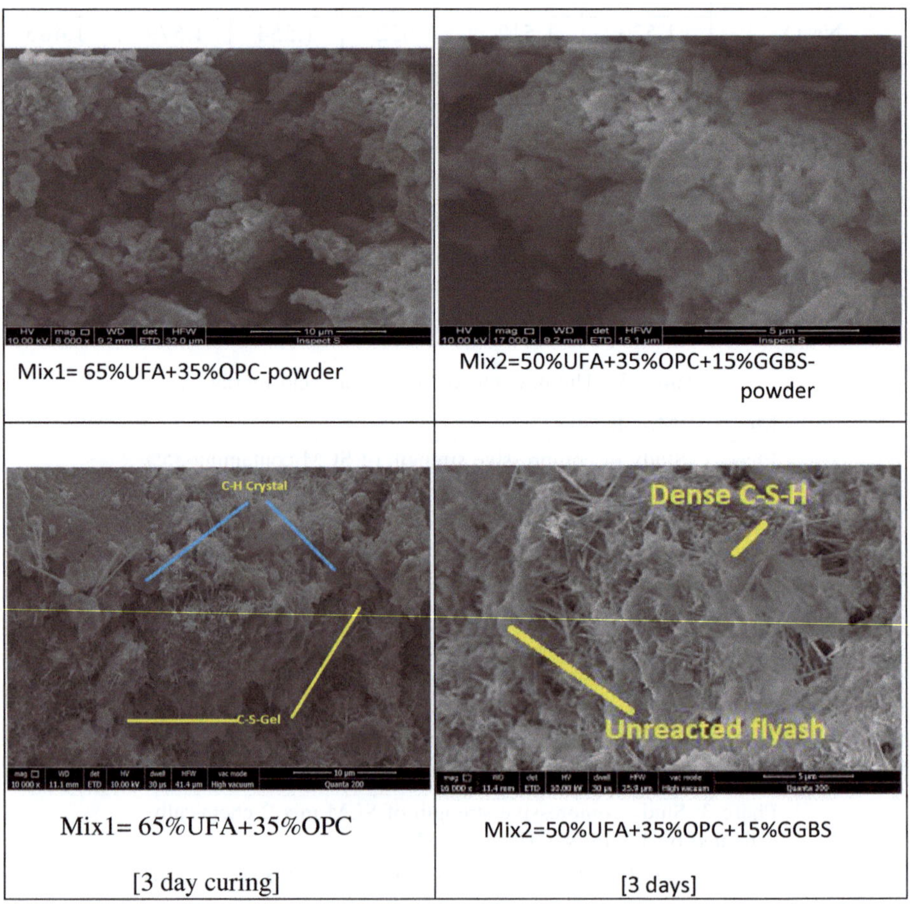

Fig. 3. SEM analysis of mixes 1-3 and mix 6

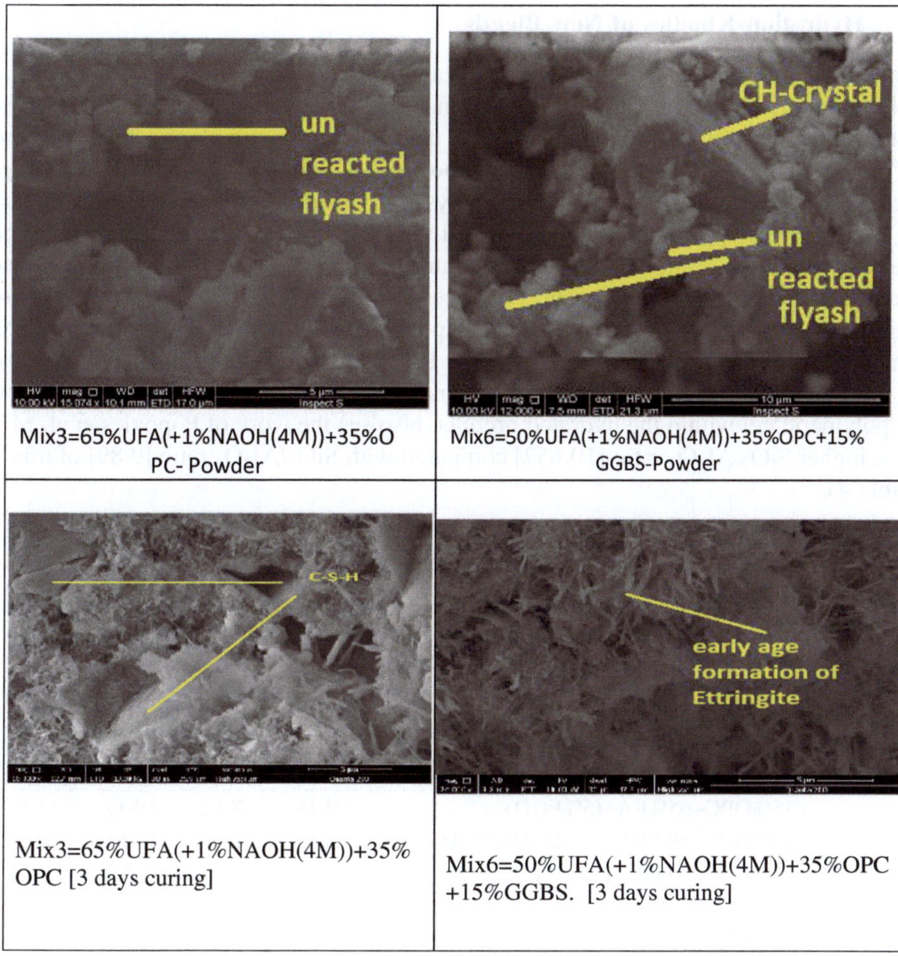

Fig. 3. *(continued)*

After discussion with the industrial client, decision was then made to apply 1% of the said chemical activation technology to mix 2 as can be seen in phase 3. The new mix is then called mix 6. The use of 1% of the chemical activation solution is purely based on meeting the client demand in reducing the cost of the cement mortar. Interestingly, the addition of 1% of the chemical said activator to mix 1 and mix two produces compressive strength of 20.59 MPa and 27.94 MPa respectively, and thus the core compressive strength target has been achieved. Mix 6 shows 26% improvement in the compressive strength compared with the compressive strength of mix 3. This is due to the present of GGBS, which with the existence of the chemical solution increases the alkalinity of the mix and thus provides peter environment for the hydration process during the curing time.

4.3 Hydration Kinetics of New Blends

The comparative strengths offered by mix 3 and mix 6 preferred SCM by the client at their optimum binder sand ratio (1:2.25) and water content (w/b of 0.45) has been demonstrated in Table 3. Though after 28 days of curing mix 1 and mix 6 attend 64.145% and 70.201% strength of control, the trend of hydration in both blends at later ages (such as 90 days) demands further analysis. Fernandez Olmo et al. suggested that the presence of Fe_2O_3 was also found retard the strength to a greater extent might be liable to produce strength lower than control at early stage. However the dense microstructure in SEM (Fig. 3) and presence of calcium silicate hydrate (C-S-H) are main parameters that contributing strength upon hydration by the new minimum OPC cement blend SCM produced in this study. It has also been found that higher SiO_2/Al_2O_3 ratio, see Table 2, confirms the simultaneous formation of C-S-H gel and geopolymeric gel within the hydrated product, also see the work of Popovics et al. Mix 6 has higher SiO_2/Al_2O_3 ratio [10.652] compared with SiO_2/Al_2O_3 ratio [9.89] of mix 3 (Table 4).

Table 4. Compressive strength at 3, 7 and 28 days of mixes

Mix No	Mix details	w/c	Compressive strength MPa at age in days of		
		0.45	3	7	28
Phase 1: compressive strength of SCM containing 35% OPC and FA activated by FGD or [FGD+GGBS] - Physio- Chemical activation					
Control	100%OPC	0.45	33.17	36.24	39.8
1	35%OPC+65%FA+(5%FGD)	0.45	8.82	10.17	13.59
2	35%OPC+50%FA+(5%FGD)+15%GGBS	0.45	12.47	13.56	19.52
Phase 2: compressive strength of SCM mix 2 chemically activated by 1% to 3% of NaOH with 4 M					
3	35%OPC+65%FA+(5%FGD)+1% NAOH	0.45	15.76	16.84	20.59
4	35%OPC+65%FA+(5%FGD)+2% NAOH	0.45	16.44	19.55	27.58
5	35%OPC+65%FA+(5%FGD)+3% NAOH	0.45	10.55	16.95	25.55
Phase 3: compressive strength of SCM mix 3 chemically activated by 1% of NaOH with 4 M					
6	35%OPC+50%FA+(5%FGD)+15%GGBS +1% NAOH	0.45	15.21	18.00	27.94

Figure 3 shows the SEM of the powder for mix 1 and 2 and the SEM for the same mixes at 3 days curing. The chemical reactions that take place between anhydrous cementitious material (Fig. 3), mix 1-powder, and mix 2-powder and water SEM - for the same mixes ar age 3 days generally complex and influenced by various factors. However, the following general features of reaction sequence and hydrates of cement concerning SEM analysis have been observed. Unreacted phases in different proportions coexist along with various reactants at different ages.

- The silicate phases lead to the formation of CH and C–S–H, the latter being amorphous or poorly crystalline.
- The reactions of the aluminate and ferrite phases in the presence of gypsum lead to the formation of C-S-H, Portlandite [$Ca(OH)_2$, ettringite [($Ca_6Al_2(SO_4)_3(OH)_{12}$-$26H_2O$], which forms during the hydration of cementatious materials and at later age converts to AFm (C_3A-$CaSO_4$-$12H_{2O}$).

5 Conclusions

Based on this research work, the following conclusions can be reported:

1. A low calcium fly ash has been activated by GDF and GGBS individually and collectively by grinding and blending with reactive NaOH 4M alkali solutions.
2. High Cao/SiO_2 and low SiO_2/Al_2O_3 ration are the main strength generating agent during the hydration process and influenced the compressive sgrength achieved for the six tested mixes.
3. NaOH alkaline solutions at a 1% to 3% and at a Molarity of 4 M were found effective in activating and improving the compressive strength of the new secondary cementations products.
4. A method of producing supplementary cementitious materials with bulk usage of waste streams has been proposed through this study. The method will not only reduce the utilization of natural minerals and fossil fuel but also omission of burning process (up to 1450 °C during OPC production) will drastically reduce CO_2 emission.

The research team is currently subjecting the new products to further development to;

- Minimise the adverse effect of Fe_2O_3 ion dioxide within the blends of the six mixes,
- Investigating the compliance of the properties of the new mixes to BS EN 197-1,
- Analysis of hydration product of all blends at age of curing after 28 days to 90 days.

References

Thomas, M.D.A., Shehata, M.H., Shashiprakash, S.G., Hopkins, D.S., Cail, K.: Use of ternary cementitious systems containing silica fume and fly ash in concrete. Cem. Concr. Res. **29**, 1207–1214 (1999a)

Sadique, M., Al Nageim, H., Atherton, W., Seton, L., Dempster, N.: Mechano-chemical activation of high-Ca fly ash by cement free blending and gypsum aided grinding. J. Constr. Build. Mater. **43**, 480–489 (2013)

Sabir, B.B.: High-strength condensed silica fume concrete. Mag. Concr. Res. **47**(172), 219–226 (1995)

Hasparyk, N.P., Monteiro, P.J.M., Carasek, H.: Effect of silica fume and rice hush ash on alkali-silica reaction. ACI Mater. J. **97**(4), 486–492 (2000)

Thomas, M.D.A., Shehata, M.H., Shashiprakash, S.G.: The use of fly ash in concrete: classification by composition. Cem. Concr. Aggreg. **21**(2), 105–110 (1999b)

Durand, B., Berard, J., Roux, R., Soles, J.A.: Alkali-silica reaction: the relation between pore solution characteristics and expansion test results. Cem. Concr. Res. **20**, 419–428 (1990)

Jones, M.R., Dhir, R.K., Csetenyi, L.J., Csetenyi, E.: Alkali Pre-activation of PFA to Maximise its use in concrete construction. Concrete Technology Unit, University of Dunde, Dunde (2004)

Rukzon, S., Chindaprasirt, P., Mahachai, R.: Effect of grinding on chemical and physical properties of rice husk ash. Int. J. Miner. Metall. Mater. **16**(2), 242 (2009)

Toutanji, H.A., El-Korchi, T.: The influence of silica fume on the compressive strength of cement paste and mortar. Cem. Concr. Res. **25**(7), 1591–1602 (1995)

Yogendran, V., Langan, B.W., Haque, M.N., Ward, M.A.: Silica fume in high strength concrete. ACI J. Mater. **84**, 124 (1987)

Rasa, E., Ketabchi, H., Afshar, M.H.: Predicting density and compressive strength of concrete cement paste containing silica fume using artificial neural networks. Trans. A Civ. Eng. **16**(1), 33–42 (2009)

Knofel, D.: Effect of alkalis on the properties of Portland cement clinker and Portland cement. Silikattechnik **22**, 262–265 (1971)

Sasidhar, K., Aruna, S., Jyothy, D.R.: An experimental investigation on behaviour of high volume alkali activated fly ash concrete. IJERMCE **2**(3), 60–66 (2017)

Stability of Concrete Containments of Nuclear Plants Under Jet Impact Loads

Mohamed Ihab S. Elmasry(✉), Nabil H. Alashkar, and Mostafa M. Hassan

Arab Academy for Science, Technology and Maritime Transport (AASTMT), Alexandria, Egypt
elmasryi@aast.edu

Abstract. Nuclear power plants are considered vital structures for generating electrical energy worldwide. In general, a nuclear power plant structure includes two containments, external and internal ones. The internal containment surrounds the main reactor as a primary shield. Meanwhile, the external containment protects the reactor from external impact loads such as jet crash, as well as being considered as a final shield between the internal of the reactor and the outer environment. This research studies the integrity of a reactor outer containment after being hit by a jet plane. Within the context of the research, an analytical model is generated using ANSYS® software to replicate the hit of the reinforced concrete containment including the circular shell and dome. An existing design of a classical nuclear reactor is used in modeling. Such design considers having the shell and the dome lined with inner steel liner plates in order to minimize the radiation flow to the outer environment in case of accidents. The studied external containment structure is assumed subjected to the impact of a jet plane, Boeing 747-200c. Riera Method is used to simulate the impact load with respect to time at a vertical level of 30 m above the upper foundation level, and at the outer surface of the external shell. Heavy weight concrete is assumed used in the shell and the dome with a compressive strength of about 60 MPa. The impact load is concentrated at 16 nodes at the outer surface of containment. The maximum deformations of the containment structure are studied and especially within the impact region of the jet plane. The containment is found to be stable after the impact of the jet plane, but with having some clear damage to some elements within the region surrounding the impact area and some secondary damages in other locations.

Keywords: Stability · Nuclear plant · Containment · Impact · Jet loads

1 Introduction

Nuclear Power Plants (NPPs) include several structures such as the internal and external containment structures and others. The external containment structure is the physical building that separates the reactor from the outer atmosphere, and houses the internal containment structure that surrounds the nuclear reactors and the associated chain reactions. In addition, NPPs are site-specific and vary in size based upon the power plant's power rating and the corresponding size of equipment. However, most

H. Rodrigues and A. Elnashai (Eds.): GeoMEast 2018, SUCI, pp. 381–398, 2019.
https://doi.org/10.1007/978-3-030-01932-7_30

typical current external containment structures range from 150 to 200 feet tall (45 to 60 m) with a diameter of approximately 150 feet (45 m) [1]. The latter structure should be airtight to prevent the release of radioactive particles into the atmosphere. Moreover, the external containment structure may be designed and constructed from steel or reinforced concrete (RC) materials.

In addition, despite the internal containment is structurally sound, yet, due to the high level of radiation and proximity to human population, a second external containment structure would be required. The latter secondary containment is thus necessary when radiation levels at the site boundary exceed the allowable limits. Accordingly, the two external and internal containments can be regarded as an enclosure building and a shield building respectively (ASCE 58) [1].

In addition, most current external containment structures have been designed as RC structures. Such concrete containment structures lined with steel are even more economical than steel containment structures for dry containment design [1]. This is attributed to the fact that a steel containment structure would require a concrete shield structure, thus concrete is required in either case, and the more economical choice is typically used.

Moreover, concrete as a construction material is commonly used for biological shielding in NPPs, medical units, and atomic research and testing facilities. This is the case where x-rays and gamma rays emissions must be considered in the design of biological shields [2]. The x-rays have a high penetration power but can be absorbed adequately by an appropriate mass of a suitable material. In contrast, gamma rays involves neutrons, which are heavy particles of atomic nuclei and do not carry an electrical charge. Polivka and Davis [2] state that concrete is an excellent shielding material that possesses the needed characteristics for both neutron and gamma ray attenuation, has satisfactory mechanical properties, and has a relatively low initial as well as maintenance costs [2].

1.1 Problem Definition

NPPs are expected to have a clear robust behavior. Accordingly, the impact of a military or civilian aircraft would not be expected to penetrate the containment of a reactor building [3]. Such challenge is of strategic significance since the damage caused by such missiles might lead to leakage of nuclear radiation. Earlier research shows that for RC containments with thicknesses greater than 1.4 m, permanent damage is negligible, whereas for thicknesses lower than 1.0 m considerable plastic strains occurred on opposite surfaces of the containment shell where different signs of strain are obtained, with the negative signs prevailing [4]. This is the case where the impact load of an aircraft on the outer nuclear containment building results from the effects of a finite amount of kinetic energy that depends upon the inertial and stiffness properties of the missile and target structure. Thereby, kinetic energy is transferred from the aircraft to the building walls and absorbed by the building components in the form of strain energy [3].

This paper studies and evaluates the effects caused by an aircraft impact, namely Boeing 747-200c, upon a typical outer RC containment of the new generation of NPPs. The step-by-step nonlinear response of such impact is studied then analyzed to identify the propagating damage in the containment structure from global and local behavior perspectives.

2 Method Statement

The response of the containment structure due to an impact of a civilian or military aircraft can be determined by equating the work done by the impacting load to the strain energy produced in the structure through applying the three basic principles of mechanics: conservation of energy, conservation of linear momentum, and virtual work principle. In addition, in the evaluation of the aircraft crash on the outer containment structure, the following steps are considered:

- Global structural response including excessive structural deformations or displacements, structure collapse or overturning, *etc*. The latter response depends clearly on the induced and propagated dynamic impact loads and stiffness characteristics of the containment subjected to impact.
- Localized structural damage due to the effects of missile impact, including penetration, perforation, scabbing, *etc*., that may lead to the failure of structural elements as a result of the effects of primary hit (aircraft) and secondary hits (engines and fuselage) that originated from an aircraft failure.

In the carried out analyses, the normal impact was chosen, according to Riera formulation [3], since it corresponds to the worst condition for the containment structure. This is the case where the structural response should be maximized for checking the reliability of the containment.

It was also assumed that the concrete may undergo gradual failure modes representing the progressive cracking and crushing of concrete itself caused by the abrupt stiffness changes (due to the propagation of the aircraft-missile kinetic energy). Moreover in the impact analysis, the damping effect was considered to be negligible as it does not affect the maximum response to impulse loading [3].

2.1 Case Study

The studied RC outer containment, as shown in Fig. 1, consists of 3 major components, the cylindrical wall, the hemispherical dome, and the cylindrical base raft foundation at which the external containment wall & inner containment structure are fixed to the foundation. In addition, based on an existing design [1], the cylindrical wall is assumed of 1.2 m thickness whereas the thickness of the RC hemispherical dome is assumed 1.05 m. The effective concrete cover is 150 mm at both faces till the reinforcement. The height of the cylindrical wall from the top of foundation till the spring line is 36.45 m. The total height of containment from the top of foundation till the apex of the dome is 60 m. The cylindrical wall and dome of containment is axisymmetric. The inner diameter of the containment is 45 m as shown in Fig. 2.

Moreover, since the containment structure must be airtight to prevent any fission emissions from bypassing, and knowing that concrete as a material is porous, then the studied containment is assumed to have a steel liner on the entire internal surface [1].

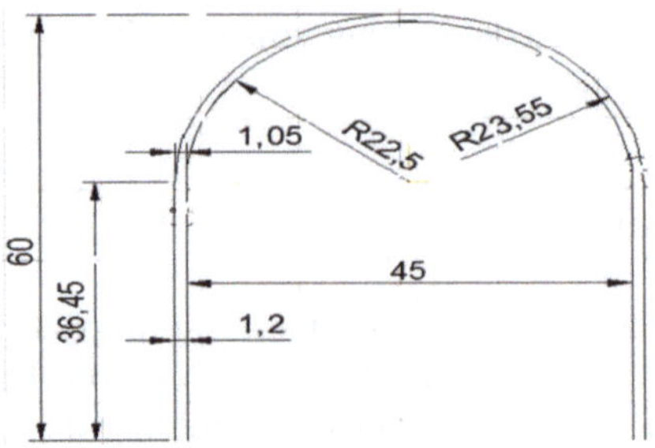

Fig. 1. Section elevation of an RC Containment wall & Dome

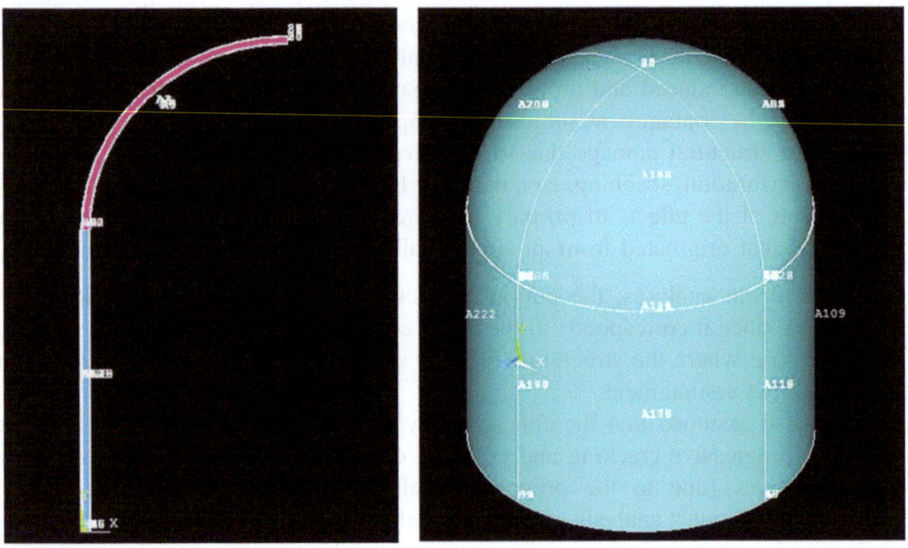

Fig. 2. General layout of the containment model in 2D and 3D

The liner steel plates are assumed anchored into the concrete by welded shear studs to the plate. This lining thus acts as a part of the system for the concrete wall [1]. Based on ASCE 58 [1], the required minimum thickness for the steel liner plates is a quarter of an inch; however, the studied existing design considers the steel liner plate of thickness three-eighths of an inch (0.9375 cm) as shown in Fig. 3.

2.2 Modeling

In the finite element model applied in ANSYS® software, the connection between the foundation and the cylindrical wall is assumed to be fixed. In addition, the local details in the RC containment structure, including penetrations, temporary openings, and embedded parts, were not considered in modeling. The meshing of the containment volume is shown in Fig. 4 where each element dimension is (2.025 × 2.025 × 1.2 m.)

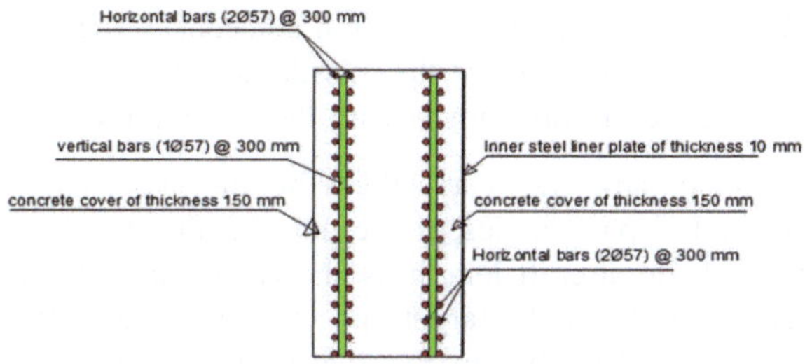

Fig. 3. Cross section plan details of shell wall reinforcement

Fig. 4. Meshing of external reinforced concrete containment building into small elements

2.2.1 Jet Impact Modeling on RC Containment

Impact loads due to the effects of a finite amount of kinetic energy that depends on the inertial and stiffness properties of missile and target structure, differs from the "impulse" load that is generally a force independent of the inertial and stiffness properties of missile and target. As a result of a jet impact, kinetic energy is transferred from the aircraft to the building walls and thereby absorbed by the building components in the form of strain energy.

From literature review, it is observed that the Riera method is efficient in evaluating damage on RC structures from severe impact loads [3]. The Riera method is an approximate method for constructing a force-time history for an aircraft striking a rigid wall based on a known distribution of mass and crushing characteristics of the aircraft along the length. This is the case where the Riera approach assumes that the aircraft can be represented like a soft missile [3, 5]. Moreover, it is assumed that as the aircraft strikes the target, a part of it close to the target gets crushed and the remaining portion of projectile undergoes elastic deformation [3].

The basic assumptions of the Riera method [5] can be summarized in:

1. The target is rigid.
2. The aircraft axis is normal to the target; angle of incidence is 90°.
3. The aircraft is separated into two parts, one being uncrushed and moving with velocity, v, and the other crushed with zero velocity.

4. All crushing takes place within a local region adjacent to the rigid target.
5. The crushing or material behavior of the missile is rigid-perfectly plastic.
6. The crushed mass of the aircraft moves with the target structure.

Consequently, some parameters clearly affect the impact scenario. Such parameters include the velocity and impact angles of the aircraft, the mass and stiffness of the aircraft as a function of the aircraft length as shown in Table 1, the aircraft loading capacity, the relative ductility of the target and that of the aircraft, and the size and location of the impact area. It is obvious that the most dangerous impact would result from a fully loaded aircraft Boeing 747-200c.

Table 1. Parameters of Boeing 747-200c aircraft

Air plane type	Boeing 747-200c
Mass at take-off (kg)	≈350000
Engine mass (kg)	≈4 × 21300
Impact speed (m/s)	≈269 (968.4 km/h)

Furthermore, it is assumed that the load versus time curve affect a constant Area equivalent to the average of total cross-sectional area of fuselage and wings. Thus, the impact contact area of the Boeing 747-200c in the analysis model is assumed 37 m^2 [3]. Moreover, in this analysis, the load is assumed to act on 16 nodes as concentrated point loads on the outer external wall of the containment. Figure 5 shows a schematic of the distribution of the loads on the loaded nodes.

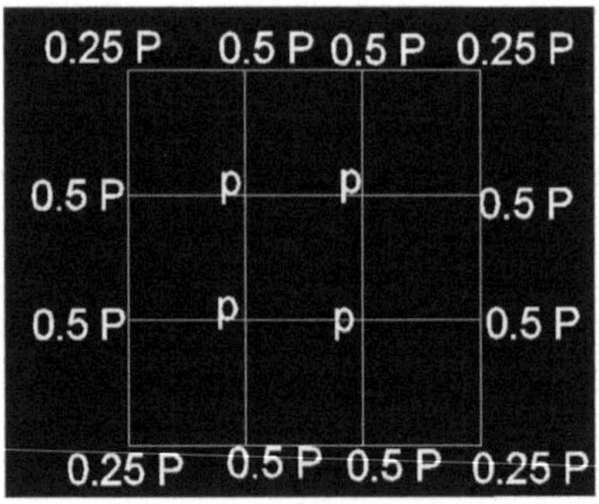

Fig. 5. Impact load distribution on nodes in the affected region

The impact force, F(t), acting on the rigid fixed containment is calculated with the Riera method, see Fig. 6, that considers one term coming from the buckling and the other one from the inertial effect of the mass releasing [5],

$$F(t) = P_c[x(t)] + \alpha m[x(t)][v(t)]^2 \quad (1)$$

where P_c represents the load necessary to crush or buckle the fuselage, $x(t)$, represents the distance from the nose of the aircraft to the point up to which crushing has progressed at time t, and m and $v(t)$ are the longitudinal mass per unit length and the velocity of the uncrushed part of the aircraft respectively. Meanwhile, α is an empirical correlation factor multiplying the effective mass on impact.

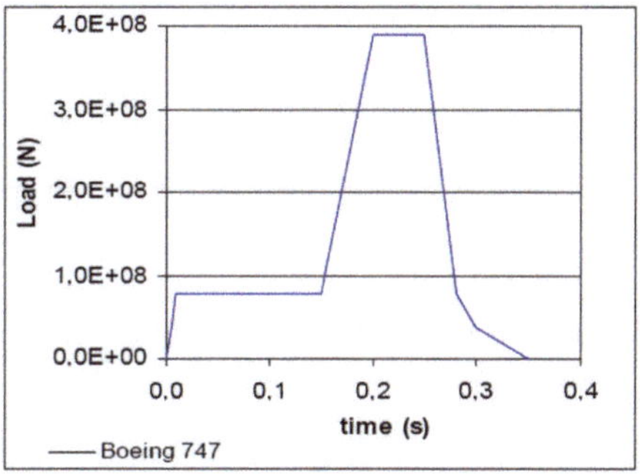

Fig. 6. Load-Time history curve of Boeing 747-200c

2.2.2 Containment Materials Modeling

<u>Concrete Modeling</u>

The concrete material is assumed to be homogenous and isotropic. The compressive uniaxial multi-linear stress-strain relationship for concrete, as defined by ANSYS®, is shown in Fig. 7 & as shown in Table 2, and calculated by the following equations:

$$f_c = E_c\mathcal{E} \quad \text{for } 0 < \mathcal{E} < \mathcal{E}_1 \quad (2)$$

$$f_c = f_c' \quad \text{for } \mathcal{E}_o < \mathcal{E} < \mathcal{E}_{cu} \quad (3)$$

$$f_c = E_c\mathcal{E}/\left(1 + (\mathcal{E}/\mathcal{E}_o)^2\right) \quad \text{for } \mathcal{E}_1 < \mathcal{E} < \mathcal{E}_o \quad (4)$$

$$\mathcal{E}_1 = (0.3 \times f_c)/E_c \quad (5)$$

$$\mathcal{E}_o = (2f_c)/E_c \quad (6)$$

where $\mathcal{E}_1$ is the strain corresponding to $0.3\, f_c$, $\mathcal{E}_o$ is the strain at peak point, and $\mathcal{E}_{cu}$ is the ultimate compressive strain [6].

Furthermore, the Poisson ratio for concrete is taken as 0.2, and the modulus of elasticity, E_c, is calculated in MPa from Eq. (7) whereas the modulus of rupture for concrete, f_r, is calculated in MPa from Eq. (8) according to ACI318-08 specifications [6].

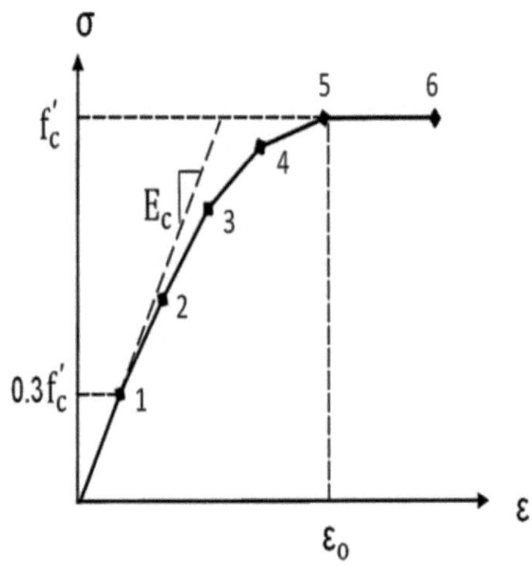

Fig. 7. Uniaxial compressive stress-strain curve for concrete

Table 2. Uniaxial compressive stress-strain for concrete in multi linear isotropic hardening

Strain	Stress (MPA)
0	0
0.00049444	18
0.001	33.337
0.002	53.217
0.003	59.73
0.003296	60

$$E_c = 4700\sqrt{f'c} \tag{7}$$

$$f_r = 0.62\sqrt{f'c} \tag{8}$$

In the analysis herein, the uniaxial compressive strength of concrete (Hi-Strength concrete) is assumed 60 MPa, whereas the uniaxial rupture stress f_r is assumed 4.8 MPa. Moreover, the shear transfer coefficients range is specified from 0.0 to 1.0, with 0.0 representing a smooth crack (complete loss of shear transfer), and 1.0 representing a rough crack (no loss of shear transfer). This specification may be made for both the closed and open crack [7, 10, 11]. When the element is cracked or crushed, a small amount of stiffness is added to the element for numerical stability as shown in Table 3. The stiffness multiplier, CSTIF, is used across a cracked face or for a crushed element, and defaults to 1E−6.

In addition, cracking is mathematically modeled at the element integration points using an approach that is called the smeared cracking model.

Table 3. Concrete strength input data

Input strength parameter	Values
Open shear transferee coefficient	0.3
Closed shear transferee coefficient	0.9
E (modulus of elasticity of concrete)	36406.043
Poisson ratio of concrete	0.2
Uniaxial cracking stress (MPA)	4.8
Uniaxial crushing stress (MPA)	60

Reinforcing Rebar Modeling

The reinforcement rebar used in the containment structure is ASTM A-615 Grade 60 steel. The stress-strain curve of the reinforcing bar is assumed to be elastic perfectly plastic as shown in Fig. 8. The latter Fig. 8 shows the bilinear material model defining the linear Isotropic hardening of the reinforcement in which the yield stress is assumed 400 MPA, the elastic modulus, E_s, assumed 2×10^5 N/mm^2, and the Poisson ratio assumed as 0.3 [8].

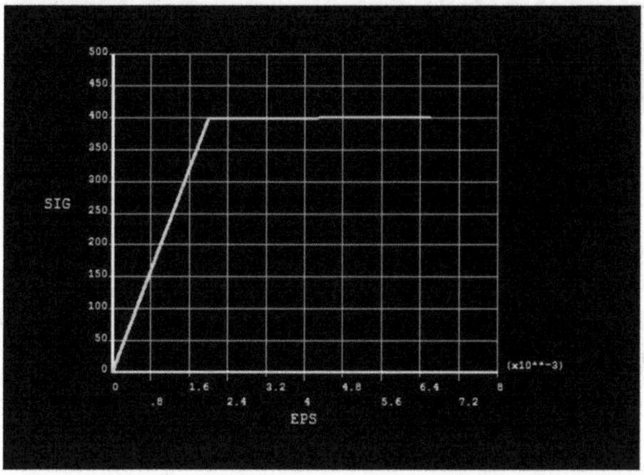

Fig. 8. Stress strain curve for steel reinforcing rebar

In addition, within ANSYS® software, rebar reinforcement is treated as an equivalent uniaxial material, which is smeared through the element section. The reinforcement layers were thus defined using the smeared reinforcement option of the SOLID65 element [8]. The reinforcement is assumed to be smeared and uniformly spread throughout the element and has uniaxial stiffness only in each direction. The amount of reinforcement was defined by material number, specifying a volume ratio and the orientation angles of the reinforcing bars.

The shell wall reinforcements (horizontal and vertical) are modeled as continuous smeared layers inside the wall section, and the same applies for the dome reinforcements. The smeared reinforcements were divided inside the shell wall and the dome into separate layers of vertical and horizontal reinforcements in order to replicate the reinforcement in each uniaxial stiffness direction [3, 9]. The thicknesses of the smeared reinforcement layers are calculated from the following equation

$$h = A/s \tag{9}$$

where h is the equivalent reinforcement layer in each direction, A is the cross section area of the reinforcement rebar, and s is the spacing between the reinforcement rebar in each direction.

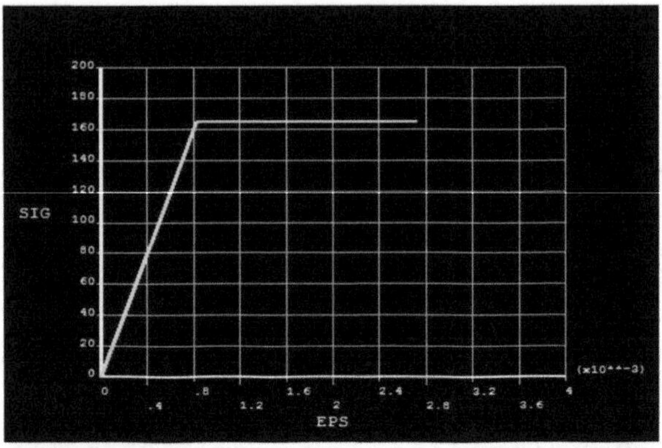

Fig. 9. Stress strain curve for the inner steel liner plate

<u>Steel Liner Modeling</u>

The 3/8″ (0.9375 cm) thick steel liner plate has a bilinear isotropic hardening plasticity model in which its yield stress is 165 MPa [1, 8] as shown in Fig. 9, and the elastic modulus, E_S, and the Poisson's ratio are assumed to be 2×10^5 N/mm^2 and 0.3 respectively.

2.2.3 Used Model Elements from ANSYS® Library

<u>SOLID65 Element</u>

The SOLID65 element [9] is used for the 3-D modelling of the RC wall and dome. The solid element is capable of cracking in tension and crushing in compression. The element is defined by eight nodes having three degrees of freedom at each node: translations in the nodal X, Y, and Z directions and the isotropic material properties. The element has one solid material and up to three rebar materials as shown in Fig. 10. The orientation of any rebar layer is defined by two angles (in degrees) from the element coordinate system.

In addition, SOLID65 allows the presence of four different materials within each element; one matrix material as concrete, and a maximum of three independent reinforcing materials. The concrete material is capable of directional integration point cracking and crushing besides incorporating plastic and creep behavior.

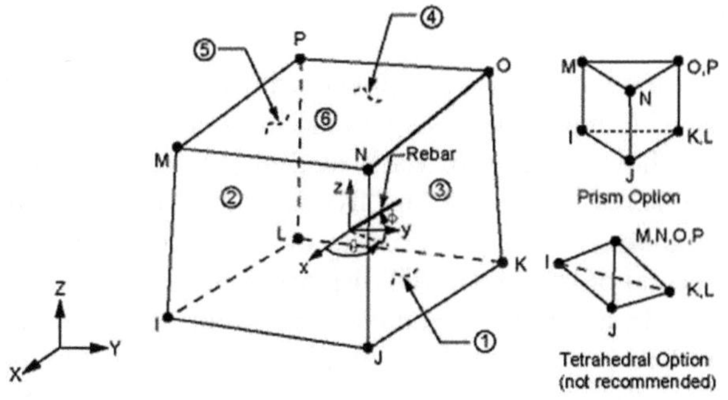

Fig. 10. Solid-65 element description

The reinforcement (which also incorporates creep and plasticity) has uniaxial stiffness only and is assumed to be smeared throughout the element.

Other assumptions and restrictions apply for the SOLID65 element. These assumptions include that the concrete material is homogenous and isotropic where cracking is permitted in three orthogonal directions at each integration point. In addition, if cracking occurs at an integration point, the cracking is modeled through an adjustment of material properties that effectively treats the cracking as a smeared band of cracks, rather than discrete cracks. Moreover, the reinforcement is assumed to be "smeared" throughout the element.

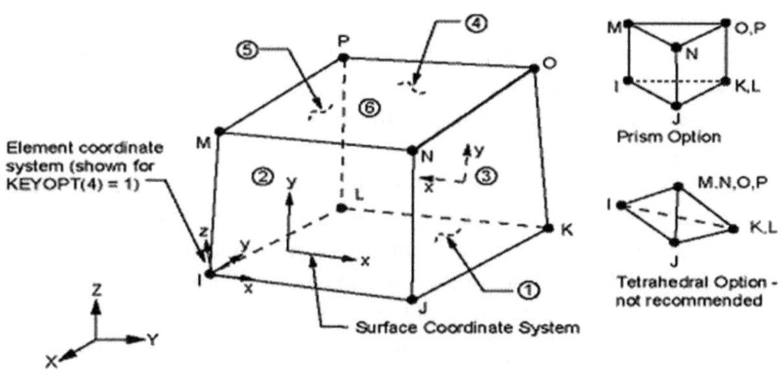

Fig. 11. Solid-45 element description

SOLID45 Element

The SOLID45 element [9] is used for the 3-D modeling of the Inner steel Liner plate, as shown in Fig. 11. The element is defined by eight nodes having three degrees of freedom at each node: translations in the nodal x, y, and z directions.

3 Analysis of Results

A full transient dynamic analysis was applied with considering a small displacement transient response. The impact analysis herein considered seven load steps in impact load-time curve as shown in Fig. 6. Each load step is either stepped or ramped. In addition, each load step was divided into five sub-steps as defined in the model. The transient dynamic analysis requires two sets of initial conditions since the dynamic equations are of second order. The initial displacement, u_o, and initial velocity ($\mathring{u}_o$) are assumed to be zero. In addition, the initial accelerations ($\ddot{u}_o$) are also assumed to be zero. Furthermore, Damping was also included in the analysis through identifying the mass damping and stiffness damping. Thus, Rayleigh damping with coefficient $\beta = 0$ was applied to the model. The coefficient α was reflected with a magnitude of 3.0 for the RC and 0.1 for the structural steel [4] such that the damping matrix, C, was identified from

$$C = \alpha m + \beta K \tag{10}$$

The outer RC containment response under the Boeing 747-200c aircraft impact load was studied at different time instances. Figure 12 shows the deformation schematic of the impact region at time 0.2 s. It can be observed that the elements within the impact region deformed in the direction of loading, which generated significant large tensile stresses at the inner surface of steel liner plates, and compressive stresses at the outer surface of concrete. The maximum displacement is found to be 46.658 mm in the region of impact at time 0.2 s. Meanwhile, other regions near the impact showed maximum displacements of 1.958 mm outside the containment in a normal 90° incidence manner, as shown in Fig. 13, and as shown in Table 4.

Plastic strains were also observed at time 0.2 s, where the maximum plastic strain in tension was 4.07E−4 in the direction of the 1st principle plastic strains as shown in Fig. 14a. The maximum plastic strain in compression was 7.94 E−4 in the direction of the 3rd principle plastic strains, as shown in Fig. 14b.

Failure or crack plotting within ANSYS® displays circles at locations of cracking or crushing in concrete elements. Cracking is shown with a circle outline in the plane of the crack, and crushing is shown with an octahedron outline [10]. If the crack has opened and then closed, the circle outline will have a "×" through it. Each integration point can crack in up to three different planes. The first crack at an integration point is shown with a red circle outline, the second crack with a green outline, and the third crack with a blue outline [9, 12].

Symbols shown at the element centroid are based on the status of all of the element's integration points. If any integration point in the element has crushed, the crushed (octahedron) symbol is shown at the centroid. If any integration point has cracked or cracked and closed, the cracked symbol is shown at the element centroid. Finally, if more than one integration point has cracked, the circle outline at the element centroid shows the average orientation of all cracked planes for that element [9, 10].

At time of 0.2 s, cracking and crushing signs appeared. Crushing occurred at the elements within the impact regions at the outer surface. These elements got fully deteriorated. In contrast, crack signs appeared at tension regions where such regions

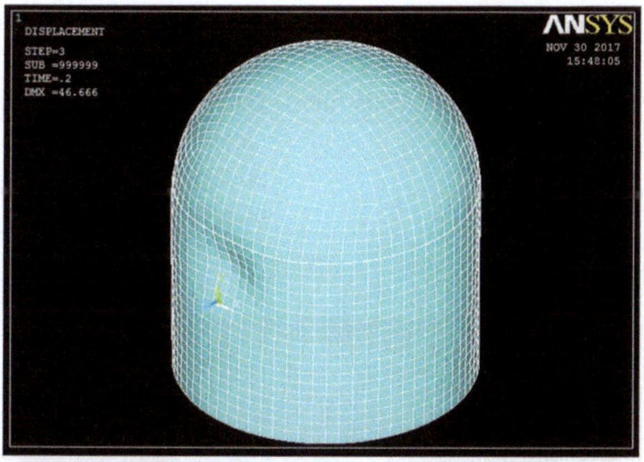

(a)

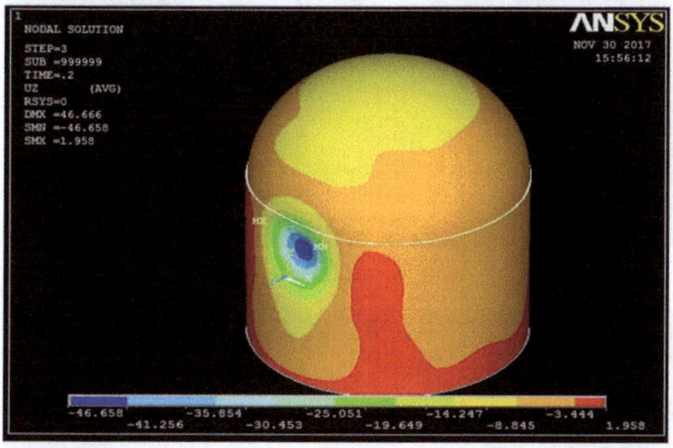

(b)

Fig. 12. (a) Deformation shape of the containment at time 0.2 s (b) Displacement distribution of the containment at time 0.2 s

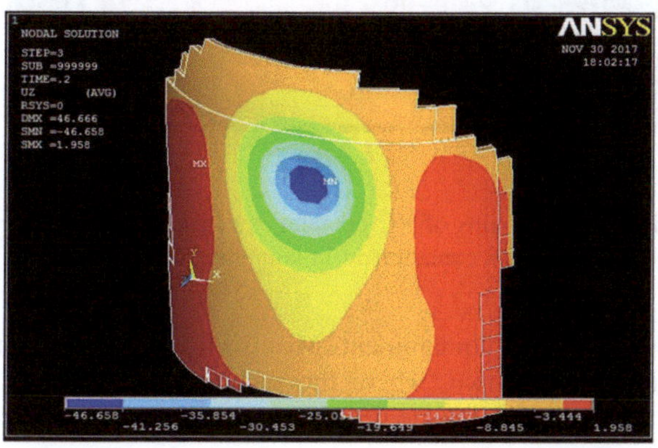

Fig. 13. Displacements of containment within the hit region in Z-axis direction

Table 4. Containment maximum displacements in Z Axis (normal to containment) within hit region

Time (s)	Displacement in +Z Axis (mm)	Displacement in −Z Axis (mm)
0.01	0.209051	5.199
0.15	1.425	30.255
0.20	1.958	46.658

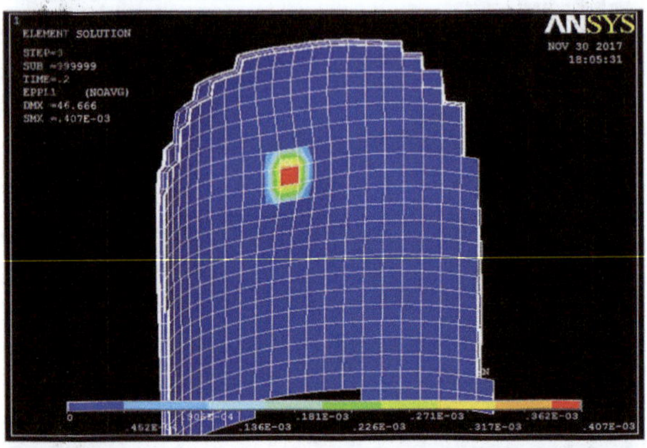

(a)

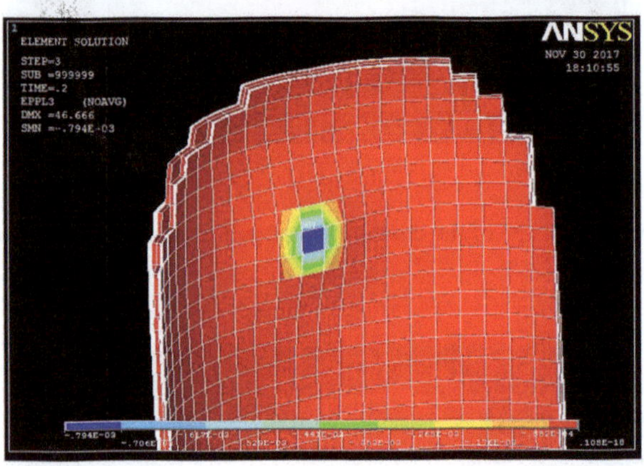

(b)

Fig. 14. (a) Principal plastic strains of RC containment in tension (1st principle plastic strain) (b) Principal plastic strains of RC containment in compression (3rd principle plastic strain)

appeared bobbing out of the containment. Meanwhile, at the bottom of the containment, flexural cracks are shown, where the flexure cracks appears as circle plane perpendicular to the maximum tensile principle stress (1st principle stresses), as shown in Fig. 15.

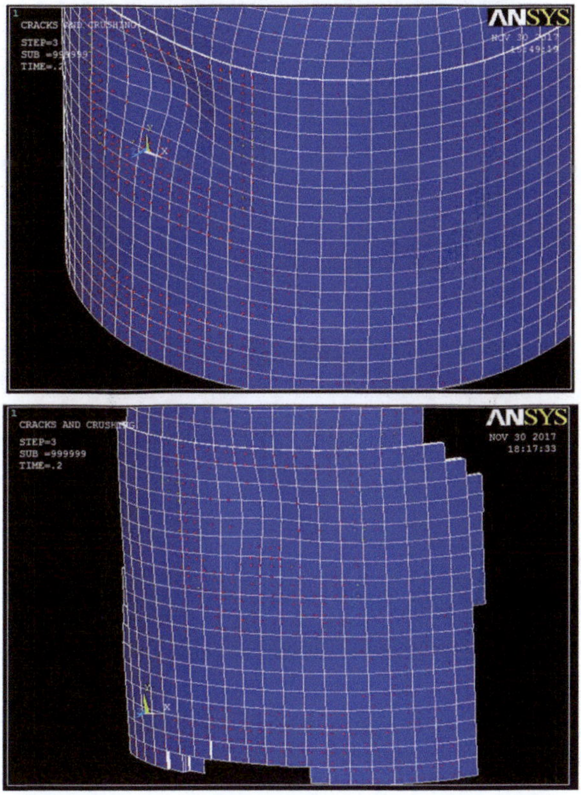

Fig. 15. Outer surface tensile cracks at time 0.2 s near the impact region

Furthermore, the response analysis indicates that the maximum tensile stress lies in the direction of 1st principal stress at time 0.2 s; the maximum principle tensile stress is 202.07 MPa towards the inner surface of impact at the steel liner plate location as shown in the highlighted elements in Fig. 16b. Other elements indicate stresses exceeding the rupture stress (4.8 MPa). Thus, these elements near the impact region suffered permanent damage and the impact region indicates failure of these elements after crashing as shown in Fig. 16a. It has to be noted though that the maximum tensile stresses within the inner surface is more significant than the compression stresses at the outer surface. The maximum Von-Misses stresses are found to be 175.578 MPa at the inner surface of the steel liner plate as shown in Fig. 17.

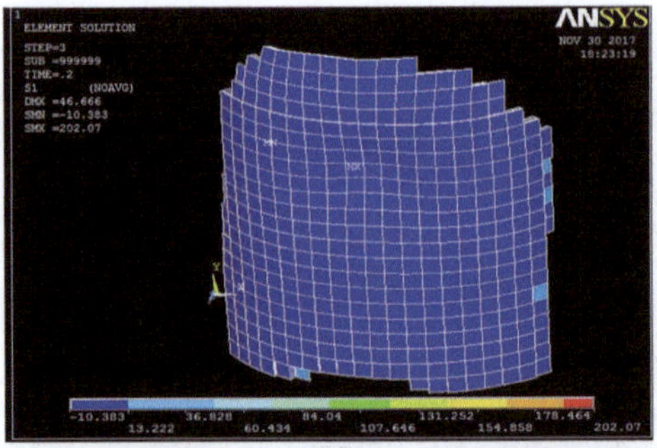

(a)

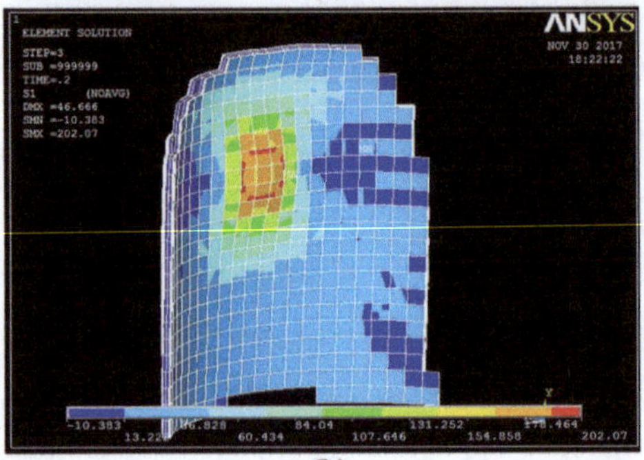

(b)

Fig. 16. (a) 1st principle stresses of outer surface of containment at time 0.2 s (b) 1st principle stresses for the inner surface of containment at time 0.2 s

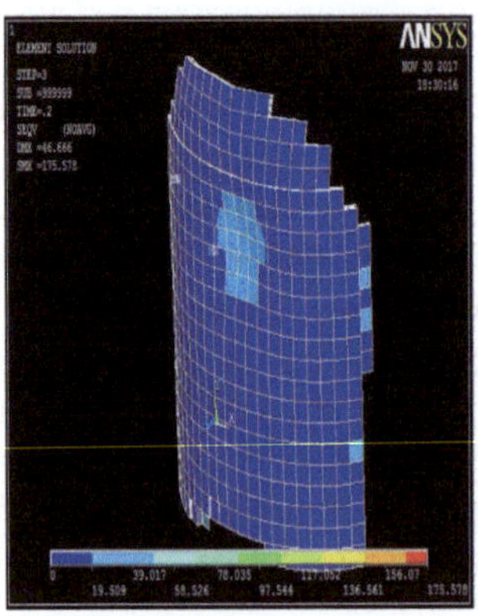

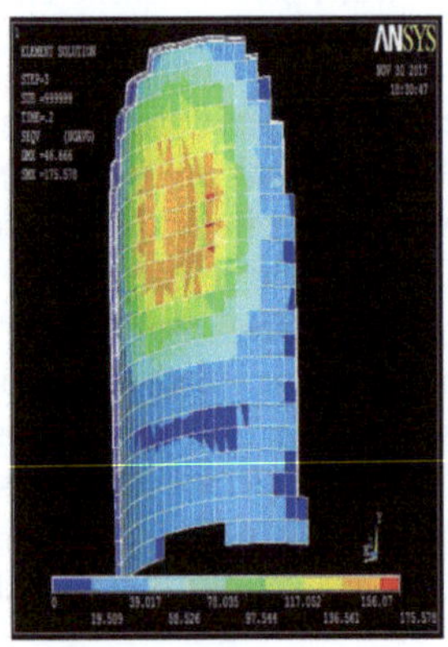

(a) Front face von mises stresses **(b) Inner face von mises stresses**

Fig. 17. Von Mises stresses at time 0.2 s

4 Conclusions and Recommendations

A study was undergone to model a jet impact load on an outer containment of an NPP. An existing design for an external containment is used as the case study. The external containment structure is subjected to normal impact load of a Boeing 747-200c aircraft using the Riera method load modeler in order to see the response of the containment, and to analyze the resulting deformations, displacements, principal stresses and strains, types and shapes of cracks that would occur in such an event. The elements within the impact region show clear deformations in the direction of loading towards the inside of the cylindrical shell wall, whereas elements in other regions away from the impact indicate deformations towards the outside of the containments. The maximum displacement in direction of loading is 46.658 mm due to the impact load of Boeing 747-200c aircraft affecting an area of 37 m^2. Permanent clear damage failure is considered in the elements near the impact region at which the tensile cracks appeared with the remaining of the containment remaining intact.

In addition, it was found that the elements at which the impact occurred suffered compressive stresses at the outer surface that reached the uniaxial compressive strength of 60 MPa. This means that concrete at the outer surface has been crushed. Rupture stress of concrete is reached at some elements at the impact region which means concrete is cracked at the inner surface of containment.

Apart from the elements in the hit region that suffered full damage, the containment remained fully intact after the impact of the aircraft with having some elements that suffered partial damage. Plastic strains were formed at time 0.2 s, the maximum plastic strains in tension reached 4.07E−4 in the direction of the 1st principle plastic strain, and the maximum plastic strains in compression reached 7.94E−4 in the direction of the 3rd principle plastic strain. The displacement and velocity of the full containment indicated its being intact except for the parts within the impact region.

References

1. Czerniewski, S.: The feasibility of modern technologies for reinforced concrete containment structures of nuclear power plants, a report of master science, chap. 3, pp. 27–31 (2009)
2. Mehta, P.K., Monteiro, P.J.M.: Concrete Microstructure, Properties, and Materials, pp. 529–531. McGraw-Hill, New York (2006)
3. Forasassi, G., Lofrano, R.: Preliminary analysis of an aircraft impact. A Report RdS, CERSE-UNIPI RL 1059/2010 (2010)
4. Sychrová, Š., Ševčík, I., Kabeláč, J., Weis, L., Němec, I.: Study of a nuclear power plant containment damage caused by impact of a plane. Int. Refer. J. Eng. Sci. (IRJES) **1**(4), 48–53 (2012)
5. James, R.J., Rashid, J.Y.R.: Severe impact dynamics of reinforced concrete structures. In: Sixth European Conference on Structural Dynamics (2005)
6. Salman, W.D.: Nonlinear behavior of RC continuous deep beam. Int. J. Eng. Res. Technol. (IJERT) (2015). ISSN 2278-0181
7. Ding, Y., Jaffer, S.: Development of finite element models for the study of ageing effects in CANDU 6 concrete containment buildings. CNL Nucl. Rev. **5**(1), 37–48 (2015). CNR.2015.00060

8. Hu, H.-T., Liang, J.I.: Ultimate analysis of BWR Mark III RC containment subjected to internal pressure. Nucl. Eng. Des. **195**(1), 1–11 (2015). https://doi.org/10.1016/s0029-5493(99)00163-6
9. ANSYS-12.1 – Theory Reference Manual
10. Dahmani, L., Khennane, A.: Crack identification in reinforced concrete beams using ANSYS software. Strength Mater. **42**(2), 232–240 (2010). https://doi.org/10.1007/s11223-010-9212-6
11. Ramadan, O.M., Abdelbaki, S.M., Saleh, A.M., Al-Khattabi, A.Y.: Modeling of reinforced concrete beams with and without opening by using ANSYS. J. Eng. Sci. (JES) Assiut Univ. **37**(4), 845–858 (2009)
12. Suanno, R.L.M, Ferrari, L.B., Prates, C.L.M.: Nonlinear analysis of reinforced concrete shells subjected to impact loads. In: Transactions of the 17th International Conference on Structural Mechanics in Reactor Technology (SMiRT 17) (2003)

Probabilistic Thickness Requirement of CFRP Plates Bonded to Reinforced Concrete Bridge Decks

A. Ahmed, J. M. Kaura, and O. S. Abejide(✉)

Department of Civil Engineering, Ahmadu Bello University, Zaria, Nigeria
abejideos@yahoo.com

Abstract. Probabilistic reinforced concrete bridge deck flexural strengthening with Carbon Fibre Reinforced Polymer (CFRP) laminates is presented. Results show that the existing and evaluated deck is in conformity with AASHTO LRFD (2010) specifications and indicate that the composite concrete deck as suggested in formulations meet the requirements for bridge decks and have the ability to sustain more live loads up to about 90% than conventional reinforced concrete decks of the same structural formulations. Also, the initial safety indices observable prove to be traffic dependent as it depends more on the live load. AASHTO design equation that corresponds to a β_{target} = 3.5 seems to be overestimated for strengthening purposes. The strengthening with a β_{target} = 4.2 as suggested herein provides a better structural reliability than β_{target} of 3.5 proposed by AASHTO LRFD (2010) provision and with no significant differences in bonded CFRP amounts required. The optimum thickness of bonded CFRP laminates required is found to be 5 mm for the upgrading of reinforced concrete bridge decks.

Keywords: Reinforced concrete bridge decks
Bonded CFRP laminates thickness · Target structural reliability

1 Introduction

The primary function of a bridge deck is to support the vehicular vertical loads and distribute these loads through the steel or reinforced superstructure to the substructure. The deterioration of bridge decks is of critical importance and it can result in bridge decks reaching the end of their service life. This degradation may be due to the lack of proper and adequate maintenance, environmental conditions or poor initial construction techniques.

The modular use of FRP composite materials is a possible solution for the rehabilitation of existing bridges and construction of new ones (Folić and Cumbo 2001; Luke 2003; Aref et al. 2005). The service life of bridges is often reduced due to the corrosion of steel reinforcing bars in bridge decks and the cracking caused by loading in excess of the original design values due to increased traffic volumes (FHWA 2002; AASHTO LRFD 2010).

H. Rodrigues and A. Elnashai (Eds.): GeoMEast 2018, SUCI, pp. 399–408, 2019.
https://doi.org/10.1007/978-3-030-01932-7_31

However, the optimized reliable thickness of CFRP plates that can be used at any time on existing reinforced concrete and steel bridges decks retrofitting needs to be obtained.

Structural reliability is concerned with the calculations and predictions of the probability of limit state violations for engineered structures (Melchers 1999). Madsen (1992) considered structural reliability as a method that attempts to treat rationally, the various sources of uncertainties. According to Tsompanakis et al. (2000), structural reliability analysis is a tool that assists the design engineer to take into account all possible uncertainties during the design and construction phases and lifetime of a structure in order to estimate its probability of failure. While reliability-based design considers the probability that the structure will last a given length of time against the agents that can induce its failure (Tsompanakis et al. 2000; Stanley et al. 1978).

2 Background of Study

Many industries, such as the aerospace and automotive industries have successfully used Fiber Reinforced Polymer Composites (FRPC). These types of composite materials offer significant advantages over conventional civil engineering materials, such as concrete and steel (Carolin et al. 2001, 2002). This is due to their chemical and corrosion resistance, lightweight, and high strength, which make them attractive for the rehabilitation of civil infrastructures (Karbhari et al. 2003; Rahman 1993).

Genetic Algorithms are adaptive heuristic search algorithms based on the evolutionary ideas of natural selection and genetics. They represent an intelligent exploitation of a random search used to solve optimization problems (Sivaraj and Ravichandran 2011). The basic techniques of the genetic algorithms are designed to simulate processes in natural systems necessary for evolution; especially those that follow the Charles Darwin's principles of "Survival of the Fittest".

This principle is similar to competition among individuals for scanty resources in nature resulting in the fittest individuals dominating over the weaker ones. Genetic algorithms simulate the survival of the fittest individuals among all the individuals in the population over successive generations for solving a problem. Each generation consists of a population of character strings that are analogous to the chromosome that are seen in DNA (Deoxyribonucleic acid). Each individual represents a point in a search space and a possible solution. The individuals in the population are then made to go through a process of evolution (Goldberg 1989; Goldberg and Deb 1991).

There are many researches that has been carried out on the experimental and numerical simulation of structural strength capacity of bonded CFRP laminates for the retrofitting and structural stability of reinforced concrete and steel structures. Various thicknesses of CFRP laminates have been used and with the results that the structural members have been adequately retrofitted with improved structural capacity in various service conditions. However, the optimum bonded CFRP thickness requirement for the formulations in proposed codes has not been fully developed. It is with this notion that a suggestion is made herein by evaluation of the bonded CFRP thickness requirement for an existing bridge along a single lane dual carriage way which is about fifty years old. This is done in order to upgrade the structural strength to sustain increased traffic loads, while sustaining and maintaining the environment with the little financial resources available instead of outright demolition and replacement of the existing bridge with a new one.

The variables considered for the structural safety of the externally bonded CFRP laminates to the bridge decks include the parameters stipulated in AASTHO LRFD (2010) and the thickness of the laminates.

3 Methodology

3.1 General

A MATLAB program was developed for the structural reliability analysis of the bridge deck before and after repairs, using genetic algorithm, and a genetic algorithm-based First Order Reliability Method (FORM). The procedure followed is outlined below.

(a) Load analysis

The FRP-Laminates bonded to reinforced concrete bridge decks are analyzed in the form of simply supported decks on steel-girders.

The theory derived for the ultimate bending capacity can be found in Täljsten (2001, 2002) and Nordin et al. (2002). The model is based on the Bernoulli´s hypothesis. The model considers the anisotropy of the composite material when used in bending. This model is also accepted by the Swedish Road Administration (2002) to be used for FRP strengthening.

Figure 1 indicates the principles of strengthening of the deck in bending as employed in this work.

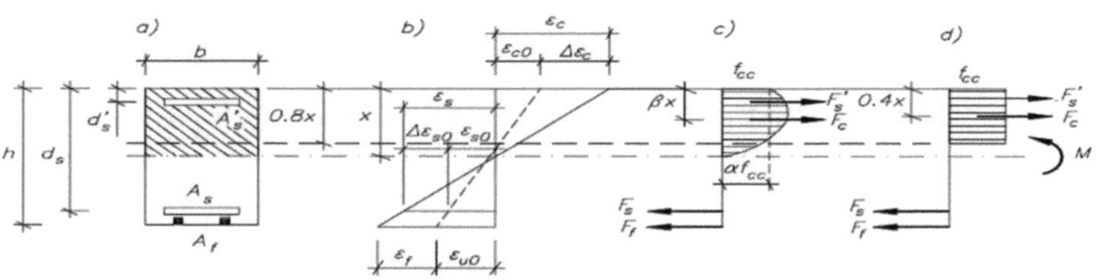

Fig. 1. Principles for strengthening in bending (Täljsten 2001, 2002; Nordin et al. 2001).

It is important to note that if there exists a strain field on the structure due to, for example, the dead load, this must be considered in design for strengthening. In Fig. 1(b), this is shown schematically, where ε_{u0} is the initial strain in the bottom face of the cross section. The influence of the creep in the concrete is taken care of with a reduced modulus of elasticity as in normal concrete design. A calculation is then made to control whether the concrete is un-cracked or not. The studied section can be considered un-cracked if the tensile capacity of the concrete is not exceeded. However, if the type of failure that can occur is assumed to be failure in the composite material without yielding

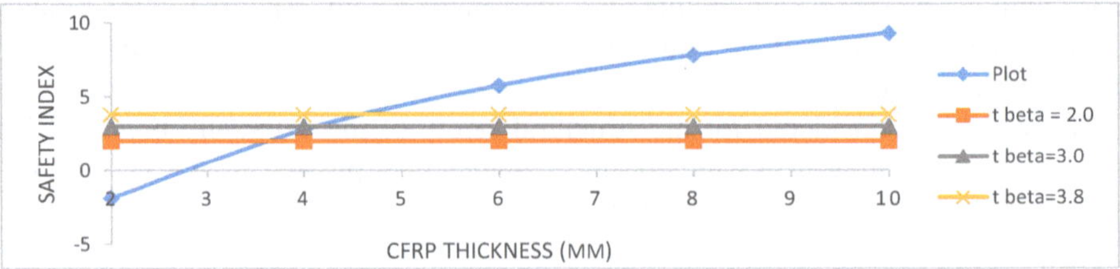

Fig. 2. Relationship of safety indices to CFRP thicknesses at α = 0.2

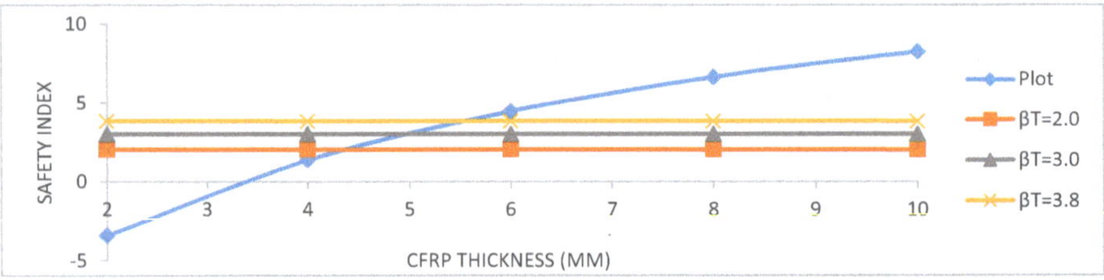

Fig. 3. Relationship of structural safety indices to CFRP thicknesses at α = 0.4

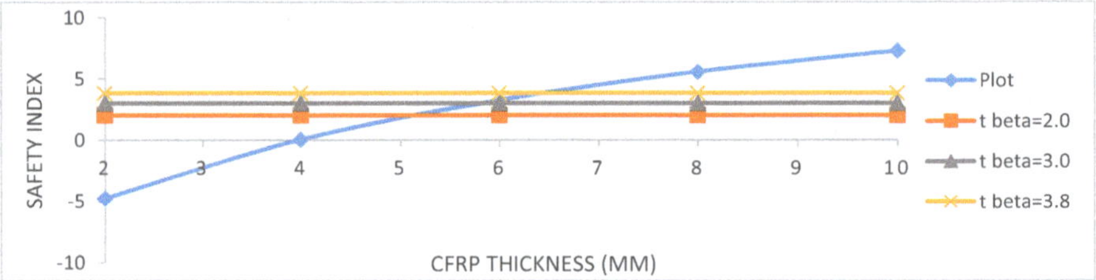

Fig. 4. Relationship of safety indices to CFRP thicknesses at α = 0.6

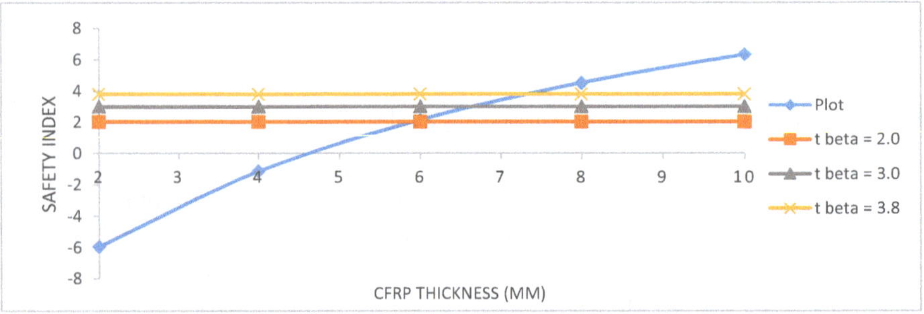

Fig. 5. Relationship of safety indices to CFRP thicknesses at α = 0.8

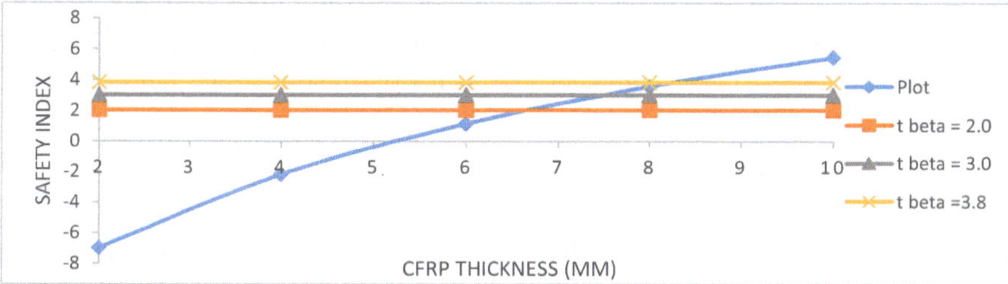

Fig. 6. Relationship of safety indices to CFRP thicknesses at $\alpha = 1.0$

in the compressive reinforcement, the bending capacity can then be expressed as, (Täljsten 2002):

$$\mathrm{M} = \frac{x - d_s'}{h - x}(\varepsilon_{fu} + \varepsilon_{u0})A_s'E_s(0.4x - d_s') + A_sf_s(d_s - 0.4x) + \mu\varepsilon_{fu}E_fA_f(h - 0.4x) \quad (1)$$

A horizontal equilibrium equation for the section (Fig. 1(d)) gives:

$$0.8f_{cc}b_x + \frac{x - d_s'}{h - x}(\varepsilon_{fu} + \varepsilon_{u0})A_s'E_s = A_sf_s + \mu\varepsilon_{fu}E_fA_f \quad (2)$$

Where x can be solved with an equation of the second degree;

$$C_1x^1 + C_2x + C_3 = 0$$

Such that,

$$\begin{aligned} C_1 &= 0.8f_{cc}b \\ C_2 &= -\ 0.8f_{cc}\mathrm{b\,h} - (\varepsilon_{fu} + \varepsilon_{u0})A_s'E_s - Asfs - \mu\,\varepsilon_{fu}\,E_fA_f \\ C_3 &= (\varepsilon_{fu} + \varepsilon_{u0})A_s'E_sd_s' + (A_sf_s + \mu\ \varepsilon_{fu}\ E_fA_f)h \end{aligned} \quad (3)$$

(b) Probability of failure

The objective of the component reliability analysis is to estimate the probability of failure as given in Eq. (4).

$$P_f = \int [g(x) \leq 0]F_x(\mathrm{X})\mathrm{dx} \quad (4)$$

where, X is random vector, $F_x(\mathrm{X})$ is joint probability density, g(x) is the limit-state function corresponding to the failure mode considered and defined such that the failure event corresponds to $g \leq 0$. Thus,

$$P_f = \int [g(x) \leq 0](2\pi)^{-\frac{n}{2}}\exp\left(-\frac{1}{2}U^TU\right)du \quad (5)$$

Where n is the number of random variables; and g(U) the limit-state function in the standard normal space. The Genetic Algorithm - based First Order Reliability Method as may be employed in FORM can be used to solve optimization problems (Sivaraj and Ravichandran 2011).

(c) Evaluation of analytical results

The level of safety specified in the design codes is determined by the target reliability, which is theoretically the optimum value considering the relationship between cost and structural reliability. In reality, it is difficult to determine such theoretical target values. Instead, the target reliability is established based on failure experiences and performances of existing structures. However, FRP structures lack such data due to the short history in the construction field. The target reliability in current codes is approximately 3.0 for building structures and 3.5 for bridges (AASHTO LRFD 2004). These levels of safety are based on the fact that failure modes are ductile, and materials are well proved to be safe through long-term uses. However, the reliabilities are much higher for brittle failure modes even in conventional materials and constructions. For instance, timber structures have reliabilities with a range of 3.5 to 6.5 (AASHTO LRFD 2004), and brittle connections may also have reliabilities over 6 or 7.

It is expected that higher target reliability indices shall be used for FRP structures, considering the brittle failure modes and the degradation of material properties for long-term uses. In order to consider the degradation effect, it has been recommended (FHWA 2002) to use 0.65 as a durability factor. Based on such evaluations, it is suggested (FHWA 2002) that the target reliability index for FRP bridge decks should be at least 7.0, which is approximately equivalent to a safety factor of 5.0.

(d) Development of flexural performance functions

The genetic algorithms consist of generating search directions in the space formed by the random variables affecting the reliability of a structural system. By following a search path until failure is reached, the reliability index 'β' corresponding to the search direction is obtained (Tsompanakis and Papadrakakis 2000).

The reliability analysis can be implemented by solving the following nonlinear optimization problem using genetic algorithms.

$$\begin{aligned} &\text{Minimise } \beta = \|\mu\|^2 = \mu^T - \mu \\ &\text{Subject to: } g(\mu) = 0 \end{aligned} \tag{6}$$

Where β is the safety index, μ is the vector of standard normal variables and $g(\mu)$ is the limit state function

Limit state function of the CFRP strengthened beam is given by:

$$G(X) = \theta_r A_s(t) f_y \left(d - \frac{0.55 A_s(t) f_y}{b f_c} \right) + \theta_{cf} E_{cf} \varepsilon_{cf} \left(h - \frac{a}{2} \right) - 0.125 \theta_w Q_k (1 + \alpha) L^2 \tag{7}$$

Where θ_r is the model uncertainty for the un-strengthened beam, $A_s(t)$ is the time dependent cross sectional area, f_y is the yield strength of the steel; reinforcement, d is the concrete cover to reinforcing steel, b is the width of the beam, f_c is the concrete compressive strength, θ_{cf} is the model uncertainty for CFRP strength, E_{cf} is the modulus of elasticity of the CFRP sheets, ε_{cf} is the strain of CFRP, h is the overall depth of the un-strengthened beam, θ_w is the model uncertainty for loading, Q_k is the applied imposed loading, α is the live to dead load ratio and L is the beam span.

(e) Statistical models of basic design variables

The evaluation of the limit state function of the CFRP strengthened beam requires the statistical models of the uncertain variables. These include the mean values, coefficient of variation as well as their probability density functions defined by the appropriate theoretical distribution models.

(e) Applications

In order to arrive at an optimum assessment of the sustainability of bridge concepts incorporating FRP bridge decks, a comparative analysis needs to be performed on a well-defined case. For this purpose, a case-study bridge was selected and evaluated with regard to maintenance optimization and long life-cycle span of the structure.

The numerical model proposed above was applied to the data obtained from the single-lane dual carriage-way steel-concrete composite deck of the Chanchagi Bridge along Takum-Wukari road in north-eastern, Nigeria. Table 1 indicates the statistical parameters used for the composite bridge deck evaluation.

Table 1. Statistical parameters of variables

S/No	Basic variables	Symbols	Distribution type	Dimension	Mean	Standard deviation	C.O. V
1	Compressive strength of concrete	f_c	Lognormal	N/mm^2	30	5	0.17
2	Yield strength	f_y	Lognormal	N/mm^2	500	30	0.06
3	Span of the slab deck	L	Deterministic	m	5	1.25	0.25
4	Reinforcement area	A_s	Deterministic	m^2	113.1	28.3	0.25
5	Slab depth	h	Normal	m	0.12	0.003	0.025
6	Distance of bars to slab bottom	A	Gamma	m	c + Ǿ/2	0.03	0.25
7	Density of concrete	γ_c	Normal	MN/m^3	0.025	0.00075	0.25
8	Live load	Q_k	Gumbel	kN/mm^2	1.5	0.315	0.21
9	Imposed long-term load	Q_{lt}	Gamma	kN/mm^2	0.5	0.75	1.5
10	Imposed short-term load	Q_{st}	Exponential	kN/mm^2	0.2	0.46	2.3
11	Uncertainty of resistance	θ_R	Lognormal	–	1.1	0.077	0.07
12	Uncertainty of load effect	θ_E	Lognormal	–	1	0.2	0.2
13	Elastic modulus of CFRP	E_{frp}	Weibull	kN/mm^2	200	6	0.03
14	Coeff. of strength reduction	ϕ	Lognormal	–	1	0.08	0.08
15	Modulus of elasticity of conc.	E_c	Lognormal	N/mm^2	31	0.90	0.03

Plate 1 shows the composite bridge before rehabilitation and the state of damage of the bridge deck. Also the procedure for application of the CFRP laminates on the deck was modelled as suggested by Folić and Cumbo 2001and Folić (1984).

Plate 1. Chanchangi bridge deck - state of damage.

3.2 Analysis and Results of Structural Reliability Evaluation

Figures 2, 3, 4, 5 and 6 give the relationship of safety indices to various CFRP laminates thicknesses for different load ratios considered. This work also extends to calibrate CFRP resistance safety factors corresponding to three target reliability levels as proposed by JCSS (2001) of $\beta = 2$, 3, and 3.8. It was observed that this complies with the AASHTO LRFD (2010) provisions. Figure 2 shows the relationship of safety indices to CFRP laminates thicknesses at 0.2 load ratio. The results show that the evaluated deck is in conformity with AASHTO LRFD (2010) specifications as regard the target safety value for bridges of 3.5, since its structural safety value is higher than the code prescription; which is 4.2 at LRF of 0.2 and with 5 mm thick bonded CFRP laminates.

Also, Fig. 6 shows the relationship of safety indices to CFRP thicknesses at LRF of 0.2; it was observed that the deck structure is stable with a safety index of 3.8 which is more than the code prescription and at 8 mm thick bonded CFRP laminates. However, this thickness is traffic dependent. The AASHTO design equation that corresponds to a $\beta_{target} = 3.5$, seems to be overestimated for strengthening purposes. Strengthening with a $\beta_{target} = 3.8$ provides a better and adequate structural reliability than the β_{target} proposed by AASHTO LRFD (2010) provisions and with no significant differences in CFRP amounts required.

4 Conclusion

This study provides a clearer understanding of CFRP-bonded composite concrete bridge decks structural behaviour while developing adequate mathematical models for obtaining optimum results for structural safety in retrofitting, upgrading and maintenance of a typical bridge deck.

The results show that the evaluated deck is in conformity with AASHTO LRFD (2010) specifications as regards the target safety value for bridges of 3.5, since the safety value obtained from results is higher than the code prescription of 4.2. Also, the initial safety index observable is proved to be traffic dependent. AASHTO design equation that corresponds to $\beta_{target} = 3.5$ seems to be overestimated for strengthening purposes. Strengthening with a $\beta_{target} = 4.2$ provides an adequate structural reliability than β_{target} of 3.5 proposed in AASHTO LRFD (2010) provisions with no significant differences in CFRP amounts required.

The result of CFRP-bonded conposite concrete deck to conventional concrete decks gave a safety index value of 11.0 for the full composite deck action at LRF of 1.0 and with 5 mm thick bonded CFRP laminates.

Thus, with an appropriate combination of structural reinforced concrete bonded CFRP laminates more efficient bridge rehabilitation, maintenance and upgrading can be realized, while sustain the environment.

References

Aref, A.J., Alampalli, S., He, Y.: performance of a fibre reinforced polymer web core skew bridge superstructure. Part I: field testing and finite element simulations. Part II: failure modes and parametric study. Compos. Struct. **69**(4), 500–509 (2005)

AASHTO LRFD : Bridge Design Specifications. American Association of State Highway and Transportation Officials, Washington, D.C. (2004)

AASHTO LRFD: Bridge Design Specifications. American Association of State Highway and Transportation Officials, Washington, D.C. (2010)

Carolin, A., Nordin, H., Täljsten, B.: Concrete beams strengthened with near surface mounted reinforcement of CFRP. In: Teng, J.-G. (ed.) International Conference on FRP Composites in Civil Engineering, vol. 2, pp. 1059–1066 (2001). ISBN: 0-08-043945-4

Carolin, A., Hejll, A., Täljsten B.: Behaviour of concrete beams strengthened with CFRP and loaded in fatigue during the strengthening process. Paper presented at ICCI, San Francisco, June 2002

Folić, R.: Analysis of the effective slab width in reinforced and prestressed concrete elements. In: Damjanić, F., Swansea, P.P. (eds.) Computer-Aided Analysis and Design of C S, Part. II, pp. 1339–1354 (1984)

Folić, R., Cumbo, A.: Application of finite elements method in analysis of composite concrete-steel beams. BAM-1770/ 2001 XCIV, TU Budapest, pp. 55–66 (2001)

Federal Highway Administration (FHWA): FRP decks and super-structures: current practice (2002). http://www.fhwa.dot.gov/bridge/frp/deckprac.htm

Goldberg, D.E.: Genetic Algorithms in Search, Optimization, and Machine Learning. Addison Wesley, New York (1989)

Goldberg, D.E., Deb, K.: A comparative analysis of selection schemes used in genetic algorithms. In: Rawlins, G. (ed.) Foundations of Genetic Algorithms. Morgan Kaufmann, San Mateo (1991)

Luke, S.: The design, installation and monitoring of an FRP bridge at West Mill, Oxford. In: Proceedings of Lightweight Bridge Decks-European Bridge Engineering Conference, Rotterdam, Netherlands, 27–28 March 2003

Karbhari, V.M.: FRP bridge decks, from design and characterization to field implementation. In: Proceedings of Lightweight Bridge Decks-European Bridge Engineering Conference, Rotterdam, Netherlands, 27–28 March 2003

Madsen, H.O.: Structural Behaviour of Timber. Timber Engineering Ltd., Vanvouver (1992)

Melchers, R.E.: Structural Reliability Analysis and Prediction, 2nd edn. Wiley, Chichester (1999)
Nordin, H., Täljsten, B., Carolin, A.: CFRP near surface mounted reinforcement (NSMR) for prestressing concrete beams. Paper presented at ICCI, San Francisco, June 2002
Rahman, A.H., Taylor, D.A., Kingsley, C.Y.: Evaluation of FRP as reinforcement for concrete bridges. In: Fiber-Reinforced-Plastic Reinforcement for Concrete Structures International Symposium, pp. 71–86. American Concrete Institute, Detroit (1993)
Sivaraj, R., Ravichandran, T.: A review of selection methods in genetic algorithm. Int. J. Eng. Sci. Technol. (IJEST) **3**(5), 3792–3793 (2011)
Stanley, K.S., William, L.G., Frank, E.W.: Differential reliability: probabilistic engineering applied to wood members in bending/tension. Forest Product Laboratory, United States Department of Agriculture, Madison, Wisconsin (1978)
Täljsten, B., Carolin, A.: CFRP - Strengthening. concrete beams strengthened with near surface mounted CFRP laminates. In: Burgoyne, C. (ed.) Fibre Reinforced Plastics for Reinforced Concrete Structures, FRPRCS-5, Cambridge, pp. 107–116 (2001)
Täljsten, B.: Strengthening of existing concrete structures with externally bonded fibre reinforced polymers, design and execution. Technical report, Division of Structural Engineering, Luleå University of Technology, Lulea (2002)
Tsompanakis, Y., Papadrakakis, M.: Robust and Efficient Methods for Reliability-based Structural Optimization (2000)
Tsompanakis, Y., Papadrakakis, M.: Efficient computational methods for large-scale structural optimization. Int. J. Comput. Eng. Sci. **1**(02), 331–354 (2000)

Author Index

H. Rodrigues and A. Elnashai (Eds.): GeoMEast 2018, SUCI, pp. 409–410, 2019.
https://doi.org/10.1007/978-3-030-01932-7

FSC
www.fsc.org
MIX
Papier aus verantwortungsvollen Quellen
Paper from responsible sources
FSC® C105338

If you have any concerns about our products,
you can contact us on
ProductSafety@springernature.com

In case Publisher is established outside the EU,
the EU authorized representative is:
Springer Nature Customer Service Center GmbH
Europaplatz 3, 69115 Heidelberg, Germany

Printed by Libri Plureos GmbH
in Hamburg, Germany